Svetlin G. Georgiev

Integral Inequalities on Time Scales

I0131943

Also of Interest

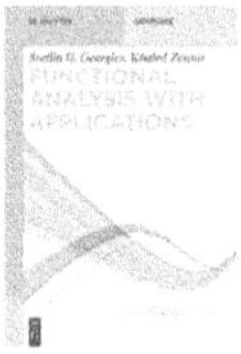

Functional Analysis with Applications
Georgiev, Svetlin G. / Zennir, Khaled, 2019
ISBN 978-3-11-065769-2, e-ISBN (PDF) 978-3-11-065772-2,
e-ISBN (EPUB) 978-3-11-065804-0

Applied Nonlinear Functional Analysis
An Introduction
Papageorgiou, Nikolaos S. / Winkert, Patrick, 2018
ISBN 978-3-11-051622-7, e-ISBN (PDF) 978-3-11-053298-2,
e-ISBN (EPUB) 978-3-11-053183-1

Variational Methods in Nonlinear Analysis
With Applications in Optimization and Partial Differential Equations
Kravvaritis, Dimitrios C. / Yannacopoulos, Athanasios N., 2020
ISBN 978-3-11-064736-5, e-ISBN (PDF) 978-3-11-064738-9,
e-ISBN (EPUB) 978-3-11-064745-7

Complex Analysis
Theory and Applications
Bulboacă, Teodor / Joshi, Santosh B. / Goswami, Pranay, 2019
ISBN 978-3-11-065782-1, e-ISBN (PDF) 978-3-11-065786-9,
e-ISBN (EPUB) 978-3-11-065803-3

Real Analysis
Measure and Integration
Markin, Marat V., 2019
ISBN 978-3-11-060097-1, e-ISBN (PDF) 978-3-11-060099-5,
e-ISBN (EPUB) 978-3-11-059882-7

Svetlin G. Georgiev

Integral Inequalities on Time Scales

DE GRUYTER

Mathematics Subject Classification 2010
39A10, 34A60, 34B37, 34C10, 34K20, 34K30

Author
Prof. Dr. Svetlin G. Georgiev
Kliment Ohridski University of Sofia
Department of Differential Equations
Faculty of Mathematics and Informatics
1126 Sofia
Bulgarien
svetlingeorgiev1@gmail.com

ISBN 978-3-11-070550-8
e-ISBN (PDF) 978-3-11-070555-3
e-ISBN (EPUB) 978-3-11-070566-9

Library of Congress Control Number: 2020939558

Bibliographic information published by the Deutsche Nationalbibliothek
The Deutsche Nationalbibliothek lists this publication in the Deutsche Nationalbibliografie; detailed bibliographic data are available on the Internet at http://dnb.dnb.de.

Cover image: Vipul Umretiya / DigitalVision Vectors / gettyimages.de
Typesetting: VTeX UAB, Lithuania
Printing and binding: CPI books GmbH, Leck

www.degruyter.com

Preface

Time scale theory was first initiated by Stefan Hilger in 1988 in his PhD thesis to unify both approaches of dynamic modeling: difference and differential equations. Similar ideas have been used before and go back in the introduction of the Riemann–Stieltjes integral which unifies sums and integrals. Many results to differential equations carry over easily to corresponding results for difference equations, while other results seem to be totally different in nature. Because of these reasons, the theory of dynamic equations is an active area of research. The time scale calculus can be applied to any fields in which dynamic processes are described by discrete or continuous time models. So, the calculus of time scales has various applications involving noncontinuous domains such as certain bug populations, phytoremediation of metals, wound healing, maximization problems in economics, and traffic problems.

This book is devoted on recent developments of some linear and nonlinear integral inequalities on time scales. The book is intended for the use in the field of dynamic calculus on time scales, dynamic and integral equations on time scales. It is also suitable for graduate courses in the above fields. The book is designed for those who have mathematical background on time scales calculus. This book contains eight chapters. The basic Gronwall's and Bellman's inequalities are considered in Chapter 1, as well as some linear Volterra-type inequalities, namely the inequalities of Gamidov and Rodrigues. We also investigate simultaneous and Pachapatte's inequalities. Chapter 2 introduces linear integro-dynamic inequalities of Pachpatte-type and linear integro-dynamic inequalities with several iterated integrals. Chapter 3 is concerned with Dragomir- and Pachpatte-type nonlinear integral inequalities. In Chapter 4, Diamond-α integral inequalities are investigated, such as Steffensen, Jensen, Radon, and Schlömilch inequalities. Chapter 5 is devoted to fractional integral inequalities on time scales. We consider the Poincaré-, Opial-, Sobolev-, and Ostrowski-type inequalities. Chapter 6 is devoted to two-dimensional linear integral inequalities. We investigate Wendroff- and Pachpatte-type two-dimensional linear integral inequalities. Chapter 7 deals with Snow-type two-dimensional linear integral inequalities. In Chapter 8, some two-dimensional Pachpatte-type linear integro-dynamic inequalities are investigated, as well as some of their modifications. Chapter 9 is devoted to Wendroff- and Pachpatte-type two-dimensional nonlinear integral inequalities. Chapter 10 introduces delay integral inequalities with one and two independent variables. Chapter 11 deals with some applications of certain linear integral inequalities and some linear integro-dynamic inequalities. We investigate the existence and uniqueness of the solutions of first order dynamic equations, as well as continuous dependence on initial conditions of the solutions of first order dynamic equations. We give applications for some second order integro-dynamic equations and Volterra integral equations. Some bounds on the solutions of delay dynamic equations are deduced.

https://doi.org/10.1515/9783110705553-201

This book is addressed to a wide audience of specialists such as mathematicians, physicists, engineers, and biologists. It can be used as a textbook at the graduate level and as a reference book for several disciplines.

The author welcomes any suggestions for the improvement of the text.

Paris
August 2019

Svetlin Georgiev

Contents

1 Linear integral inequalities on time scales

In this chapter are investigated Gronwall's and Bellman's inequalities, some linear Volterra type inequalities, the inequalities of Gamidov and Rodrigues. They are investigated simultaneous inequalities and Pachapatte's inequalities. The presentation in this chapter follows some results in [1, 2, 5–7, 16, 24–26, 29].

Suppose that $\mathbb{T}$ is a time scale with forward jump operator and delta differentiation operator σ and Δ, respectively. Let also, $a, b \in \mathbb{T}$, $a < b$, and $J = [a, b]$ be a time scale interval. The set of all rd-continuous functions on $\mathbb{T}$ will be denoted with $\mathcal{C}_{rd}$. The set of all rd-continuous and regressive functions on $\mathbb{T}$ will be denoted by $\mathcal{R}$, i. e., $\mathcal{R}$ is the set of all rd-continuous functions $f : \mathbb{T} \to \mathbb{R}$ such that

$$1 + \mu(t)f(t) \neq 0, \quad t \in \mathbb{T}.$$

If $A \subset \mathbb{T}$, by $\mathcal{R}(A)$ we will denote the set of all functions $f : A \to \mathbb{R}$ which are rd-continuous on A and

$$1 + \mu(t)f(t) \neq 0 \quad \text{for any } t \in A.$$

We define the sets

$$\mathcal{R}^+ = \{f \in \mathcal{R} : 1 + \mu(t)f(t) > 0 \text{ for all } t \in \mathbb{T}\},$$
$$\mathcal{R}^+(A) = \{f \in \mathcal{R}(A) : 1 + \mu(t)f(t) > 0 \text{ for all } t \in A\}.$$

1.1 The inequalities of Gronwall and Bellman

Theorem 1.1.1 (Gronwall's Inequality). *Let $x \in \mathcal{C}_{rd}(J)$ be a nonnegative function and c, p be nonnegative constants. Then the inequality*

$$x(t) \leq \int_a^t (px(s) + c)\Delta s, \quad t \in J,$$

implies the inequality

$$x(t) \leq c \int_a^t e_{\ominus p}(\sigma(s), t)\Delta s, \quad t \in J.$$

Proof. Let

$$g(t) = \int_a^t (px(s) + c)\Delta s, \quad t \in J.$$

https://doi.org/10.1515/9783110705553-001

Then

$$\begin{aligned} g(a) &= 0, \\ x(t) &\le g(t), \quad t \in J, \end{aligned}$$

and

$$\begin{aligned} g^{\Delta}(t) &= px(t) + c \\ &\le pg(t) + c, \quad t \in J, \end{aligned}$$

or

$$g^{\Delta}(t) - pg(t) \le c, \quad t \in J.$$

We have

$$\begin{aligned} (g(\cdot)e_{\ominus p}(\cdot, a))^{\Delta}(t) &= g^{\Delta}(t)e_{\ominus p}(\sigma(t), a) \\ &\quad + (\ominus p)(t)g(t)e_{\ominus p}(t, a) \\ &= g^{\Delta}(t)e_{\ominus p}(\sigma(t), a) \\ &\quad - \frac{pg(t)}{1 + p\mu(t)}e_{\ominus p}(t, a) \\ &= g^{\Delta}(t)e_{\ominus p}(\sigma(t), a) \\ &\quad - \frac{pg(t)}{1 + p\mu(t)}(1 + p\mu(t))e_{\ominus p}(\sigma(t), a) \\ &= g^{\Delta}(t)e_{\ominus p}(\sigma(t), a) \\ &\quad - pg(t)e_{\ominus p}(\sigma(t), a) \\ &= (g^{\Delta}(t) - pg(t))e_{\ominus p}(\sigma(t), a) \\ &\le ce_{\ominus p}(\sigma(t), a), \quad t \in J. \end{aligned}$$

Hence,

$$g(t)e_{\ominus p}(t, a) \le c\int_a^t e_{\ominus p}(\sigma(s), a)\Delta s, \quad t \in J,$$

and

$$\begin{aligned} x(t) &\le g(t) \\ &\le c\int_a^t \frac{e_{\ominus p}(\sigma(s), a)}{e_{\ominus p}(t, a)}\Delta s \\ &= c\int_a^t e_{\ominus p}(\sigma(s), t)\Delta s, \quad t \in J. \end{aligned}$$

This completes the proof. □

Example 1.1.2. Let $\mathbb{T} = l\mathbb{Z}$, $l > 0$, x, c and p be as in Theorem 1.1.1. Then

$$\sigma(t) = t + l, \quad t \in \mathbb{T},$$

and

$$\begin{aligned}
\int_a^t (px(s) + c)\Delta s &= l \sum_{n=a}^{t-l} (px(n) + c), \quad t \in J,\ t > a,\\
(\ominus p)(t) &= -\frac{p}{1 + \mu(t)p}\\
&= -\frac{p}{1 + lp}, \ t \in \mathbb{T},\\
e_{\ominus p}(\sigma(s), t) &= e_{\ominus p}(s + l, t)\\
&= \prod_{n=s+l}^{t-l} (1 + l(\ominus p))^{-1}\\
&= \prod_{n=s+l}^{t-l} \left(1 - \frac{lp}{1 + lp}\right)^{-1}\\
&= \prod_{n=s+l}^{t-l} (1 + lp)\\
&= (1 + lp)^{\frac{t-s-l}{l}}, \quad t, s \in J,\ t - l \geq s + l.
\end{aligned}$$

Hence, by Theorem 1.1.1, it follows that the inequality

$$x(t) \leq l \sum_{n=a}^{t-l} (px(n) + c), \quad t \in J,\ t > a,$$

implies the inequality

$$x(t) \leq cl \sum_{s=a}^{t-l} (1 + lp)^{\frac{t-s-l}{l}}, \quad t \in J,\ t \geq a + l.$$

Example 1.1.3. Let $\mathbb{T} = q^{\mathbb{N}_0}$, $q > 0$. Let also, x, p and c be as in Theorem 1.1.1. Suppose that

$$a = q^{r_1}, \quad t = q^{r_2}, \quad s = q^{r_3}, \quad r_2 > r_3 \geq r_1.$$

We have

$$\sigma(t) = qt, \quad t \in \mathbb{T},$$

and

$$\int_a^t (px(s) + c)\Delta s = \int_{q^{r_1}}^{q^{r_2}} (px(s) + c)\Delta s$$

$$
\begin{aligned}
&= \sum_{s=q^{r_1}}^{q^{r_2-1}} (q-1)s(px(s)+c),\\
(\ominus p)(t) &= -\frac{p}{1+\mu(t)p}\\
&= -\frac{p}{1+(q-1)tp},\\
e_{\ominus p}(\sigma(s),t) &= e_{\ominus p}(qs,t)\\
&= e_{\ominus p}(q^{r_3+1},q^{r_2})\\
&= \prod_{z=q^{r_3+1}}^{q^{r_2-1}} \left(1-(q-1)z\left(\frac{p}{1+(q-1)sp}\right)\right)^{-1}\\
&= \prod_{z=q^{r_3+1}}^{q^{r_2-1}} (1+(q-1)zp), \quad r_2 \ge r_3+2.
\end{aligned}
$$

Then, by Theorem 1.1.1, the inequality

$$
x(t) \le \sum_{s=q^{r_1}}^{q^{r_2-1}} (q-1)s(px(s)+c), \quad r_2 \ge r_3+1,
$$

implies the inequality

$$
x(t) \le c \sum_{s=q^{r_1}}^{q^{r_2-1}} (q-1)s \prod_{z=qs}^{q^{r_2-1}} (1+(q-1)zp), \quad r_2 \ge r_1+2.
$$

Theorem 1.1.4 **(Bellman's Inequality).** *Let $x, f \in \mathcal{C}_{rd}(J)$ be nonnegative functions and c be a nonnegative constant. Then the inequality*

$$
x(t) \le c + \int_a^t f(s)x(s)\Delta s, \quad t \in J,
$$

implies the inequality

$$
x(t) \le c e_f(t,a), \quad t \in J.
$$

Proof. Let

$$
y(t) = c + \int_a^t f(s)x(s)\Delta s, \quad t \in J.
$$

Then

$$
\begin{aligned}
y(a) &= c,\\
x(t) &\le y(t), \quad t \in J,
\end{aligned}
$$

and

$$\begin{aligned} y^{\Delta}(t) &= \left(c + \int_a^t f(s)x(s)\Delta s\right)^{\Delta} \\ &= f(t)x(t) \\ &\le f(t)y(t), \quad t \in J. \end{aligned} \tag{1.1}$$

Hence,

$$\begin{aligned} (ye_{\ominus f}(\cdot, a))^{\Delta}(t) &= y^{\Delta}(t)e_{\ominus f}(\sigma(t), a) \\ &\quad + y(t)(\ominus f)(t)e_{\ominus f}(t, a) \\ &= y^{\Delta}(t)e_{\ominus f}(\sigma(t), a) \\ &\quad - \frac{y(t)f(t)}{1 + \mu(t)f(t)}e_{\ominus f}(t, a) \\ &= y^{\Delta}(t)e_{\ominus f}(\sigma(t), a) \\ &\quad - y(t)f(t)e_{\ominus f}(\sigma(t), a) \\ &= (y^{\Delta}(t) - y(t)f(t))e_{\ominus f}(\sigma(t), a). \end{aligned} \tag{1.2}$$

Since $f \in \mathcal{R}^+(J)$, then

$$1 + \mu(t)f(t) > 0, \quad t \in J,$$

and

$$\begin{aligned} 1 + \mu(t)(\ominus f)(t) &= 1 - \frac{\mu(t)f(t)}{1 + \mu(t)f(t)} \\ &= \frac{1}{1 + \mu(t)f(t)} \\ &> 0, \quad t \in J, \end{aligned}$$

i. e., $\ominus f \in \mathcal{R}^+(J)$. Therefore

$$e_{\ominus f}(t, a) > 0, \quad e_{\ominus f}(\sigma(t), a) > 0, \quad t \in J.$$

Hence, applying inequality (1.1) to inequality (1.2), we get

$$(ye_{\ominus f}(\cdot, a))^{\Delta}(t) \le 0, \quad t \in J,$$

whereupon, integrating from a to t, $t \in J$, we get

$$\begin{aligned} y(t)e_{\ominus f}(t, a) - y(a)e_{\ominus f}(a, a) &= y(t)e_{\ominus f}(t, a) - c \\ &\le 0, \quad t \in J. \end{aligned}$$

From here,

$$y(t)e_{\ominus f}(t,a) \le c, \quad t \in J,$$

or

$$\begin{aligned} y(t) &\le \frac{c}{e_{\ominus f}(t,a)} \\ &= ce_f(t,a), \quad t \in J. \end{aligned}$$

This completes the proof. □

Example 1.1.5. Let $\mathbb{T} = l\mathbb{Z}$, $l > 0$, x, f and c be as in Theorem 1.1.4. We have

$$\sigma(t) = t + l, \quad t \in \mathbb{T},$$

and

$$\begin{aligned} \int_a^t f(s)x(s)\Delta s &= l\sum_{s=a}^{t-l} f(s)x(s), \\ e_f(t,a) &= \prod_{s=a}^{t-l}(1 + lf(s)), \quad t \in J,\ t > a. \end{aligned}$$

Then, by Theorem 1.1.4, it follows that the inequality

$$x(t) \le c + l\sum_{s=a}^{t-l} f(s)x(s), \quad t \in J,\ t > a,$$

implies the inequality

$$x(t) \le c\prod_{s=a}^{t-l}(1 + lf(s)), \quad t \in J,\ t > a.$$

Example 1.1.6. Let $\mathbb{T} = q^{\mathbb{N}_0}$, $q > 1$, x, f and c be as in Theorem 1.1.4. Let also

$$a = q^{r_1}, \quad t = q^{r_2}, \quad r_1, r_2 \in \mathbb{N}_0, \quad r_2 > r_1.$$

We have

$$\sigma(t) = qt, \quad t \in \mathbb{T},$$

and

$$\int_a^t f(s)x(s)\Delta s = \sum_{s=q^{r_1}}^{q^{r_2-1}} \mu(s)f(s)x(s)$$

$$= (q-1)\sum_{s=q^{r_1}}^{q^{r_2-1}} sf(s)x(s),$$

$$e_f(t,a) = \prod_{s=q^{r_1}}^{q^{r_2-1}} (1+(q-1)sf(s)).$$

Then, by Theorem 1.1.4, it follows that the inequality

$$x(q^{r_2}) \le c\sum_{s=q^{r_1}}^{q^{r_2-1}} sf(s)x(s), \quad r_2 > r_1,$$

implies the inequality

$$x(q^{r_2}) \le c\prod_{s=q^{r_1}}^{q^{r_2-1}} (1+(q-1)sf(s)), \quad r_2 > r_1.$$

Theorem 1.1.7. *Let $x, f \in \mathcal{C}_{rd}(J)$ be nonnegative functions. Let also $g \in \mathcal{C}_{rd}(J)$ be a positive and nondecreasing function. Then the inequality*

$$x(t) \le g(t) + \int_a^t x(s)f(s)\Delta s, \quad t \in J, \tag{1.3}$$

implies the inequality

$$x(t) \le g(t)e_f(t,a), \quad t \in J.$$

Proof. Let

$$y(t) = \frac{x(t)}{g(t)}, \quad t \in J.$$

Then, using (1.3), we get

$$\begin{aligned}\frac{x(t)}{g(t)} &\le 1 + \int_a^t \frac{x(s)f(s)}{g(t)}\Delta s\\ &\le 1 + \int_a^t \frac{x(s)f(s)}{g(s)}\Delta s, \quad t \in J,\end{aligned}$$

whereupon

$$y(t) \le 1 + \int_a^t f(s)y(s)\Delta s, \quad t \in J.$$

Hence, by Theorem 1.1.4, we obtain

$$y(t) \leq e_f(t,a), \quad t \in J.$$

Therefore

$$\frac{x(t)}{g(t)} \leq e_f(t,a), \quad t \in J,$$

or

$$x(t) \leq g(t)e_f(t,a), \quad t \in J.$$

This completes the proof. □

Example 1.1.8. Let $\mathbb{T} = l\mathbb{Z}$, $l > 0$, x, f and g be as in Theorem 1.1.4. Then

$$\sigma(t) = t + l, \quad t \in \mathbb{T},$$

and

$$\int_a^t f(s)x(s)\Delta s = l \sum_{s=a}^{t-l} f(s)x(s),$$

$$e_f(t,a) = \prod_{s=a}^{t-l}(1 + lf(s)), \quad t \in J,\ t > a.$$

Then, by Theorem 1.1.4, it follows that the inequality

$$x(t) \leq g(t) + l \sum_{s=a}^{t-l} f(s)x(s), \quad t \in J,\ t > a,$$

implies the inequality

$$x(t) \leq g(t) \prod_{s=a}^{t-l}(1 + lf(s)), \quad t \in J,\ t > a.$$

Example 1.1.9. Let $\mathbb{T} = q^{\mathbb{N}_0}$, $q > 1$, x, f and g be as in Theorem 1.1.4. Let also

$$a = q^{r_1}, \quad t = q^{r_2}, \quad r_1, r_2 \in \mathbb{N}_0, \quad r_2 > r_1.$$

We have

$$\sigma(t) = qt, \quad t \in \mathbb{T},$$

and

$$\int_a^t f(s)x(s)\Delta s = \sum_{s=q^{r_1}}^{q^{r_2-1}} \mu(s)f(s)x(s)$$

$$= (q-1)\sum_{s=q^{r_1}}^{q^{r_2-1}} sf(s)x(s),$$

$$e_f(t,a) = \prod_{s=q^{r_1}}^{q^{r_2-1}} (1+(q-1)sf(s)).$$

Then, by Theorem 1.1.4, it follows that the inequality

$$x(q^{r_2}) \le g(q^{r_2}) + \sum_{s=q^{r_1}}^{q^{r_2-1}} sf(s)x(s), \quad r_2 > r_1,$$

implies the inequality

$$x(q^{r_2}) \le g(q^{r_2}) \prod_{s=q^{r_1}}^{q^{r_2-1}} (1+(q-1)sf(s)), \quad r_2 > r_1.$$

Theorem 1.1.10. *Let $x, f, g, h \in \mathcal{C}_{rd}(J)$ be nonnegative functions. Then the inequality*

$$x(t) \le f(t) + g(t)\int_a^t h(s)x(s)\Delta s, \quad t \in J, \tag{1.4}$$

implies the inequality

$$x(t) \le f(t) + g(t)\int_a^t h(s)f(s)e_{\ominus(hg)}(\sigma(s),t)\Delta s, \quad t \in J.$$

Proof. Let

$$y(t) = \int_a^t h(s)x(s)\Delta s, \quad t \in J.$$

Then, using (1.4), we get

$$x(t) \le f(t) + g(t)y(t), \quad t \in J. \tag{1.5}$$

We have

$$\begin{aligned} y^\Delta(t) &= h(t)x(t) \\ &\le h(t)(f(t) + g(t)y(t)), \quad t \in J, \end{aligned}$$

or

$$y^\Delta(t) - y(t)g(t)h(t) \le h(t)f(t), \quad t \in J,$$

and

$$\begin{aligned}
y(a) &= 0,\\
(y(\cdot)e_{\ominus(hg)}(\cdot,a))^{\Delta}(t) &= y^{\Delta}(t)e_{\ominus(hg)}(\sigma(t),a)\\
&\quad + y(t)(\ominus(hg))(t)e_{\ominus(hg)}(t,a)\\
&= y^{\Delta}(t)e_{\ominus(hg)}(\sigma(t),a)\\
&\quad - \frac{y(t)h(t)g(t)}{1+\mu(t)h(t)g(t)}e_{\ominus(hg)}(t,a)\\
&= y^{\Delta}(t)e_{\ominus(hg)}(\sigma(t),a)\\
&\quad - y(t)h(t)g(t)e_{\ominus(hg)}(\sigma(t),a)\\
&= (y^{\Delta}(t) - y(t)h(t)g(t))e_{\ominus(hg)}(\sigma(t),a)\\
&\le h(t)f(t)e_{\ominus(hg)}(\sigma(t),a), \quad t \in J.
\end{aligned}$$

We integrate the last inequality from a to t and get

$$y(t)e_{\ominus(hg)}(t,a) \le \int_a^t h(s)f(s)e_{\ominus(hg)}(\sigma(s),a)\Delta s, \quad t \in J.$$

Since $hg \in \mathcal{R}^+(J)$, we have

$$e_{\ominus(hg)}(t,a) > 0, \quad t \in J.$$

Therefore

$$\begin{aligned}
y(t) &\le \frac{1}{e_{\ominus(hg)}(t,a)}\int_a^t h(s)f(s)e_{\ominus(hg)}(\sigma(s),a)\Delta s\\
&= \int_a^t h(s)f(s)e_{\ominus(hg)}(a,t)e_{\ominus(hg)}(\sigma(s),a)\Delta s\\
&= \int_a^t h(s)f(s)e_{\ominus(hg)}(\sigma(s),t)\Delta s, \quad t \in J.
\end{aligned}$$

Hence, using (1.5), we obtain

$$\begin{aligned}
x(t) &\le f(t) + g(t)y(t)\\
&\le f(t) + g(t)\int_a^t h(s)f(s)e_{\ominus(hg)}(\sigma(s),t)\Delta s, \quad t \in J.
\end{aligned}$$

This completes the proof. □

Example 1.1.11. Let $\mathbb{T} = l\mathbb{Z}$, $l > 0$, x, f, g and h be as in Theorem 1.1.10. Then

$$\sigma(t) = t + l, \quad t \in \mathbb{T},$$

and

$$\int_a^t h(s)x(s)\Delta s = l \sum_{s=a}^{t-l} h(s)x(s), \quad t \in J,\ t > a,$$

$$\ominus(hg)(t) = -\frac{h(t)g(t)}{1 + lh(t)g(t)}, \quad t \in J,$$

$$\begin{aligned} e_{\ominus(hg)}(\sigma(s), t) &= e_{\ominus(hg)}(s + l, t) \\ &= \prod_{z=s+l}^{t-l} \left(1 + l\left(-\frac{h(z)g(z)}{1 + lh(z)g(z)}\right)\right)^{-1} \\ &= \prod_{z=s+l}^{t-l} (1 + lh(z)g(z)), \quad t \geq s + 2l, \end{aligned}$$

$$\int_a^t h(s)f(s)e_{\ominus(hg)}(\sigma(s), t)\Delta s = l \sum_{s=a}^{t-l} h(s)f(s) \prod_{z=s+l}^{t-l} (1 + lh(z)g(z)), \quad t \geq s + 2l.$$

Then, by Theorem 1.1.10, it follows that the inequality

$$x(t) \leq f(t) + lg(t) \sum_{s=a}^{t-l} h(s)x(s), \quad t > a,$$

implies the inequality

$$x(t) \leq f(t) + lg(t) \sum_{s=a}^{t-l} h(s)f(s) \prod_{z=s+l}^{t-l} (1 + lh(z)g(z)), \quad t \in J,\ t \geq a + 2l.$$

Theorem 1.1.12. *Let $x, g, h \in \mathcal{C}_{rd}(J)$ be nonnegative functions, $f \in \mathcal{C}_{rd}(J)$ be a positive nondecreasing function. Then the inequality*

$$x(t) \leq f(t) + g(t) \int_a^t h(s)x(s)\Delta s, \quad t \in J, \tag{1.6}$$

implies the inequality

$$x(t) \leq f(t)\left(1 + g(t) \int_a^t h(s)e_{\ominus(hg)}(\sigma(s), t)\Delta s\right), \quad t \in J.$$

Proof. Let

$$y(t) = \frac{x(t)}{f(t)}, \quad t \in J.$$

Then, by inequality (1.6), we get

$$\begin{aligned}\frac{x(t)}{f(t)} &\le 1+\frac{g(t)}{f(t)}\int_a^t h(s)x(s)\Delta s\\ &\le 1+g(t)\int_a^t h(s)\frac{x(s)}{f(s)}\Delta s\\ &= 1+g(t)\int_a^t y(s)h(s)\Delta s, \quad t\in J,\end{aligned}$$

or

$$y(t)\le 1+g(t)\int_a^t y(s)h(s)\Delta s, \quad t\in J.$$

Hence, using Theorem 1.1.10, we get

$$y(t)\le 1+g(t)\int_a^t h(s)e_{\ominus(hg)}(\sigma(s),t)\Delta s, \quad t\in J.$$

Therefore

$$\frac{x(t)}{f(t)}\le 1+g(t)\int_a^t h(s)e_{\ominus(hg)}(\sigma(s),t)\Delta s, \quad t\in J.$$

This completes the proof. □

Example 1.1.13. Let $\mathbb{T} = l\mathbb{Z}$, $l > 0$. Then

$$\sigma(t) = t + l, \quad \mu(t) = l, \quad t\in\mathbb{T}.$$

Suppose that $x, g, h \in \mathcal{C}_{rd}(J)$ are nonnegative functions and $f \in \mathcal{C}_{rd}(J)$ is a positive nondecreasing function. Then

$$\begin{aligned}\ominus(hg)(t) &= -\frac{h(t)g(t)}{1+\mu(t)h(t)g(t)}\\ &= -\frac{h(t)g(t)}{1+lh(t)g(t)},\\ e_{\ominus(hg)}(\sigma(s),t) &= \prod_{z=s+l}^{t-l}\left(1-\frac{lh(z)g(z)}{1+lh(z)g(z)}\right)^{-1}\\ &= \prod_{z=s+l}^{t-l}(1+lh(z)g(z)).\end{aligned}$$

Then the inequality

$$x(t) \le f(t) + g(t) \sum_{s=a}^{t-l} lh(s)x(s), \quad t \in J,$$

implies the inequality

$$\begin{aligned} x(t) &\le f(t)\left(1 + g(t) \sum_{s=a}^{t-l} lh(s)e_{\ominus(hg)}(\sigma(s), t)\right) \\ &= f(t)\left(1 + lg(t) \sum_{s=a}^{t-l} h(s) \prod_{z=s+l}^{t-l} (1 + lh(z)g(z))\right), \end{aligned}$$

$t \in J, t \ge a + 2l$.

Example 1.1.14. Let $a = 1$, $\mathbb{T} = q^{\mathbb{N}_0}$, $q > 1$. Then

$$\sigma(t) = qt, \quad \mu(t) = (q-1)t, \quad t \in \mathbb{T}.$$

Suppose that $x, g, h \in \mathcal{C}_{rd}(J)$ are nonnegative functions and $f \in \mathcal{C}_{rd}(J)$ is a positive nondecreasing function. Then the inequality

$$x(q^t) \le f(q^t) + g(q^t) \sum_{s=q^r}^{q^{t-1}} (q-1)sh(s)x(s), \quad t \in \mathbb{N}_0,\ t \ge r+1,\ q^t \in J,$$

implies the inequality

$$\begin{aligned} x(q^t) \le f(q^t)\Bigg(1 + g(q^t) &\sum_{s=q^r}^{q^{t-1}} (q-1)sh(s) \\ &\times \prod_{z=q^{r+1}}^{q^{t-1}} (1 + (q-1)zh(z)g(z))\Bigg) \end{aligned}$$

$t \in \mathbb{N}, t \ge r + 2, q^t \in J$.

Theorem 1.1.15. *Let $f, g, h, p \in \mathcal{C}_{rd}(J)$ be nonnegative functions. Then the inequality*

$$x(t) \le f(t) + g(t) \int_a^t (h(s)x(s) + p(s))\Delta s, \quad t \in J, \tag{1.7}$$

implies the inequality

$$x(t) \le f(t) + g(t) \int_a^t (h(s)f(s) + p(s))e_{\ominus(hg)}(\sigma(s), t)\Delta s, \quad t \in J.$$

Proof. Let

$$y(t) = \int_a^t (h(s)x(s) + p(s))\Delta s, \quad t \in J.$$

Then $y(a) = 0$ and

$$x(t) \le f(t) + g(t)y(t), \quad t \in J, \tag{1.8}$$

and

$$\begin{aligned} y^\Delta(t) &= h(t)x(t) + p(t) \\ &\le h(t)(f(t) + g(t)y(t)) + p(t) \\ &= h(t)f(t) + h(t)g(t)y(t) + p(t), \quad t \in J. \end{aligned}$$

Therefore

$$y^\Delta(t) - h(t)g(t)y(t) \le h(t)f(t) + p(t), \quad t \in J.$$

Hence,

$$\begin{aligned} \left(y(\cdot)e_{\ominus(hg)}(\cdot, a)\right)^\Delta(t) &= y^\Delta(t)e_{\ominus(hg)}(\sigma(t), a) \\ &\quad + (\ominus(hg))(t)y(t)e_{\ominus(hg)}(t, a) \\ &= y^\Delta(t)e_{\ominus(hg)}(\sigma(t), a) \\ &\quad - h(t)g(t)y(t)e_{\ominus(hg)}(\sigma(t), a) \\ &= (y^\Delta(t) - h(t)g(t)y(t))e_{\ominus(hg)}(\sigma(t), a) \\ &\le (h(t)f(t) + p(t))e_{\ominus(hg)}(\sigma(t), a), \quad t \in J. \end{aligned}$$

Integrating the last inequality from a to t, we get

$$y(t)e_{\ominus(hg)}(t, a) \le \int_a^t (h(s)f(s) + p(s))e_{\ominus(hg)}(\sigma(s), a)\Delta s, \quad t \in J.$$

From here,

$$\begin{aligned} y(t) &\le e_{\ominus(hg)}(a, t) \int_a^t (h(s)f(s) + p(s))e_{\ominus(hg)}(\sigma(s), a)\Delta s \\ &= \int_a^t (h(s)f(s) + p(s))e_{\ominus(hg)}(\sigma(s), t)\Delta s, \quad t \in J. \end{aligned}$$

Now, using (1.8), we obtain

$$x(t) \le f(t) + g(t)y(t)$$

$$\leq f(t) + g(t)\int_a^t (h(s)f(s) + p(s))e_{\ominus(hg)}(\sigma(s), t)\Delta s, \quad t \in J.$$

This completes the proof. □

Theorem 1.1.16. *Let $x, f, g, h \in \mathcal{C}_{rd}(J)$ be nonnegative functions. Then the inequality*

$$x(t) \geq f(z) + g(t)\int_z^t h(s)x(s)\Delta s, \quad a \leq z \leq t \leq b,$$

implies the inequality

$$x(t) \geq f(z)e_{\ominus p}(z, t), \quad a \leq z \leq t \leq b,$$

where

$$p(z) = g(t)h(z), \quad (\ominus p)(z) = -\frac{g(t)h(z)}{1 + \mu(z)g(t)h(z)}, \quad a \leq z \leq t \leq b.$$

Proof. Let

$$y(z) = x(t) - g(t)\int_z^t h(s)x(s)\Delta s, \quad a \leq z \leq t \leq b.$$

Then

$$\begin{aligned} y(t) &= x(t), \quad t \in J, \\ y(t) &\geq f(z), \\ y(t) &\leq x(t), \quad a \leq z \leq t \leq b, \end{aligned}$$

and

$$\begin{aligned} y^\Delta(z) &= g(t)h(z)x(z) \\ &\geq g(t)h(z)y(z) \\ &= p(z)y(z), \quad a \leq z \leq t \leq b. \end{aligned}$$

Then

$$y^\Delta(z) - p(z)y(z) \geq 0, \quad a \leq z \leq t \leq b.$$

We have

$$\begin{aligned} (y(\cdot)e_{\ominus p}(\cdot, a))^\Delta(z) &= y^\Delta(z)e_{\ominus p}(\sigma(z), a) \\ &\quad + y(z)(\ominus p)(z)e_{\ominus p}(z, a) \end{aligned}$$

$$\begin{aligned}&= y^{\Delta}(z)e_{\ominus p}(\sigma(z),a)\\ &\quad - y(z)p(z)e_{\ominus p}(\sigma(z),a)\\ &= (y^{\Delta}(z)-p(z)y(z))e_{\ominus p}(\sigma(z),a)\\ &\geq 0, \quad a\leq z\leq t\leq b.\end{aligned}$$

Hence,

$$\begin{aligned}y(t)e_{\ominus p}(t,a) &\geq y(z)e_{\ominus p}(z,a)\\ &\geq f(z)e_{\ominus p}(z,a), \quad a\leq z\leq t\leq b,\end{aligned}$$

or

$$\begin{aligned}x(t) &\geq f(z)\frac{e_{\ominus p}(z,a)}{e_{\ominus p}(t,a)}\\ &= f(z)e_{\ominus p}(z,t), \quad a\leq z\leq t\leq b.\end{aligned}$$

This completes the proof. □

Example 1.1.17. Let $\mathbb{T} = l\mathbb{Z}$, $l > 0$, $a, b, z, t \in \mathbb{T}$, $a \leq z \leq t \leq b$. Suppose that x, f, g, h and p are as in Theorem 1.1.16. Then

$$\begin{aligned}\int_z^t h(s)x(s)\Delta s &= \sum_{n=z}^{t-l} lh(n)x(n),\\ (\ominus p)(z) &= -\frac{g(t)h(z)}{1+lg(t)h(z)},\\ e_{\ominus p}(z,t) &= \prod_{n=z}^{t-l}\frac{1}{1-l\frac{g(t)h(n)}{1+lg(t)h(n)}}\\ &= \prod_{n=z}^{t-1}(1+lg(t)h(n)).\end{aligned}$$

Then, by Theorem 1.1.16, it follows that the inequality

$$x(t) \geq f(z) + lg(t)\sum_{n=z}^{t-l} h(n)x(n), \quad a\leq z < t\leq b,$$

implies the inequality

$$x(t) \geq f(z)\prod_{n=z}^{t-l}(1+lg(t)h(z)), \quad a\leq z < t\leq b.$$

Example 1.1.18. Let $\mathbb{T} = q^{\mathbb{N}_0}$, $q > 0$, $a, b, z, t \in \mathbb{T}$, $a \leq z \leq t \leq b$, $z = q^{r_1}$, $t = q^{r_2}$, $r_1 < r_2$, $r_1, r_2 \in \mathbb{N}$. Let also, x, f, g, h, and p be as in Theorem 1.1.16. Then

$$\int_z^t h(s)x(s)\Delta s = \int_{q^{r_1}}^{q^{r_2}} h(s)x(s)\Delta s$$

$$
\begin{aligned}
&= \sum_{n=q^{r_1}}^{q^{r_2-1}} (q-1)nh(n)x(n), \\
(\ominus p)(q^{r_1}) &= -\frac{g(q^{r_2})h(q^{r_1})}{1-(q-1)ng(q^{r_2})h(q^{r_1})}, \\
e_{\ominus p}(z,t) &= \prod_{n=q^{r_1}}^{q^{r_2-1}} \left(1+(q-1)n\left(-\frac{g(q^{r_2})h(n)}{1+(q-1)ng(q^{r_2})h(n)}\right)\right)^{-1} \\
&= \prod_{n=q^{r_1}}^{q^{r_2-1}} (1+(q-1)ng(q^{r_2})h(n)).
\end{aligned}
$$

Then, by Theorem 1.1.16, it follows that the inequality

$$x(q^{r_2}) \geq f(q^{r_1}) + g(q^{r_2}) \sum_{n=q^{r_1}}^{q^{r_2-1}} (q-1)nh(n)x(n),$$

implies the inequality

$$x(q^{r_2}) \geq f(q^{r_1}) \prod_{n=q^{r_1}}^{q^{r_2-1}} (1+(q-1)ng(q^{r_2})h(n)).$$

1.2 Volterra-type integral inequalities

Theorem 1.2.1. *Let $x, f \in C_{rd}(J)$, $k \in C_{rd}(J \times J)$ be nonnegative functions on $J \times J$, and*

$$x(t) \leq f(t) + \int_a^t k(t,s)x(s)\Delta s, \quad t \in J. \tag{1.9}$$

Let also

$$
\begin{aligned}
k_1(t,s) &= k(t,s), \\
k_l(t,s) &= \int_s^t k(t,s_1)k_{l-1}(s_1,s)\Delta s_1, \quad l \in \mathbb{N},\ l \geq 2,\ a \leq s \leq t \leq b.
\end{aligned}
$$

Suppose that

$$H(t,s) = \sum_{n=1}^{\infty} k_n(t,s)$$

is an uniformly convergent series on $a \leq s \leq t \leq b$. Then

$$x(t) \leq f(t) + \int_a^t H(t,s)f(s)\Delta s, \quad t \in J. \tag{1.10}$$

Proof. We have

$$
\begin{aligned}
x(t) &\le f(t) + \int_a^t k(t,s)x(s)\Delta s \\
&\le f(t) + \int_a^t k(t,s)\left(f(s) + \int_a^s k(s,s_1)x(s_1)\Delta s_1 \right)\Delta s \\
&= f(t) + \int_a^t k(t,s)f(s)\Delta s \\
&\quad + \int_a^t \int_a^s k(t,s)k(s,s_1)x(s_1)\Delta s_1 \Delta s \\
&= f(t) + \int_a^t k(t,s)f(s)\Delta s \\
&\quad + \int_a^t \left(\int_{s_1}^t k(t,s)k(s,s_1)\Delta s \right) x(s_1)\Delta s_1 \\
&= f(t) + \int_a^t k_1(t,s)f(s)\Delta s + \int_a^t k_2(t,s)x(s)\Delta s, \quad t \in J.
\end{aligned}
$$

Hence, for $t \in J$, we get

$$
\begin{aligned}
x(t) &\le f(t) + \int_a^t k(t,s)x(s)\Delta s \\
&\le f(t) + \int_a^t k(t,s)\left(f(s) + \int_a^s k_1(s,s_1)f(s_1)\Delta s_1 + \int_a^s k_2(s,s_1)x(s_1)\Delta s_1 \right)\Delta s \\
&= f(t) + \int_a^t k(t,s)f(s)\Delta s \\
&\quad + \int_a^t \int_a^s k(t,s)k_1(s,s_1)f(s_1)\Delta s_1 \Delta s \\
&\quad + \int_a^t \int_a^s k(t,s)k_2(s,s_1)x(s_1)\Delta s_1 \Delta s \\
&= f(t) + \int_a^t k_1(t,s)f(s)\Delta s
\end{aligned}
$$

$$
\begin{aligned}
&+\int_a^t\left(\int_{s_1}^t k(t,s)k_1(s,s_1)\Delta s\right)f(s_1)\Delta s_1\\
&+\int_a^t\left(\int_{s_1}^t k(t,s)k_2(s,s_1)\Delta s\right)x(s_1)\Delta s_1\\
&= f(t)+\int_a^t k_1(t,s)f(s)\Delta s+\int_a^t k_2(t,s)f(s)\Delta s\\
&+\int_a^t k_3(t,s)x(s)\Delta s\\
&= f(t)+\int_a^t (k_1(t,s)+k_2(t,s))f(s)\Delta s\\
&+\int_a^t k_3(t,s)x(s)\Delta s.
\end{aligned}
$$

Assume that

$$
x(t)\le f(t)+\sum_{l=1}^{n}\int_a^t k_l(t,s)f(s)\Delta s+\int_a^t k_{n+1}(t,s)x(s)\Delta s,\quad t\in J, \tag{1.11}
$$

for some $n\in\mathbb{N}$. We will prove that

$$
x(t)\le f(t)+\sum_{l=1}^{n+1}\int_a^t k_l(t,s)f(s)\Delta s+\int_a^t k_{n+2}(t,s)x(s)\Delta s,\quad t\in J.
$$

Indeed, for $t\in J$, we have

$$
\begin{aligned}
x(t)&\le f(t)+\int_a^t k(t,s)x(s)\Delta s\\
&\le f(t)+\int_a^t k(t,s)\left(f(s)+\sum_{l=1}^{n}\int_a^s k_l(s,s_1)f(s_1)\Delta s_1+\int_a^s k_{n+1}(s,s_1)x(s_1)\Delta s_1\right)\Delta s\\
&= f(t)+\int_a^t k(t,s)f(s)\Delta s\\
&+\sum_{l=1}^{n}\int_a^t\int_a^s k(t,s)k_l(s,s_1)f(s_1)\Delta s_1\Delta s\\
&+\int_a^t\int_a^s k(t,s)k_{n+1}(s,s_1)x(s_1)\Delta s_1\Delta s
\end{aligned}
$$

$$
\begin{aligned}
&= f(t) + \int_a^t k(t,s)f(s)\Delta s \\
&\quad + \sum_{l=1}^{n} \int_a^t \left(\int_{s_1}^t k(t,s)k_l(s,s_1)\Delta s \right) f(s_1)\Delta s_1 \\
&\quad + \int_a^t \left(\int_{s_1}^t k(t,s)k_{n+1}(s,s_1)\Delta s \right) x(s_1)\Delta s_1 \\
&= f(t) + \int_a^t k_1(t,s)f(s)\Delta s + \sum_{l=1}^{n} \int_a^t k_{l+1}(t,s)f(s)\Delta s \\
&\quad + \int_a^t k_{n+2}(t,s)x(s)\Delta s \\
&= f(t) + \int_a^t k_1(t,s)f(s)\Delta s \\
&\quad + \sum_{l=2}^{n+1} \int_a^t k_l(t,s)f(s)\Delta s \\
&\quad + \int_a^t k_{n+2}(t,s)x(s)\Delta s \\
&= f(t) + \sum_{l=1}^{n+1} \int_a^t k_l(t,s)f(s)\Delta s + \int_a^t k_{n+2}(t,s)x(s)\Delta s.
\end{aligned}
$$

Therefore (1.11) holds for any $n \in \mathbb{N}$. Letting $n \to \infty$ in (1.11) and using that $H(t,s) = \sum_{n=1}^{\infty} k_n(t,s)$ is a uniformly convergent series on $a \le s \le t \le b$, we get inequality (1.10). This completes the proof. □

Theorem 1.2.2. *Let $x \in \mathcal{C}_{rd}(J)$ be a nonnegative function, $k \in \mathcal{C}_{rd}(J \times J)$ be a nonnegative function and $k(t,s)$ be nondecreasing in t for each $s \in J$. If*

$$x(t) \le c + \int_a^t k(t,s)x(s)\Delta s, \quad t \in J, \tag{1.12}$$

for a nonnegative constant c, then

$$x(t) \le c e_{k(t,\cdot)}(t,a), \quad t \in J. \tag{1.13}$$

Proof. Fix a $T \in J$. Then for $a \le t \le T$, using (1.12), we get

$$x(t) \le c + \int_a^t k(T,s)x(s)\Delta s.$$

We apply Theorem 1.1.4 for $t \in [a, T]$ and get

$$x(t) \le ce_{k(T,\cdot)}(t,a), \quad t \in [a,T].$$

In particular,

$$x(T) \le ce_{k(T,\cdot)}(T,a).$$

Since $T \in J$ was arbitrarily chosen, we get inequality (1.13). This completes the proof. □

Theorem 1.2.3. *Let $x \in \mathcal{C}_{rd}(J)$ be a nonnegative function, $k \in \mathcal{C}_{rd}(J \times J)$ be a nonnegative function and $k(t,s)$ be nondecreasing in t for each $s \in J$. Let also, $g \in \mathcal{C}_{rd}(J)$ be a positive and nondecreasing function. If*

$$x(t) \le g(t) + \int_a^t k(t,s)x(s)\Delta s, \quad t \in J, \tag{1.14}$$

then

$$x(t) \le g(t)e_{k(t,\cdot)}(t,a), \quad t \in J. \tag{1.15}$$

Proof. Let $T \in J$ be arbitrarily chosen and fixed. Then, by (1.14), we obtain

$$x(t) \le g(t) + \int_a^t k(T,s)x(s)\Delta s, \quad a \le t \le T.$$

We apply Theorem 1.1.7 for $a \le t \le T$ and get

$$x(t) \le g(t)e_{k(T,\cdot)}(t,a).$$

In particular, when $t = T$, we obtain

$$x(T) \le g(T)e_{k(T,\cdot)}(T,a).$$

Because $T \in J$ was arbitrarily chosen, we obtain inequality (1.15). This completes the proof. □

Theorem 1.2.4. *Let $x, f, g, h, p \in \mathcal{C}_{rd}(J)$ be nonnegative functions, $k \in \mathcal{C}_{rd}(J \times J)$ be a nonnegative function such that $k_t^\Delta(t,s)$ exists for any $t,s \in J$, and $k_t^\Delta(t,s) \ge 0$ for any $t,s \in J$. Let also*

$$a(t) = k(\sigma(t),t)h(t)g(t) + \int_a^t k_t^\Delta(t,s)h(s)g(s)\Delta s,$$

$$b(t) = k(\sigma(t), t)(h(t)f(t) + p(t)) + \int_a^t k_t^\Delta(t,s)(h(s)f(s) + p(s))\Delta s,$$

$$B(t) = \int_a^t b(s)\Delta s, \quad t \in J.$$

Suppose that $a \in \mathcal{R}^+(J)$ and

$$x(t) \le f(t) + g(t)\int_a^t k(t,s)(h(s)x(s) + p(s))\Delta s, \quad t \in J. \tag{1.16}$$

Then

$$x(t) \le B(t) + \int_a^t a(s)B(s)e_{\ominus a}(\sigma(s), t)\Delta s, \quad t \in J. \tag{1.17}$$

Proof. Let

$$z(t) = \int_a^t k(t,s)(h(s)x(s) + p(s))\Delta s, \quad t \in J.$$

Then

$$z(a) = 0$$

and

$$z^\Delta(t) = k(\sigma(t), t)(h(t)x(t) + p(t)) + \int_a^t k_t^\Delta(t,s)(h(s)x(s) + p(s))\Delta s \ge 0, \quad t \in J. \tag{1.18}$$

Therefore z is nonnegative and nondecreasing on J. By (1.16), we obtain

$$x(t) \le f(t) + g(t)z(t), \quad t \in J.$$

From (1.18), we get

$$z^\Delta(t) \le k(\sigma(t), t)(h(t)f(t) + h(t)g(t)z(t) + p(t)) + \int_a^t k_t^\Delta(t,s)(h(s)f(s) + h(s)g(s)z(s) + p(s))\Delta s$$

$$\begin{aligned}
&\leq k(\sigma(t),t)(h(t)f(t)+p(t)) \\
&\quad + k(\sigma(t),t)h(t)g(t)z(t) \\
&\quad + \int_a^t k_t^\Delta(t,s)(h(s)f(s)+p(s))\Delta s \\
&\quad + \left(\int_a^t k_t^\Delta(t,s)h(s)g(s)\Delta s\right)z(t) \\
&= \left(k(\sigma(t),t)h(t)g(t) + \int_a^t k_t^\Delta(t,s)h(s)g(s)\Delta s\right)z(t) \\
&\quad + k(\sigma(t),t)(h(t)f(t)+p(t)) \\
&\quad + \int_a^t k_t^\Delta(t,s)(h(s)f(s)+p(s))\Delta s \\
&= a(t)z(t) + b(t), \quad t \in J.
\end{aligned}$$

Integrating the last inequality, we get

$$\begin{aligned}
z(t) &\leq \int_a^t b(s)\Delta s + \int_a^t a(s)z(s)\Delta s \\
&= B(t) + \int_a^t a(s)z(s)\Delta s, \quad t \in J.
\end{aligned}$$

Hence, using Theorem 1.1.10, we obtain

$$x(t) \leq B(t) + \int_a^t a(s)B(s)e_{\ominus a}(\sigma(s),t)\Delta s, \quad t \in J,$$

i. e., we get (1.17). This completes the proof. □

1.3 The inequalities of Gamidov and Rodrigues

Theorem 1.3.1 (Gamidov's Inequality). *Let $x, f, g_i, h_i \in C_{rd}(J)$, $i \in \{1,\dots,n\}$, be nonnegative functions, and*

$$g(t) = \sup_{i\in\{1,\dots,n\}} g_i(t), \quad h(t) = \sum_{i=1}^n h_i(t), \quad t \in J.$$

Then the inequality

$$x(t) \leq f(t) + \sum_{i=1}^n g_i(t)\int_a^t h_i(s)x(s)\Delta s, \quad t \in J, \tag{1.19}$$

implies the inequality

$$x(t) \le f(t) + g(t) \int_a^t h(s)f(s)e_{\ominus(hg)}(\sigma(s), t)\Delta s, \quad t \in J. \tag{1.20}$$

Proof. By inequality (1.19), we get

$$\begin{aligned} x(t) &\le f(t) + g(t) \int_a^t \sum_{i=1}^n h_i(s)x(s)\Delta s \\ &= f(t) + g(t) \int_a^t h(s)x(s)\Delta s, \quad t \in J. \end{aligned}$$

Hence, from Theorem 1.1.10, we get inequality (1.20). This completes the proof. □

Theorem 1.3.2 (Gamidov's Inequality). *Let* $n \in \mathbb{N}$, $n \ge 2$, $x, f, g_1, g_2, h_i \in \mathcal{C}_{rd}(J)$, $i \in \{1, \ldots, n\}$, *be nonnegative functions,*

$$a = t_1 \le t_2 \le \cdots \le t_n = b,$$

c_i, $i \in \{1, \ldots, n\}$, *be nonnegative constants,*

$$\begin{aligned} m_i &= c_i \int_{t_1}^{t_i} h_i(s)x(s)\Delta s, \quad i \in \{2, \ldots, n\}, \\ F(t) &= f(t) + g_2(t) \sum_{i=2}^n m_i, \\ p_1(t) &= f(t) + g_1(t) \int_a^t h_1(s)f(s)e_{\ominus(h_1 g_1)}(\sigma(s), t)\Delta s, \\ p_2(t) &= g_2(t) + g_1(t) \int_a^t h_1(s)g_2(s)e_{\ominus(h_2 g_1)}(\sigma(s), t)\Delta s. \end{aligned}$$

Suppose that

$$M^{-1} = 1 - \sum_{j=2}^n c_j \int_{t_1}^{t_j} h_j(s)p_2(s)\Delta s > 0.$$

Then the inequality

$$x(t) \le f(t) + g_1(t) \int_{t_1}^t h_1(s)x(s)\Delta s, \quad t \in J, \tag{1.21}$$

implies the inequality

$$x(t) \le p_1(t) + Mp_2(t)\sum_{i=2}^{n} c_i \int_{t_1}^{t_i} h_i(s)p_1(s)\Delta s, \quad t \in J. \tag{1.22}$$

Proof. Using inequality (1.21), we obtain

$$\begin{aligned} x(t) &\le \left(f(t) + g_2(t)\sum_{i=2}^{n} m_i \right) + g_1(t)\int_{t_1}^{t} h_1(s)x(s)\Delta s \\ &= F(t) + g_1(t)\int_{t_1}^{t} h_1(s)x(s)\Delta s, \quad t \in J. \end{aligned}$$

Hence, by Theorem 1.1.10, we obtain

$$\begin{aligned} x(t) &\le F(t) + g_1(t)\int_{t_1}^{t} h_1(s)F(s)e_{\ominus(h_1g_1)}(\sigma(s), t)\Delta s \\ &= f(t) + g_2(t)\sum_{i=2}^{n} m_i \\ &\quad + g_1(t)\int_{t_1}^{t} h_1(s)\left(f(s) + g_2(s)\sum_{i=2}^{n} m_i \right)e_{\ominus(h_1g_1)}(\sigma(s), t)\Delta s \\ &= f(t) + g_1(t)\int_{t_1}^{t} h_1(s)f(s)e_{\ominus(h_1g_1)}(\sigma(s), t)\Delta s \\ &\quad + \sum_{i=2}^{n} m_i\left(g_2(t) + g_1(t)\int_{t_1}^{t} h_1(s)g_2(s)e_{\ominus(h_1g_1)}(\sigma(s), t)\Delta s \right) \\ &= p_1(t) + p_2(t)\sum_{i=2}^{n} m_i, \quad t \in J. \end{aligned} \tag{1.23}$$

Note that

$$\begin{aligned} \sum_{i=2}^{n} m_i &= \sum_{i=2}^{n} c_i \int_{t_1}^{t_i} h_i(s)x(s)\Delta s \\ &\le \sum_{i=2}^{n} c_i \int_{t_1}^{t_i} h_i(s)\left(p_1(s) + p_2(s)\sum_{i=2}^{n} m_i \right)\Delta s \\ &= \sum_{i=2}^{n} c_i \int_{t_1}^{t_i} h_i(s)p_1(s)\Delta s \end{aligned}$$

$$+\sum_{i=2}^{n} m_i\left(\sum_{j=2}^{n} c_j \int_{t_1}^{t_j} h_j(s)p_2(s)\Delta s\right),$$

or

$$\sum_{i=2}^{n} m_i\left(1-\sum_{j=2}^{n} c_j \int_{t_1}^{t_j} h_j(s)p_2(s)\Delta s\right) \le \sum_{i=2}^{n} c_i \int_{t_1}^{t_i} h_i(s)p_1(s)\Delta s.$$

Therefore

$$\sum_{i=2}^{n} m_i \le M\sum_{i=2}^{n} c_i \int_{t_1}^{t_i} h_i(s)p_1(s)\Delta s.$$

Hence from (1.23), we get inequality (1.22). This completes the proof. □

We denote $\mathbb{R}_+ = [0,\infty)$. Below we suppose that $\mathbb{R}_+ \cap \mathbb{T} \neq \emptyset$ and $\sup \mathbb{T} = \infty$.

Theorem 1.3.3. *Let $x, f \in C_{rd}(\mathbb{R}_+ \cap \mathbb{T})$ be nonnegative functions such that*

$$1-\mu(t)f(t) > 0, \quad t \in \mathbb{R}_+ \cap \mathbb{T},$$

and

$$\left|\int_0^\infty \log(1-\mu(t)f(t))\Delta t\right| < \infty.$$

Let also c be a nonnegative constant. Then the inequality

$$x(t) \le c + \int_t^\infty f(s)x(s)\Delta s, \quad t \in \mathbb{R}_+ \cap \mathbb{T}, \tag{1.24}$$

implies the inequality

$$x(t) \le cf(t)e_g(\infty, t), \quad t \in \mathbb{R}_+ \cap \mathbb{T},$$

where

$$g(t) = \frac{f(t)}{1-\mu(t)f(t)}, \quad t \in \mathbb{R}_+ \cap \mathbb{T}.$$

Proof. Let

$$y(t) = c + \int_t^\infty f(s)x(s)\Delta s, \quad t \in \mathbb{R}_+ \cap \mathbb{T}.$$

Then

$$\lim_{t\to\infty} y(t) = c$$

and, using (1.24), we have

$$x(t) \le y(t), \quad t \in \mathbb{R}_+ \cap \mathbb{T}. \tag{1.25}$$

Also, we have

$$\begin{aligned} y^\Delta(t) &= -f(t)x(t) \\ &\ge -f(t)y(t), \end{aligned}$$

or

$$y^\Delta(t) + f(t)y(t) \ge 0, \quad t \in \mathbb{R}_+ \cap \mathbb{T}.$$

Using the last inequality, we get

$$\begin{aligned} (y(t)e_g(t,\infty))^\Delta &= y^\Delta(t)e_g(\sigma(t),\infty) \\ &\quad + y(t)g(t)e_g(t,\infty) \\ &= y^\Delta(t)e_g(\sigma(t),\infty) \\ &\quad + \frac{y(t)g(t)}{1+\mu(t)g(t)}(1+\mu(t)g(t))e_g(t,\infty) \\ &= y^\Delta(t)e_g(\sigma(t),\infty) \\ &\quad + \frac{y(t)g(t)}{1+\mu(t)g(t)}e_g(\sigma(t),\infty) \\ &= \left(y^\Delta(t) + \frac{g(t)}{1+\mu(t)g(t)}y(t)\right)e_g(\sigma(t),\infty) \\ &= \left(y^\Delta(t) + \frac{\frac{f(t)}{1-\mu(t)f(t)}}{1+\frac{\mu(t)f(t)}{1-\mu(t)f(t)}}y(t)\right)e_g(\sigma(t),\infty) \\ &= (y^\Delta(t) + f(t)y(t))e_g(\sigma(t),\infty). \end{aligned} \tag{1.26}$$

Observe that

$$\begin{aligned} e_g(\sigma(t),\infty) &= e^{\int_\infty^{\sigma(t)} \frac{1}{\mu(s)}\log(1+\mu(s)g(s))\Delta s} \\ &= e^{\int_\infty^{\sigma(t)} \frac{1}{\mu(s)}\log(1+\frac{\mu(s)f(s)}{1-\mu(s)f(s)})\Delta s} \\ &= e^{\int_\infty^{\sigma(t)} \frac{1}{\mu(s)}\log(\frac{1}{1-\mu(s)f(s)})\Delta s} \\ &= e^{-\int_\infty^{\sigma(t)} \frac{1}{\mu(s)}\log(1-\mu(s)f(s))\Delta s} \\ &= e^{\int_{\sigma(t)}^\infty \frac{1}{\mu(s)}\log(1-\mu(s)f(s))\Delta s} \\ &\ge 0, \quad t \in \mathbb{R}_+ \cap \mathbb{T}. \end{aligned}$$

Hence, by (1.26), we obtain that

$$(y(t)e_g(t,\infty))^\Delta \geq 0, \quad t \in \mathbb{R}_+ \cap \mathbb{T}.$$

We integrate the last inequality from t to ∞ and get

$$\lim_{t\to\infty}(y(t)e_g(t,\infty)) - y(t)e_g(t,\infty) \geq 0,$$

or

$$y(t)e_g(t,\infty) \leq c,$$

or

$$y(t) \leq ce_g(\infty,t), \quad t \in \mathbb{R}_+ \cap \mathbb{T}.$$

From here and from (1.25), we obtain

$$x(t) \leq ce_g(\infty,t), \quad t \in \mathbb{R}_+ \cap \mathbb{T}.$$

This completes the proof. □

Theorem 1.3.4. *Let $x, g \in \mathcal{C}_{rd}(\mathbb{R}_+ \cap \mathbb{T})$ be nonnegative functions. Let also $f \in \mathcal{C}_{rd}(\mathbb{R}_+ \cap \mathbb{T})$ be a positive decreasing function and*

$$1 - \mu(t)g(t) > 0, \quad t \in \mathbb{R}_+ \cap \mathbb{T},$$

$$\left|\int_t^\infty \log(1-\mu(s)g(s))\Delta s\right| < \infty.$$

Then the inequality

$$x(t) \leq f(t) + \int_t^\infty g(s)x(s)\Delta s, \quad t \in \mathbb{R}_+ \cap \mathbb{T},$$

implies the inequality

$$x(t) \leq f(t)g(t)e_h(\infty,t), \quad t \in \mathbb{R}_+ \cap \mathbb{T},$$

where

$$h(t) = \frac{g(t)}{1-\mu(t)g(t)}, \quad t \in \mathbb{R}_+ \cap \mathbb{T}.$$

Proof. Using that f is decreasing and nonnegative on $\mathbb{R}_+ \cap \mathbb{T}$, we have

$$\frac{x(t)}{f(t)} \leq 1 + \int_t^\infty \frac{g(s)}{f(t)}x(s)\Delta s$$

$$\le 1 + \int_t^\infty \frac{g(s)}{f(s)} x(s)\Delta s, \quad t \in \mathbb{R}_+ \cap \mathbb{T}.$$

Hence, by Theorem 1.3.3, we obtain

$$\frac{x(t)}{f(t)} \le g(t) e_h(\infty, t),$$

or

$$x(t) \le f(t) g(t) e_h(\infty, t), \quad t \in \mathbb{R}_+ \cap \mathbb{T}.$$

This completes the proof. □

Theorem 1.3.5. *Let $x, f, g \in \mathcal{C}_{rd}(\mathbb{R}_+ \cap \mathbb{T})$ be nonnegative functions, $y \in \mathcal{C}_{rd}(\mathbb{R}_+ \cap \mathbb{T})$ be a positive decreasing function, c be a nonnegative constant,*

$$\begin{aligned} 1 - \mu(t)g(t) &> 0, \\ 1 - \beta - \mu(t)g(t) &> 0, \\ \left| \int_t^\infty \log\Big(1 - \beta - \mu(s)g(s)\Big)\Delta s \right| &< \infty, \quad t \in \mathbb{R}_+ \cap \mathbb{T}, \\ \left| \int_0^\infty \log(1 - \mu(t)g(t))\Delta t \right| &< \infty, \end{aligned}$$

and yx be a bounded function on $\mathbb{R}_+ \cap \mathbb{T}$. Suppose also that there exists a sufficiently large $a \in \mathbb{R}_+ \cap \mathbb{T}$ such that

$$\beta = \int_a^\infty f(s)\Delta s + \int_a^\infty g(s)\Delta s < 1.$$

Then the inequality

$$x(t) \le c + \int_a^t f(s)x(s)\Delta s + \frac{1}{y(t)} \int_t^\infty y(s)g(s)x(s)\Delta s, \quad t \in \mathbb{R}_+ \cap \mathbb{T},$$

implies the inequality

$$x(t) \le c \frac{g(t)}{(1-\beta)^2} e_a(\infty, t), \quad t \in \mathbb{R}_+ \cap \mathbb{T},\ t \ge a,$$

where

$$h(t) = \frac{g(t)}{1 - \beta - \mu(t)g(t)}, \quad t \in \mathbb{R}_+ \cap \mathbb{T},\ t \ge a,$$

and

$$x(t) \le cg(t)e_{h_1}(\infty, t), \quad t \in \mathbb{R}_+ \cap \mathbb{T},\ t \le a,$$

where

$$h_1(t) = \frac{g(t)}{1 - \mu(t)g(t)}, \quad t \in \mathbb{R}_+ \cap \mathbb{T},\ t \le a.$$

Proof.

1. Let $t \ge a$, $t \in \mathbb{R}_+ \cap \mathbb{T}$. We set

$$y(t) = \max_{a \le s \le t} x(s).$$

Then y is an increasing function on $[a, \infty) \cap \mathbb{T}$, $y \in \mathcal{C}_{rd}([a, \infty) \cap \mathbb{T})$ and

$$x(t) \le y(t), \quad t \in \mathbb{R}_+ \cap \mathbb{T},\ t \ge a.$$

Also, yy is a bounded function on $(\mathbb{R}_+ \cap \mathbb{T}) \cap [a, \infty)$. For a given $t \ge a$, there exists $t_1 \in [a, t]$ such that

$$y(t) = x(t_1).$$

Hence,

$$\begin{aligned}
y(t) &= x(t_1) \\
&\le c + \int_a^{t_1} f(s)x(s)\Delta s + \frac{1}{y(t_1)} \int_{t_1}^{\infty} y(s)g(s)x(s)\Delta s \\
&\le c + \int_a^{t_1} f(s)y(s)\Delta s + \frac{1}{y(t_1)} \int_{t_1}^{\infty} y(s)g(s)y(s)\Delta s \\
&= c + \int_a^{t_1} f(s)y(s)\Delta s + \frac{1}{y(t_1)} \int_{t_1}^{t} y(s)g(s)y(s)\Delta s \\
&\quad + \frac{1}{y(t_1)} \int_t^{\infty} y(s)g(s)y(s)\Delta s \\
&\le c + \int_a^{t_1} f(s)y(s)\Delta s + \frac{1}{y(t_1)} \int_{t_1}^{t} y(t_1)g(s)y(s)\Delta s \\
&\quad + \frac{1}{y(t_1)} \int_t^{\infty} y(s)g(s)y(s)\Delta s
\end{aligned}$$

$$
\begin{aligned}
&\le c + \int_a^{t_1} f(s)y(s)\Delta s + \int_{t_1}^{t} g(s)y(s)\Delta s \\
&\quad + \frac{1}{y(t)} \int_t^{\infty} y(s)g(s)y(s)\Delta s \\
&\le c + \int_a^{t_1} f(s)y(s)\Delta s + y(t) \int_{t_1}^{t} g(s)\Delta s \\
&\quad + \frac{1}{y(t)} \int_t^{\infty} y(s)g(s)y(s)\Delta s \\
&\le c + \int_a^{t_1} f(s)y(s)\Delta s + y(t) \int_a^{\infty} g(s)\Delta s \\
&\quad + \frac{1}{y(t)} \int_t^{\infty} y(s)g(s)y(s)\Delta s \\
&\le c + y(t) \int_a^{t_1} f(s)\Delta s + y(t) \int_a^{\infty} g(s)\Delta s \\
&\quad + \frac{1}{y(t)} \int_t^{\infty} y(s)g(s)y(s)\Delta s \\
&\le c + y(t)\left(\int_a^{\infty} f(s)\Delta s + \int_a^{\infty} g(s)\Delta s \right) + \frac{1}{y(t)} \int_t^{\infty} y(s)g(s)y(s)\Delta s \\
&= c + \beta y(t) + \frac{1}{y(t)} \int_t^{\infty} y(s)g(s)y(s)\Delta s, \quad t \in \mathbb{R}_+ \cap \mathbb{T},\ t \ge a.
\end{aligned}
$$

Hence,

$$
(1-\beta)y(t) \le c + \frac{1}{y(t)} \int_t^{\infty} y(s)g(s)y(s)\Delta s
$$

and

$$
y(t)y(t) \le \frac{c}{1-\beta}y(t) + \frac{1}{1-\beta} \int_t^{\infty} y(s)g(s)y(s)\Delta s, \quad t \in \mathbb{R}_+ \cap \mathbb{T},\ t \ge a.
$$

Hence, by Theorem 1.3.4, it follows that

$$
y(t)y(t) \le cy(t)\frac{g(t)}{(1-\beta)^2} e_h(\infty, t),
$$

or

$$y(t) \leq c\frac{g(t)}{(1-\beta)^2}e_h(\infty, t), \quad t \in \mathbb{R}_+ \cap \mathbb{T},\ t \geq a.$$

Therefore

$$x(t) \leq c\frac{g(t)}{(1-\beta)^2}e_h(\infty, t), \quad t \in \mathbb{R}_+ \cap \mathbb{T},\ t \geq a.$$

2. Let $t \in \mathbb{R}_+ \cap \mathbb{T}$, $t \leq a$. Then

$$\int_a^t f(s)x(s)\Delta s \leq 0$$

and

$$\begin{aligned} x(t) &\leq c + \frac{1}{y(t)}\int_t^\infty y(s)g(s)x(s)\Delta s \\ &\leq c + \int_t^\infty g(s)x(s)\Delta s, \quad t \in \mathbb{R}_+ \cap \mathbb{T},\ t \leq a. \end{aligned}$$

Hence, by Theorem 1.3.3, it follows that

$$x(t) \leq cg(t)e_{h_1}(\infty, t), \quad t \in \mathbb{R}_+ \cap \mathbb{T},\ t \leq a.$$

This completes the proof. □

1.4 Simultaneous inequalities

In this section we suppose that $0 \in \mathbb{T}$.

Theorem 1.4.1. *Let $x, y, h_1, h_2, h_3, h_4 \in C_{rd}(\mathbb{R}_+ \cap \mathbb{T})$ be nonnegative functions, λ, k_1, and k_2 be nonnegative constants, and*

$$\begin{aligned} x(t) &\leq k_1 + \int_0^t h_1(s)x(s)\Delta s + \int_0^t e_\lambda(s,0)h_2(s)y(s)\Delta s, \\ y(t) &\leq k_2 + \int_0^t e_{\ominus\lambda}(s,0)h_3(s)x(s)\Delta s + \int_0^t h_4(s)y(s)\Delta s, \quad t \in \mathbb{R}_+ \cap \mathbb{T}. \end{aligned} \tag{1.27}$$

Then

$$\begin{aligned} x(t) &\leq (k_1M + k_2)e_{\lambda\oplus h}(t, 0), \\ y(t) &\leq (k_1M + k_2)e_h(t, 0), \quad t \in \mathbb{R}_+ \cap \mathbb{T}, \end{aligned}$$

where

$$h(t) = \max\{h_1(t) + h_3(t), \quad h_2(t) + h_4(t)\},$$
$$M = \max_{t\geq 0} e_{\ominus\lambda}(t,0), \quad t \in \mathbb{R}_+ \cap \mathbb{T}.$$

Proof. Since $\lambda \geq 0$, we have

$$e_\lambda(s,0) \leq e_\lambda(t,0), \quad t \geq s,\ s,t \in \mathbb{R}_+ \cap \mathbb{T},$$

and

$$e_{\ominus\lambda}(t,0) \leq e_{\ominus\lambda}(s,0), \quad t \geq s,\ s,t \in \mathbb{R}_+ \cap \mathbb{T}.$$

Hence,

$$\begin{aligned} x(t) &\leq k_1 + \int_0^t h_1(s)x(s)\Delta s + \int_0^t e_\lambda(t,0)h_2(s)y(s)\Delta s \\ &= k_1 + \int_0^t h_1(s)x(s)\Delta s + e_\lambda(t,0)\int_0^t h_2(s)y(s)\Delta s, \end{aligned}$$

whereupon

$$\begin{aligned} e_{\ominus\lambda}(t,0)x(t) &\leq k_1 e_{\ominus\lambda}(t,0) + e_{\ominus\lambda}(t,0)\int_0^t h_1(s)x(s)\Delta s \\ &\quad + \int_0^t h_2(s)y(s)\Delta s \\ &\leq k_1 e_{\ominus\lambda}(t,0) + \int_0^t e_{\ominus\lambda}(s,0)h_1(s)x(s)\Delta s \\ &\quad + \int_0^t h_2(s)y(s)\Delta s \\ &\leq Mk_1 + \int_0^t e_{\ominus\lambda}(s,0)h_1(s)x(s)\Delta s \\ &\quad + \int_0^t h_2(s)y(s)\Delta s, \quad t \in \mathbb{R}_+ \cap \mathbb{T}. \end{aligned} \tag{1.28}$$

Let

$$F(t) = e_{\ominus\lambda}(t,0)x(t) + y(t), \quad t \in \mathbb{R}_+ \cap \mathbb{T}.$$

Hence, after adding (1.28) and the second inequality of (1.27), we obtain

$$\begin{aligned}
F(t) &\leq (k_1 M + k_2) + \int_0^t e_{\ominus\lambda}(s,0)(h_1(s) + h_3(s))x(s)\Delta s \\
&\quad + \int_0^t (h_2(s) + h_4(s))y(s)\Delta s \\
&\leq (k_1 M + k_2) + \int_0^t e_{\ominus\lambda}(s,0)x(s)h(s)\Delta s \\
&\quad + \int_0^t h(s)y(s)\Delta s \\
&= (k_1 M + k_2) + \int_0^t F(s)h(s)\Delta s, \quad t \in \mathbb{R}_+ \cap \mathbb{T}.
\end{aligned}$$

From the last inequality and Theorem 1.1.4, we get

$$F(t) \leq (k_1 M + k_2)e_h(t,0), \quad t \in \mathbb{R}_+ \cap \mathbb{T}.$$

Therefore

$$e_{\ominus\lambda}(t,0)x(t) \leq (k_1 M + k_2)e_h(t,0),$$

or

$$\begin{aligned}
x(t) &\leq (k_1 M + k_2)e_\lambda(t,0)e_h(t,0) \\
&= (k_1 M + k_2)e_{\lambda\oplus h}(t,0), \quad t \in \mathbb{R}_+ \cap \mathbb{T},
\end{aligned}$$

and

$$y(t) \leq (k_1 M + k_2)e_h(t,0), \quad t \in \mathbb{R}_+ \cap \mathbb{T}.$$

This completes the proof. □

Theorem 1.4.2. *Let $x, y, a, b, h_1, h_2, h_3, h_4 \in C_{rd}(\mathbb{R}_+ \cap \mathbb{T})$ be nonnegative functions, λ be a nonnegative constant, and*

$$\begin{aligned}
x(t) &\leq a(t) + p(t)\int_0^t h_1(s)x(s)\Delta s + p(t)\int_0^t e_\lambda(s,0)h_2(s)y(s)\Delta s, \\
y(t) &\leq b(t) + p(t)\int_0^t e_{\ominus\lambda}(s,0)h_3(s)x(s)\Delta s + p(t)\int_0^t h_4(s)y(s)\Delta s, \quad t \in \mathbb{R}_+ \cap \mathbb{T}.
\end{aligned} \tag{1.29}$$

Then

$$x(t) \leq \left(a(t) + b(t)e_\lambda(t,0)\right) + p(t)e_\lambda(t,0)\int_0^t h(s)\left(a(s)e_{\ominus\lambda}(s,0) + b(s)\right)e_{\ominus(hp)}(\sigma(s),t)\Delta s,$$

and

$$y(t) \leq \left(a(t)e_{\ominus\lambda}(t,0) + b(t)\right) + p(t)\int_0^t h(s)\left(a(s)e_{\ominus\lambda}(s,0) + b(s)\right)e_{\ominus(hp)}(\sigma(s),t)\Delta s,$$

$t \in \mathbb{R}_+ \cap \mathbb{T}$, *where*

$$h(t) = \max\{h_1(t) + h_3(t), \quad h_2(t) + h_4(t)\}, \quad t \in \mathbb{R}_+ \cap \mathbb{T}.$$

Proof. Since $\lambda \geq 0$, we have

$$e_\lambda(s,0) \leq e_\lambda(t,0), \quad t \geq s, \ s,t \in \mathbb{R}_+ \cap \mathbb{T},$$

and

$$e_{\ominus\lambda}(t,0) \leq e_{\ominus\lambda}(s,0), \quad t \geq s, \ s,t \in \mathbb{R}_+ \cap \mathbb{T}.$$

Hence,

$$\begin{aligned} x(t) &\leq a(t) + p(t)\int_0^t h_1(s)x(s)\Delta s + p(t)\int_0^t e_\lambda(t,0)h_2(s)y(s)\Delta s \\ &= a(t) + p(t)\int_0^t h_1(s)x(s)\Delta s + e_\lambda(t,0)p(t)\int_0^t h_2(s)y(s)\Delta s, \end{aligned}$$

whereupon

$$\begin{aligned} e_{\ominus\lambda}(t,0)x(t) &\leq a(t)e_{\ominus\lambda}(t,0) + e_{\ominus\lambda}(t,0)p(t)\int_0^t h_1(s)x(s)\Delta s \\ &\quad + p(t)\int_0^t h_2(s)y(s)\Delta s \\ &\leq a(t)e_{\ominus\lambda}(t,0) + p(t)\int_0^t e_{\ominus\lambda}(s,0)h_1(s)x(s)\Delta s \\ &\quad + p(t)\int_0^t h_2(s)y(s)\Delta s, \quad t \in \mathbb{R}_+ \cap \mathbb{T}. \end{aligned} \tag{1.30}$$

Let

$$F(t) = e_{\ominus\lambda}(t,0)x(t) + y(t), \quad t \in \mathbb{R}_+ \cap \mathbb{T}.$$

Hence, after adding (1.30) and the second inequality of (1.29), we obtain

$$\begin{aligned}
F(t) &\le (a(t)e_{\ominus\lambda}(t,0) + b(t)) + p(t)\int_0^t e_{\ominus\lambda}(s,0)(h_1(s) + h_3(s))x(s)\Delta s \\
&\quad + p(t)\int_0^t (h_2(s) + h_4(s))y(s)\Delta s \\
&\le (a(t)e_{\ominus\lambda}(t,0) + b(t)) + p(t)\int_0^t e_{\ominus\lambda}(s,0)x(s)h(s)\Delta s \\
&\quad + p(t)\int_0^t h(s)y(s)\Delta s \\
&= (a(t)e_{\ominus\lambda}(t,0) + b(t)) + p(t)\int_0^t F(s)h(s)\Delta s, \quad t \in \mathbb{R}_+ \cap \mathbb{T}.
\end{aligned}$$

From the last inequality and Theorem 1.1.10, we get

$$F(t) \le (a(t)e_{\ominus\lambda}(t,0)+b(t))+p(t)\int_0^t h(s)(a(s)e_{\ominus\lambda}(s,0)+b(s))e_{\ominus(hp)}(\sigma(s),t)\Delta s, \quad t \in \mathbb{R}_+\cap\mathbb{T}.$$

Therefore

$$e_{\ominus\lambda}(t,0)x(t) \le (a(t)e_{\ominus\lambda}(t,0) + b(t)) + p(t)\int_0^t h(s)(a(s)e_{\ominus\lambda}(s,0) + b(s))e_{\ominus(hp)}(\sigma(s),t)\Delta s,$$

or

$$x(t) \le (a(t) + b(t)e_{\lambda}(t,0)) + p(t)e_{\lambda}(t,0)\int_0^t h(s)(a(s)e_{\ominus\lambda}(s,0) + b(s))e_{\ominus(hp)}(\sigma(s),t)\Delta s,$$

and

$$y(t) \le (a(t)e_{\ominus\lambda}(t,0) + b(t)) + p(t)\int_0^t h(s)(a(s)e_{\ominus\lambda}(s,0) + b(s))e_{\ominus(hp)}(\sigma(s),t)\Delta s,$$

$t \in \mathbb{R}_+ \cap \mathbb{T}$. This completes the proof. □

1.5 Pachpatte's inequalities

Theorem 1.5.1 (Pachpatte's Inequality). *Let $x, f, g \in \mathcal{C}_{rd}(\mathbb{R}_+ \cap \mathbb{T})$ be nonnegative functions and*

$$x(t) \le c + \int_0^t f(s)x(s)\Delta s + \int_0^t f(s)\left(\int_0^s g(y)x(y)\Delta y\right)\Delta s,$$

$t \in \mathbb{R}_+ \cap \mathbb{T}$, where c is a nonnegative constant. Then

$$x(t) \le c e_{f+g}(t,0)$$

and

$$x(t) \le c\left(1 + \int_0^t f(s)e_{f+g}(s,0)\Delta s\right), \quad t \in \mathbb{R}_+ \cap \mathbb{T}.$$

Proof. Let

$$\begin{aligned} z(t) &= c + \int_0^t f(s)x(s)\Delta s \\ &\quad + \int_0^t f(s)\left(\int_0^s g(y)x(y)\Delta y\right)\Delta s, \quad t \in \mathbb{R}_+ \cap \mathbb{T}. \end{aligned}$$

Then

$$z(0) = c$$

and

$$\begin{aligned} x(t) &\le z(t), \\ z^\Delta(t) &= f(t)x(t) + f(t)\int_0^t g(y)x(y)\Delta y \\ &= f(t)\left(x(t) + \int_0^t g(y)x(y)\Delta y\right) \\ &\le f(t)\left(z(t) + \int_0^t g(y)z(y)\Delta y\right), \quad t \in \mathbb{R}_+ \cap \mathbb{T}. \end{aligned} \tag{1.31}$$

Let

$$m(t) = z(t) + \int_0^t g(y)z(y)\Delta y, \quad t \in \mathbb{R}_+ \cap \mathbb{T}.$$

Then

$$\begin{aligned} z(t) &\le m(t),\\ z^{\Delta}(t) &\le f(t)m(t),\\ m(0) &= z(0)\\ &= c,\\ m^{\Delta}(t) &= z^{\Delta}(t) + g(t)z(t)\\ &\le f(t)m(t) + g(t)m(t)\\ &= (f(t) + g(t))m(t), \quad t \in \mathbb{R}_+ \cap \mathbb{T}. \end{aligned}$$

Hence,

$$\begin{aligned} m(t) &\le m(0) + \int_0^t (f(y) + g(y))m(y)\Delta y\\ &= c + \int_0^t (f(y) + g(y))m(y)\Delta y, \quad t \in \mathbb{R}_+ \cap \mathbb{T}. \end{aligned}$$

From here and Theorem 1.1.4, we conclude that

$$\begin{aligned} m(t) &\le c e_{f+g}(t, 0),\\ x(t) &\le z(t)\\ &\le m(t)\\ &\le c e_{f+g}(t, 0), \quad t \in \mathbb{R}_+ \cap \mathbb{T}. \end{aligned}$$

Hence, by (1.31), we get

$$z^{\Delta}(t) \le c f(t) e_{f+g}(t, 0), \quad t \in \mathbb{R}_+ \cap \mathbb{T}.$$

Then

$$\begin{aligned} z(t) - z(0) &\le c \int_0^t f(s) e_{f+g}(s, 0)\Delta s,\\ z(t) &\le c\left(1 + \int_0^t f(s) e_{f+g}(s, 0)\Delta s\right), \quad t \in \mathbb{R}_+ \cap \mathbb{T}. \end{aligned}$$

Therefore

$$\begin{aligned} x(t) &\le z(t)\\ &\le c\left(1 + \int_0^t f(s) e_{f+g}(s, 0)\Delta s\right), \quad t \in \mathbb{R}_+ \cap \mathbb{T}. \end{aligned}$$

This completes the proof. □

Theorem 1.5.2 (Pachpatte's Inequality). *Let $x, f, p, g \in \mathcal{C}_{rd}(\mathbb{R}_+ \cap \mathbb{T})$ be nonnegative functions, c be a nonnegative constant, and*

$$x(t) \le c + \int_0^t (f(s)x(s) + p(s))\Delta s + \int_0^t f(s)\left(\int_0^s g(y)x(y)\Delta y\right)\Delta s,$$

$t \in \mathbb{R}_+ \cap \mathbb{T}$. Then

$$\begin{aligned} x(t) &\le c + \int_0^t (p(s) + q(s)f(s))\Delta s \\ &\quad + \int_0^t f(s)\int_0^s (f(y) + g(y))q(y)e_{\ominus(f+g)}(\sigma(y), s)\Delta y\Delta s, \end{aligned}$$

$t \in \mathbb{R}_+ \cap \mathbb{T}$, where

$$q(t) = c + \int_0^t p(s)\Delta s, \quad t \in \mathbb{R}_+ \cap \mathbb{T}.$$

Proof. Let

$$z(t) = c + \int_0^t (f(s)x(s) + p(s))\Delta s + \int_0^t f(s)\left(\int_0^s g(y)x(y)\Delta y\right)\Delta s,$$

$t \in \mathbb{R}_+ \cap \mathbb{T}$. Then

$$\begin{aligned} z(0) &= c, \\ x(t) &\le z(t), \\ z^\Delta(t) &= f(t)x(t) + p(t) + f(t)\int_0^t g(y)x(y)\Delta y \\ &\le z(t)f(t) + p(t) + f(t)\int_0^t g(y)z(y)\Delta y \\ &= p(t) + f(t)\left(z(t) + \int_0^t g(y)z(y)\Delta y\right), \quad t \in \mathbb{R}_+ \cap \mathbb{T}. \end{aligned}$$

Let

$$v(t) = z(t) + \int_0^t g(y)z(y)\Delta y, \quad t \in \mathbb{R}_+ \cap \mathbb{T}.$$

Then

$$\begin{aligned} z(t) &\le v(t), \\ z^\Delta(t) &\le p(t) + f(t)v(t), \\ v(0) &= z(0) \\ &= c, \\ v^\Delta(t) &= z^\Delta(t) + g(t)z(t) \\ &\le p(t) + f(t)v(t) + g(t)v(t) \\ &\le p(t) + (f(t) + g(t))v(t), \quad t \in \mathbb{R}_+ \cap \mathbb{T}. \end{aligned}$$

Hence,

$$\begin{aligned} v(t) &\le c + \int_0^t p(s)\Delta s + \int_0^t (f(s) + g(s))v(s)\Delta s \\ &= q(t) + \int_0^t (f(s) + g(s))v(s)\Delta s, \quad t \in \mathbb{R}_+ \cap \mathbb{T}. \end{aligned}$$

From here and Theorem 1.1.10, we get

$$v(t) \le q(t) + \int_0^t (f(s) + g(s))q(s)e_{\ominus(f+g)}(\sigma(s), t)\Delta s,$$

$t \in \mathbb{R}_+ \cap \mathbb{T}$. Therefore

$$\begin{aligned} z^\Delta(t) &\le p(t) + q(t)f(t) \\ &\quad + f(t)\int_0^t (f(s) + g(s))q(s)e_{\ominus(f+g)}(\sigma(s), t)\Delta s \end{aligned}$$

and

$$\begin{aligned} z(t) &\le c + \int_0^t (p(s) + q(s)f(s))\Delta s \\ &\quad + \int_0^t f(s)\int_0^s (f(y) + g(y))q(y)e_{\ominus(f+g)}(\sigma(y), s)\Delta y\Delta s, \end{aligned}$$

$t \in \mathbb{R}_+ \cap \mathbb{T}$. Consequently,

$$x(t) \le z(t)$$

$$\leq c + \int_0^t (p(s) + q(s)f(s))\Delta s$$
$$+ \int_0^t f(s) \int_0^s (f(y) + g(y))q(y)e_{\ominus(f+g)}(\sigma(y), s)\Delta y\Delta s,$$

$t \in \mathbb{R}_+ \cap \mathbb{T}$. This completes the proof. □

Corollary 1.5.3. *Let $x, f, g, p \in \mathcal{C}_{rd}(\mathbb{R}_+ \cap \mathbb{T})$ be nonnegative functions, c be a nonnegative constant, and*

$$x(t) \leq c + \int_0^t f(s)x(s)\Delta s + \int_0^t f(s)\left(\int_0^s (g(y)x(y) + p(y))\Delta y\right)\Delta s, \quad t \in \mathbb{R}_+ \cap \mathbb{T}.$$

Then

$$x(t) \leq c + \int_0^t (q(s) + l(s)f(s))\Delta s$$
$$+ \int_0^t f(s) \int_0^s (f(y) + g(y))l(y)e_{\ominus(f+g)}(\sigma(y), s)\Delta y\Delta s, \quad t \in \mathbb{R}_+ \cap \mathbb{T},$$

where

$$q(t) = f(t) \int_0^t p(y)\Delta y,$$
$$l(t) = c + \int_0^t q(s)\Delta s, \quad t \in \mathbb{R}_+ \cap \mathbb{T}.$$

Proof. We have

$$x(t) \leq c + \int_0^t f(s)x(s)\Delta s + \int_0^t f(s) \int_0^s p(y)\Delta y\Delta s$$
$$+ \int_0^t f(s)\left(\int_0^s g(y)x(y)\Delta y\right)\Delta s$$
$$= c + \int_0^t (f(s)x(s) + q(s))\Delta s$$
$$+ \int_0^t f(s)\left(\int_0^s g(y)x(y)\Delta y\right)\Delta s, \quad t \in \mathbb{R}_+ \cap \mathbb{T}.$$

Hence, by Theorem 1.5.2, we get the desired result. This completes the proof. □

Corollary 1.5.4. *Let $x, f, g, h \in \mathcal{C}_{rd}(\mathbb{R}_+ \cap \mathbb{T})$ be nonnegative functions, c be a nonnegative constant, and*

$$x(t) \le c + \int_0^t f(s)x(s)\Delta s + \int_0^t g(s)\left(x(s) + \int_0^s h(y)x(y)\Delta y\right)\Delta s, \quad t \in \mathbb{R}_+ \cap \mathbb{T}.$$

Then

$$x(t) \le c + c\int_0^t (f(s) + g(s))\Delta s + c\int_0^t (f(s) + g(s))\int_0^s (f(y) + g(y) + h(y))e_{\ominus(f+g+h)}(\sigma(y), s)\Delta y\Delta s,$$

$t \in \mathbb{R}_+ \cap \mathbb{T}$.

Proof. We have

$$\begin{aligned} x(t) &\le c + \int_0^t (f(s) + g(s))x(s)\Delta s + \int_0^t g(s)\left(\int_0^s h(y)x(y)\Delta y\right)\Delta s \\ &\le c + \int_0^t (f(s) + g(s))x(s)\Delta s + \int_0^t (f(s) + g(s))\left(\int_0^s h(y)x(y)\Delta y\right)\Delta s, \quad t \in \mathbb{R}_+ \cap \mathbb{T}. \end{aligned}$$

Hence, by Theorem 1.5.2, we get the desired result. This completes the proof. □

Theorem 1.5.5 (Pachpatte's Inequality). *Let $x, f, g, h, p \in \mathcal{C}_{rd}(\mathbb{R}_+ \cap \mathbb{T})$ be nonnegative functions, and*

$$x(t) \le h(t) + p(t)\int_0^t f(s)x(s)\Delta s + p(t)\int_0^t f(s)p(s)\left(\int_0^s g(y)x(y)\Delta y\right)\Delta s, \quad t \in \mathbb{R}_+ \cap \mathbb{T}.$$

Then

$$x(t) \le h(t) + p(t)l(t) + p(t)\int_0^t q(s)h(s)e_{\ominus q}(\sigma(s),t)\Delta s, \quad t \in \mathbb{R}_+ \cap \mathbb{T},$$

where

$$l(t) = \int_0^t (f(s) + g(s))h(s)\Delta s,$$

$$q(t) = p(t)(f(t) + g(t)), \quad t \in \mathbb{R}_+ \cap \mathbb{T}.$$

Proof. Let

$$z(t) = \int_0^t f(s)x(s)\Delta s + \int_0^t f(s)p(s)\left(\int_0^s g(y)x(y)\Delta y\right)\Delta s, \quad t \in \mathbb{R}_+ \cap \mathbb{T}.$$

Then

$$z(0) = 0,$$

$$x(t) \le h(t) + p(t)z(t),$$

and

$$\begin{aligned} z^\Delta(t) &= f(t)x(t) + f(t)p(t)\int_0^t g(s)x(s)\Delta s \\ &\le f(t)h(t) + f(t)p(t)z(t) \\ &\quad + f(t)p(t)\int_0^t g(s)(h(s) + p(s)z(s))\Delta s \\ &= f(t)\Bigg(h(t) + p(t)\Bigg(z(t) \\ &\quad + \int_0^t g(s)(h(s) + p(s)z(s))\Delta s\Bigg)\Bigg), \quad t \in \mathbb{R}_+ \cap \mathbb{T}. \end{aligned}$$

We set

$$m(t) = z(t) + \int_0^t g(s)(h(s) + p(s)z(s))\Delta s, \quad t \in \mathbb{R}_+ \cap \mathbb{T}.$$

Then

$$\begin{aligned} z(t) &\le m(t), \\ m(0) &= z(0) \\ &= 0, \end{aligned}$$

and

$$z^{\Delta}(t) \le f(t)(h(t) + p(t)m(t)), \quad t \in \mathbb{R}_+ \cap \mathbb{T}.$$

Also,

$$\begin{aligned} m^{\Delta}(t) &= z^{\Delta}(t) + g(t)h(t) + g(t)p(t)z(t) \\ &\le f(t)(h(t) + p(t)m(t)) \\ &\quad + g(t)h(t) + g(t)p(t)m(t) \\ &= (f(t) + g(t))h(t) \\ &\quad + p(t)(f(t) + g(t))m(t), \quad t \in \mathbb{R}_+ \cap \mathbb{T}. \end{aligned}$$

Hence,

$$\begin{aligned} m(t) &\le \int_0^t (f(s) + g(s))h(s)\Delta s \\ &\quad + \int_0^t p(s)(f(s) + g(s))m(s)\Delta s \\ &= l(t) + \int_0^t q(s)m(s)\Delta s, \quad t \in \mathbb{R}_+ \cap \mathbb{T}. \end{aligned}$$

Hence, by Theorem 1.1.10, we get

$$m(t) \le l(t) + \int_0^t q(s)l(s)e_{\ominus q}(\sigma(s), t)\Delta s, \quad t \in \mathbb{R}_+ \cap \mathbb{T}.$$

Therefore

$$\begin{aligned} x(t) &\le h(t) + p(t)z(t) \\ &\le h(t) + p(t)m(t) \\ &\le h(t) + p(t)l(t) \\ &\quad + p(t)\int_0^t q(s)l(s)e_{\ominus q}(\sigma(s), t)\Delta s, \quad t \in \mathbb{R}_+ \cap \mathbb{T}. \end{aligned}$$

This completes the proof. □

Theorem 1.5.6 (Pachpatte's Inequality). *Let $x, f, g, h \in \mathcal{C}_{rd}(\mathbb{R}_+ \cap \mathbb{T})$ be nonnegative functions, c be a nonnegative constant, and*

$$\begin{aligned} x(t) \le c &+ \int_0^t f(s)x(s)\Delta s + \int_0^t f(s)\left(\int_0^s g(y)x(y)\Delta y\right)\Delta s \\ &+ \int_0^t f(s)\left(\int_0^s g(y)\left(\int_0^y h(\tau)x(\tau)\Delta\tau\right)\Delta y\right)\Delta s, \end{aligned}$$

$t \in \mathbb{R}_+ \cap \mathbb{T}$. Then

$$x(t) \le c\left(1 + \int_0^t (f(s) + g(s))e_{f+g+h}(s,0)\Delta s\right), \quad t \in \mathbb{R}_+ \cap \mathbb{T}.$$

Proof. Let

$$\begin{aligned} z(t) = c &+ \int_0^t f(s)x(s)\Delta s + \int_0^t f(s)\left(\int_0^s g(y)x(y)\Delta y\right)\Delta s \\ &+ \int_0^t f(s)\left(\int_0^s g(y)\left(\int_0^y h(\tau)x(\tau)\Delta\tau\right)\Delta y\right)\Delta s, \end{aligned}$$

$t \in \mathbb{R}_+ \cap \mathbb{T}$. Then

$$\begin{aligned} x(t) &\le z(t), \\ z(0) &= c, \\ z^\Delta(t) &= f(t)x(t) + f(t)\int_0^t g(y)x(y)\Delta y \\ &\quad + f(t)\int_0^t g(y)\left(\int_0^y h(s)x(s)\Delta s\right)\Delta y \\ &= f(t)\left(x(t) + \int_0^t g(y)x(y)\Delta y \right.\\ &\quad \left. + \int_0^t g(y)\left(\int_0^y h(s)x(s)\Delta s\right)\Delta y\right) \\ &\le f(t)\left(z(t) + \int_0^t g(y)z(y)\Delta y \right.\\ &\quad \left. + \int_0^t g(y)\left(\int_0^y h(s)z(s)\Delta s\right)\Delta y\right), \quad t \in \mathbb{R}_+ \cap \mathbb{T}. \end{aligned}$$

Let

$$v(t) = z(t) + \int_0^t g(y)z(y)\Delta y + \int_0^t g(y)\left(\int_0^y h(s)z(s)\Delta s\right)\Delta y, \quad t \in \mathbb{R}_+ \cap \mathbb{T}.$$

Then

$$\begin{aligned} v(0) &= z(0) \\ &= c, \\ z^\Delta(t) &\le f(t)v(t), \\ z(t) &\le v(t), \\ v^\Delta(t) &= z^\Delta(t) + g(t)z(t) \\ &\quad + g(t)\int_0^t h(s)z(s)\Delta s \\ &\le f(t)v(t) + g(t)v(t) \\ &\quad + g(t)\int_0^t h(s)v(s)\Delta s \\ &= (f(t)+g(t))v(t) + g(t)\int_0^t h(s)v(s)\Delta s \\ &\le (f(t)+g(t))v(t) + (f(t)+g(t))\int_0^t h(s)v(s)\Delta s, \quad t \in \mathbb{R}_+ \cap \mathbb{T}. \end{aligned}$$

Hence,

$$v(t) \le c + \int_0^t (f(s)+g(s))v(s)\Delta s + \int_0^t (f(s)+g(s))\left(\int_0^s h(y)v(y)\Delta y\right)\Delta s, \quad t \in \mathbb{R}_+ \cap \mathbb{T}.$$

Hence, by Theorem 1.5.1, we get

$$v(t) \le c\left(1 + \int_0^t (f(s)+g(s))e_{f+g+h}(s,0)\Delta s\right), \quad t \in \mathbb{R}_+ \cap \mathbb{T}.$$

Therefore

$$x(t) \le z(t)$$

$$\leq v(t)$$
$$\leq c\left(1+\int_0^t (f(s)+g(s))e_{f+g+h}(s,0)\Delta s\right), \quad t\in\mathbb{R}_+\cap\mathbb{T}.$$

This completes the proof. □

Theorem 1.5.7. *Let $x, k, p, f, g, h \in \mathcal{C}_{rd}(\mathbb{R}_+\cap\mathbb{T})$ be nonnegative functions, c be a nonnegative constant, $f(t)\leq g(t)$, $t\in\mathbb{R}_+\cap\mathbb{T}$, and*

$$x(t)\leq k(t)+p(t)\left(\int_0^t f(s)x(s)\Delta s+\int_0^t f(s)p(s)\left(\int_0^s g(\tau)x(\tau)\Delta\tau\right)\Delta s\right.$$
$$\left.+\int_0^t f(s)p(s)\left(\int_0^s g(\tau)p(\tau)\left(\int_0^\tau h(y)x(y)\Delta y\right)\Delta\tau\right)\Delta s\right),$$

$t\in\mathbb{R}_+\cap\mathbb{T}$. Then

$$x(t)\leq k(t)+p(t)l(t)+2p(t)r(t)$$
$$+2p(t)\int_0^t c(s)l(s)e_{\ominus c}(\sigma(s),t)\Delta s, \quad t\in\mathbb{R}_+\cap\mathbb{T},$$

where

$$q(t)=f(t)k(t)+f(t)p(t)\int_0^t g(s)k(s)\Delta s$$
$$+f(t)p(t)\int_0^t g(\tau)p(\tau)\left(\int_0^\tau h(y)k(y)\Delta y\right)\Delta\tau,$$
$$r(t)=\int_0^t (p(s)g(s)+h(s)p(s))l(s)\Delta s,$$
$$c(t)=2(p(t)g(t)+h(t)p(t)),$$
$$l(t)=\int_0^t q(s)\Delta s, \quad t\in\mathbb{R}_+\cap\mathbb{T}.$$

Proof. Let

$$z(t)=\int_0^t f(s)x(s)\Delta s+\int_0^t f(s)p(s)\left(\int_0^s g(\tau)x(\tau)\Delta\tau\right)\Delta s$$
$$+\int_0^t f(s)p(s)\left(\int_0^s g(\tau)p(\tau)\left(\int_0^\tau h(y)x(y)\Delta y\right)\Delta\tau\right)\Delta s,$$

$t \in \mathbb{R}_+ \cap \mathbb{T}$. Then

$$
\begin{aligned}
z(0) &= 0,\\
x(t) &\le k(t) + p(t)z(t),\\
z^{\Delta}(t) &= f(t)x(t) + f(t)p(t)\int_0^t g(s)x(s)\Delta s\\
&\quad + f(t)p(t)\int_0^t g(\tau)p(\tau)\left(\int_0^{\tau} h(y)x(y)\Delta y\right)\Delta\tau\\
&\le f(t)k(t) + f(t)p(t)z(t)\\
&\quad + f(t)p(t)\int_0^t g(s)k(s)\Delta s\\
&\quad + f(t)p(t)\int_0^t p(s)g(s)z(s)\Delta s\\
&\quad + f(t)p(t)\int_0^t g(\tau)p(\tau)\left(\int_0^{\tau} h(y)k(y)\Delta y\right)\Delta\tau\\
&\quad + f(t)p(t)\int_0^t g(\tau)p(\tau)\left(\int_0^{\tau} h(y)p(y)z(y)\Delta y\right)\Delta\tau\\
&= q(t) + f(t)p(t)z(t)\\
&\quad + f(t)p(t)\int_0^t p(s)g(s)z(s)\Delta s\\
&\quad + f(t)p(t)\int_0^t g(\tau)p(\tau)\left(\int_0^{\tau} h(y)p(y)z(y)\Delta y\right)\Delta\tau\\
&= q(t) + f(t)p(t)\left(z(t) + \int_0^t p(s)g(s)z(s)\Delta s\right.\\
&\quad \left. + \int_0^t g(\tau)p(\tau)\left(\int_0^{\tau} h(y)p(y)z(y)\Delta y\right)\Delta\tau\right),
\end{aligned}
$$

$t \in \mathbb{R}_+ \cap \mathbb{T}$. Let

$$
\begin{aligned}
v(t) &= z(t) + \int_0^t p(s)g(s)z(s)\Delta s\\
&\quad + \int_0^t g(\tau)p(\tau)\left(\int_0^{\tau} h(y)p(y)z(y)\Delta y\right)\Delta\tau, \quad t \in \mathbb{R}_+ \cap \mathbb{T}.
\end{aligned}
$$

Then

$$\begin{aligned}
v(0) &= z(0) \\
&= 0, \\
z^{\Delta}(t) &\le q(t) + f(t)p(t)v(t), \\
z(t) &\le v(t), \\
v^{\Delta}(t) &= z^{\Delta}(t) + p(t)g(t)z(t) \\
&\quad + p(t)g(t)\int_0^t h(y)p(y)z(y)\Delta y \\
&\le q(t) + f(t)p(t)v(t) + p(t)g(t)v(t) \\
&\quad + p(t)g(t)\int_0^t h(y)p(y)v(y)\Delta y \\
&= q(t) + g(t)p(t)\left(2v(t) + \int_0^t h(y)p(y)v(y)\Delta y\right), \quad t \in \mathbb{R}_+ \cap \mathbb{T}.
\end{aligned}$$

Hence,

$$\begin{aligned}
v(t) &\le \int_0^t q(s)\Delta s + 2\int_0^t p(s)g(s)v(s)\Delta s \\
&\quad + \int_0^t g(s)p(s)\int_0^s h(y)p(y)v(y)\Delta y\Delta s \\
&\le l(t) + 2\int_0^t p(s)g(s)v(s)\Delta s \\
&\quad + 4\int_0^t g(s)p(s)\int_0^s h(y)p(y)v(y)\Delta y\Delta s, \quad t \in \mathbb{R}_+ \cap \mathbb{T}.
\end{aligned}$$

From here and Theorem 1.5.5, we obtain

$$v(t) \le l(t) + 2r(t) + 2\int_0^t c(s)l(s)e_{\ominus c}(\sigma(s), t)\Delta s, \quad t \in \mathbb{R}_+ \cap \mathbb{T}.$$

Therefore

$$\begin{aligned}
x(t) &\le k(t) + p(t)z(t) \\
&\le k(t) + p(t)v(t)
\end{aligned}$$

$$\leq k(t) + p(t)l(t) + 2p(t)r(t)$$
$$+ 2p(t)\int_0^t c(s)l(s)e_{\ominus c}(\sigma(s), t)\Delta s, \quad t \in \mathbb{R}_+ \cap \mathbb{T}.$$

This completes the proof. □

Theorem 1.5.8. *Let $x, f, g \in \mathcal{C}_{rd}(\mathbb{R}_+ \cap \mathbb{T})$ be nonnegative functions, $h \in \mathcal{C}_{rd}(\mathbb{R}_+ \cap \mathbb{T})$ be a positive and nondecreasing function, and*

$$x(t) \leq h(t) + \int_0^t f(s)x(s)\Delta s + \int_0^t f(s)\left(\int_0^s g(y)x(y)\Delta y\right)\Delta s,$$

$t \in \mathbb{R}_+ \cap \mathbb{T}$. Then

$$x(t) \leq h(t)\left(1 + \int_0^t f(s)e_{f+g}(s, 0)\Delta s\right), \quad t \in \mathbb{R}_+ \cap \mathbb{T}.$$

Proof. Since h is positive and nondecreasing on $\mathbb{R}_+ \cap \mathbb{T}$, we get

$$\frac{x(t)}{h(t)} \leq 1 + \frac{1}{h(t)}\int_0^t f(s)x(s)\Delta s + \frac{1}{h(t)}\int_0^t f(s)\left(\int_0^s g(y)x(y)\Delta y\right)\Delta s$$
$$\leq 1 + \int_0^t f(s)\frac{x(s)}{h(s)}\Delta s + \int_0^t f(s)\left(\int_0^s g(y)\frac{x(y)}{h(y)}\Delta y\right)\Delta s,$$

$t \in \mathbb{R}_+ \cap \mathbb{T}$. Hence, by Theorem 1.5.1, we get

$$\frac{x(t)}{h(t)} \leq 1 + \int_0^t f(s)e_{f+g}(s, 0)\Delta s, \quad t \in \mathbb{R}_+ \cap \mathbb{T},$$

or

$$x(t) \leq h(t)\left(1 + \int_0^t f(s)e_{f+g}(s, 0)\Delta s\right), \quad t \in \mathbb{R}_+ \cap \mathbb{T}.$$

This completes the proof. □

Theorem 1.5.9. *Let $x, h, p \in \mathcal{C}_{rd}(\mathbb{R}_+ \cap \mathbb{T})$ be nonnegative functions, $f \in \mathcal{C}^2_{rd}(\mathbb{R}_+ \cap \mathbb{T})$ and $g \in \mathcal{C}^1_{rd}(\mathbb{R}_+ \cap \mathbb{T})$ be positive functions, c be a positive constant, and*

$$x(t) \leq c + \int_0^t f(s)\left(h(s) + \int_0^s g(\tau)\left(\int_0^\tau p(y)x(y)\Delta y\right)\Delta\tau\right)\Delta s,$$

$t \in \mathbb{R}_+ \cap \mathbb{T}$. *Then*

$$x(t) \leq \left(c + \int_0^t f(s)h(s)\Delta s\right)\left(1 + \int_0^t q(s)e_{\ominus q}(\sigma(s), t)\Delta s\right),$$

$t \in \mathbb{R}_+ \cap \mathbb{T}$, *where*

$$q(t) = f(t)\int_0^t g(y)\int_0^y p(\tau)\Delta\tau\Delta y, \quad t \in \mathbb{R}_+ \cap \mathbb{T}.$$

Proof. Let

$$l(t) = c + \int_0^t f(s)h(s)\Delta s, \quad t \in \mathbb{R}_+ \cap \mathbb{T}.$$

Then

$$x(t) \leq l(t) + \int_0^t f(s)\int_0^s g(\tau)\left(\int_0^\tau p(y)x(y)\Delta y\right)\Delta\tau\Delta s, \quad t \in \mathbb{R}_+ \cap \mathbb{T}.$$

Note that l is a positive and nondecreasing function on $\mathbb{R}_+ \cap \mathbb{T}$. Then

$$\begin{aligned}
\frac{x(t)}{l(t)} &\leq 1 + \frac{1}{l(t)}\int_0^t f(s)\int_0^s g(\tau)\left(\int_0^\tau p(y)x(y)\Delta y\right)\Delta\tau\Delta s \\
&\leq 1 + \int_0^t f(s)\int_0^s g(\tau)\left(\int_0^\tau p(y)\frac{x(y)}{l(y)}\Delta y\right)\Delta\tau\Delta s, \quad t \in \mathbb{R}_+ \cap \mathbb{T}.
\end{aligned}$$

Let

$$z(t) = 1 + \int_0^t f(s)\int_0^s g(\tau)\left(\int_0^\tau p(y)\frac{x(y)}{l(y)}\Delta y\right)\Delta\tau\Delta s, \quad t \in \mathbb{R}_+ \cap \mathbb{T}.$$

Then

$$\begin{aligned}
&\frac{x(t)}{l(t)} \leq z(t), \\
&z(0) = 1, \\
&z^\Delta(t) = f(t)\int_0^t g(\tau)\left(\int_0^\tau p(y)\frac{x(y)}{l(y)}\Delta y\right)\Delta\tau,
\end{aligned}$$

$t \in \mathbb{R}_+ \cap \mathbb{T}$. Hence,

$$\begin{aligned} \frac{z^\Delta(t)}{f(t)} &= \int_0^t g(\tau)\left(\int_0^\tau p(y)\frac{x(y)}{l(y)}\Delta y\right)\Delta\tau, \\ \left(\frac{z^\Delta(t)}{f(t)}\right)^\Delta &= g(t)\int_0^t p(y)\frac{x(y)}{l(y)}\Delta y, \\ \frac{1}{g(t)}\left(\frac{z^\Delta(t)}{f(t)}\right)^\Delta &= \int_0^t p(y)\frac{x(y)}{l(y)}\Delta y, \\ \left(\frac{1}{g(t)}\left(\frac{z^\Delta(t)}{f(t)}\right)^\Delta\right)^\Delta &= p(t)\frac{x(t)}{l(t)} \\ &\le p(t)z(t), \quad t \in \mathbb{R}_+ \cap \mathbb{T}. \end{aligned}$$

Therefore

$$\frac{(\frac{1}{g(t)}(\frac{z^\Delta(t)}{f(t)})^\Delta)^\Delta}{z(t)} \le p(t), \quad t \in \mathbb{R}_+ \cap \mathbb{T}.$$

Note that

$$z^\Delta(t) \ge 0, \quad z(t) \ge 0, \quad \frac{1}{g(t)}\left(\frac{z^\Delta(t)}{f(t)}\right)^\Delta \ge 0.$$

Therefore

$$\frac{(\frac{1}{g(t)}(\frac{z^\Delta(t)}{f(t)})^\Delta)^\Delta}{z(t)} \le p(t) + \frac{\frac{1}{g(t)}(\frac{z^\Delta(t)}{f(t)})^\Delta z^\Delta(t)}{z(t)z(\sigma(t))},$$

$t \in \mathbb{R}_+ \cap \mathbb{T}$. Since z is nondecreasing on $\mathbb{R}_+ \cap \mathbb{T}$, we have

$$\frac{(\frac{1}{g(t)}(\frac{z^\Delta(t)}{f(t)})^\Delta)^\Delta z(t)}{(z(t))^2} \ge \frac{(\frac{1}{g(t)}(\frac{z^\Delta(t)}{f(t)})^\Delta)^\Delta z(t)}{z(t)z(\sigma(t))}, \quad t \in \mathbb{R}_+ \cap \mathbb{T}.$$

Therefore

$$\frac{(\frac{1}{g(t)}(\frac{z^\Delta(t)}{f(t)})^\Delta)^\Delta z(t)}{z(t)z(\sigma(t))} \le p(t) + \frac{\frac{1}{g(t)}(\frac{z^\Delta(t)}{f(t)})^\Delta z^\Delta(t)}{z(t)z(\sigma(t))},$$

or

$$\left(\frac{\frac{1}{g(t)}(\frac{z^\Delta(t)}{f(t)})^\Delta}{z(t)}\right)^\Delta \le p(t), \quad t \in \mathbb{R}_+ \cap \mathbb{T}.$$

From here,

$$\frac{\frac{1}{g(t)}\left(\frac{z^\Delta(t)}{f(t)}\right)^\Delta}{z(t)} \le \int_0^t p(s)\Delta s,$$

or

$$\frac{\left(\frac{z^\Delta(t)}{f(t)}\right)^\Delta}{z(t)} \le g(t)\int_0^t p(s)\Delta s,$$

or

$$\frac{\left(\frac{z^\Delta(t)}{f(t)}\right)^\Delta z(t)}{(z(t))^2} \le g(t)\int_0^t p(s)\Delta s, \quad t \in \mathbb{R}_+ \cap \mathbb{T}.$$

Since z is nondecreasing, we obtain

$$\begin{aligned}\frac{\left(\frac{z^\Delta(t)}{f(t)}\right)^\Delta z(t)}{z(t)z(\sigma(t))} &\le \frac{\left(\frac{z^\Delta(t)}{f(t)}\right)^\Delta z(t)}{(z(t))^2}\\ &\le g(t)\int_0^t p(s)\Delta s + \frac{\frac{z^\Delta(t)}{f(t)}z^\Delta(t)}{z(t)z(\sigma(t))},\end{aligned}$$

or

$$\left(\frac{\frac{z^\Delta(t)}{f(t)}}{z(t)}\right)^\Delta \le g(t)\int_0^t p(s)\Delta s, \quad t \in \mathbb{R}_+ \cap \mathbb{T}.$$

From here,

$$\frac{\frac{z^\Delta(t)}{f(t)}}{z(t)} \le \int_0^t g(s)\int_0^s p(y)\Delta y\Delta s,$$

or

$$\frac{z^\Delta(t)}{f(t)} \le z(t)\int_0^t g(s)\int_0^s p(y)\Delta y\Delta s,$$

or

$$z^\Delta(t) \le z(t)f(t)\int_0^t g(s)\int_0^s p(y)\Delta y\Delta s,$$

or

$$z(t) \le 1 + \int_0^t \left(f(s) \int_0^s g(y) \int_0^y p(\tau)\Delta\tau\Delta y \right) z(s)\Delta s, \quad t \in \mathbb{R}_+ \cap \mathbb{T}.$$

Hence, by Theorem 1.1.10, we get

$$z(t) \le 1 + \int_0^t q(s)e_{\ominus q}(\sigma(s), t)\Delta s, \quad t \in \mathbb{R}_+ \cap \mathbb{T}.$$

Consequently,

$$\begin{aligned} x(t) &\le l(t)\left(1 + \int_0^t q(s)e_{\ominus q}(\sigma(s), t)\Delta s\right) \\ &= \left(c + \int_0^t f(s)h(s)\Delta s\right)\left(1 + \int_0^t q(s)e_{\ominus q}(\sigma(s), t)\Delta s\right), \end{aligned}$$

$t \in \mathbb{R}_+ \cap \mathbb{T}$. This completes the proof. □

Theorem 1.5.10. *Let $x, h, p \in \mathcal{C}_{rd}(\mathbb{R}_+ \cap \mathbb{T})$ be nonnegative functions, $f \in \mathcal{C}^1_{rd}(\mathbb{R}_+ \cap \mathbb{T})$ be a positive function, c be a positive constant, and*

$$x(t) \le c + \int_0^t f(s)\left(h(s) + \int_0^s p(y)x(y)\Delta y \right)\Delta s, \quad t \in \mathbb{R}_+ \cap \mathbb{T}.$$

Then

$$x(t) \le g(t) + \int_0^t l(s)g(s)e_{\ominus l}(\sigma(s), t)\Delta s, \quad t \in \mathbb{R}_+ \cap \mathbb{T},$$

where

$$\begin{aligned} g(t) &= c + \int_0^t f(s)h(s)\Delta s, \\ l(t) &= f(t)\int_0^t p(y)\Delta y, \quad t \in \mathbb{R}_+ \cap \mathbb{T}. \end{aligned}$$

Proof. Let

$$z(t) = c + \int_0^t f(s)\left(h(s) + \int_0^s p(y)x(y)\Delta y \right)\Delta s, \quad t \in \mathbb{R}_+ \cap \mathbb{T}.$$

Then

$$
\begin{aligned}
x(t) &\le z(t),\\
z(0) &= c,\\
z^{\Delta}(t) &= f(t)h(t) + f(t)\int_0^t p(y)x(y)\Delta y, \quad t \in \mathbb{R}_+ \cap \mathbb{T}.
\end{aligned}
$$

Hence,

$$
\begin{aligned}
z^{\Delta}(t) &\ge 0,\\
\frac{z^{\Delta}(t) - f(t)h(t)}{f(t)} &= \int_0^t p(y)x(y)\Delta y\\
&\ge 0, \quad t \in \mathbb{R}_+ \cap \mathbb{T},
\end{aligned}
$$

and

$$
\begin{aligned}
\left(\frac{z^{\Delta}(t) - f(t)h(t)}{f(t)}\right)^{\Delta} &= p(t)x(t)\\
&\le p(t)z(t), \quad t \in \mathbb{R}_+ \cap \mathbb{T}.
\end{aligned}
$$

Therefore

$$
\begin{aligned}
\frac{\left(\frac{z^{\Delta}(t)-f(t)h(t)}{f(t)}\right)^{\Delta}}{z(t)} &\le p(t)\\
&\le p(t) + \frac{\frac{z^{\Delta}(t)-f(t)h(t)}{f(t)} z^{\Delta}(t)}{z(t)z(\sigma(t))}, \quad t \in \mathbb{R}_+ \cap \mathbb{T}.
\end{aligned}
$$

Consequently,

$$
\begin{aligned}
\frac{\left(\frac{z^{\Delta}(t)-f(t)h(t)}{f(t)}\right)^{\Delta} z(t)}{z(t)z(\sigma(t))} &\le \frac{\left(\frac{z^{\Delta}(t)-f(t)h(t)}{f(t)}\right)^{\Delta} z(t)}{(z(t))^2}\\
&\le p(t) + \frac{\frac{z^{\Delta}(t)-f(t)h(t)}{f(t)} z^{\Delta}(t)}{z(t)z(\sigma(t))}, \quad t \in \mathbb{R}_+ \cap \mathbb{T}.
\end{aligned}
$$

Hence,

$$
\left(\frac{\frac{z^{\Delta}(t)-f(t)h(t)}{f(t)}}{z(t)}\right)^{\Delta} \le p(t)
$$

and

$$
\frac{z^{\Delta}(t) - f(t)h(t)}{f(t)z(t)} \le \int_0^t p(s)\Delta s,
$$

or

$$\frac{z^{\Delta}(t) - f(t)h(t)}{z(t)} \le f(t) \int_0^t p(s)\Delta s,$$

or

$$z^{\Delta}(t) \le f(t)h(t) + z(t)f(t) \int_0^t p(s)\Delta s, \quad t \in \mathbb{R}_+ \cap \mathbb{T}.$$

Therefore

$$\begin{aligned} z(t) &\le c + \int_0^t f(s)h(s)\Delta s + \int_0^t f(s)\left(\int_0^s p(y)\Delta y \right) z(s)\Delta s \\ &= g(t) + \int_0^t l(s)z(s)\Delta s, \quad t \in \mathbb{R}_+ \cap \mathbb{T}. \end{aligned}$$

Hence, by Theorem 1.1.10, we get

$$z(t) \le g(t) + \int_0^t l(s)g(s)e_{\ominus l}(\sigma(s), t)\Delta s, \quad t \in \mathbb{R}_+ \cap \mathbb{T},$$

whereupon

$$x(t) \le g(t) + \int_0^t l(s)g(s)e_{\ominus l}(\sigma(s), t)\Delta s, \quad t \in \mathbb{R}_+ \cap \mathbb{T}.$$

This completes the proof. □

Theorem 1.5.11. *Let $x, v, p, f, g \in \mathcal{C}_{rd}(J)$ be nonnegative functions, p be decreasing on J, $1 - \mu(s)h(s) > 0$ for $a \le s \le t$, and*

$$x(t) \ge v(s) - p(t)\left(\int_s^t f(y)v(y)\Delta y + \int_s^t f(y)\left(\int_y^t g(\tau)v(\tau)\Delta\tau \right)\Delta y \right),$$

$a \le s \le t \le b$. Then

$$x(t) \ge v(s)\left(1 + p(t) \int_s^t f(y)e_{\frac{h}{1-\mu h}}(t, y)\Delta y \right)^{-1}, \quad a \le s \le t \le b,$$

where

$$h(s) = p(s)f(s) + g(s), \quad a \le s \le t \le b.$$

Proof. Let $t \in J$ be fixed and define, for $a \le s \le t \le b$,

$$z(s) = x(t) + p(t)\left(\int_s^t f(y)v(y)\Delta y + \int_s^t f(y)\left(\int_y^t g(\tau)v(\tau)\Delta\tau\right)\Delta y\right),$$

$a \le s \le t \le b$. Then

$$\begin{aligned} z(t) &= x(t),\\ v(s) &\le z(s),\\ z^\Delta(s) &= -p(t)f(s)\left(v(s) + \int_s^t g(y)v(y)\Delta y\right)\\ &\ge -p(t)f(s)\left(z(s) + \int_s^t g(y)z(y)\Delta y\right), \end{aligned}$$

$a \le s \le t \le b$. Let

$$r(s) = z(s) + \int_s^t g(y)z(y)\Delta y, \quad a \le s \le t \le b.$$

Then

$$\begin{aligned} z(s) &\le r(s),\\ r^\Delta(s) &= z^\Delta(s) - g(s)z(s)\\ &\ge -p(s)f(s)r(s) - g(s)z(s)\\ &\ge -p(s)f(s)r(s) - g(s)r(s), \end{aligned}$$

or

$$r^\Delta(s) + (p(s)f(s) + g(s))r(s) \ge 0,$$

or

$$r^\Delta(s) + h(s)r(s) \ge 0, \quad a \le s \le t \le b.$$

Note that

$$\begin{aligned} \left(r(s)e_{\frac{h}{1-\mu h}}(s,a)\right)^\Delta &= r^\Delta(s)e_{\frac{h}{1-\mu h}}(\sigma(s),a)\\ &\quad + r(s)\frac{h(s)}{1-\mu(s)h(s)}e_{\frac{h}{1-\mu h}}(s,a)\\ &= r^\Delta(s)\left(1 + \frac{\mu(s)h(s)}{1-\mu(s)h(s)}\right)e_{\frac{h}{1-\mu h}}(s,a)\\ &\quad + r(s)\frac{h(s)}{1-\mu(s)h(s)}e_{\frac{h}{1-\mu h}}(s,a) \end{aligned}$$

$$
\begin{aligned}
&= \frac{1}{1-\mu(s)h(s)}(r^{\Delta}(s) + h(s)r(s))e_{\frac{h}{1-\mu h}}(s,a) \\
&\geq 0, \quad a \leq s \leq t \leq b.
\end{aligned}
$$

Therefore

$$
r(t)e_{\frac{h}{1-\mu h}}(t,a) \geq r(s)e_{\frac{h}{1-\mu h}}(s,a),
$$

or

$$
r(t) \geq r(s)e_{\frac{h}{1-\mu h}}(s,t),
$$

or

$$
\begin{aligned}
r(s) &\leq e_{\frac{h}{1-\mu h}}(t,s)r(t) \\
&\leq x(t)e_{\frac{h}{1-\mu h}}(t,s), \quad a \leq s \leq t \leq b.
\end{aligned}
$$

Therefore

$$
z^{\Delta}(s) \geq -p(t)f(s)x(t)e_{\frac{h}{1-\mu h}}(t,s)
$$

and

$$
z(t) - z(s) \geq -p(t)x(t)\int_s^t f(y)e_{\frac{h}{1-\mu h}}(t,y)\Delta y,
$$

or

$$
\begin{aligned}
z(t) &\geq z(s) - p(t)x(t)\int_s^t f(y)e_{\frac{h}{1-\mu h}}(t,y)\Delta y \\
&\geq v(s) - p(t)x(t)\int_s^t f(y)e_{\frac{h}{1-\mu h}}(t,y)\Delta y,
\end{aligned}
$$

or

$$
x(t) \geq v(s) - p(t)x(t)\int_s^t f(y)e_{\frac{h}{1-\mu h}}(t,y)\Delta y,
$$

or

$$
\left(1 + p(t)\int_s^t f(y)e_{\frac{h}{1-\mu h}}(t,y)\Delta y\right)x(t) \geq v(s),
$$

or

$$
x(t) \geq v(s)\left(1 + p(t)\int_s^t f(y)e_{\frac{h}{1-\mu h}}(t,y)\Delta y\right)^{-1},
$$

$a \leq s \leq t \leq b$. This completes the proof. □

2 Linear integro-dynamic inequalities

This chapter introduces linear integro-dynamic inequalities of Pachpatte type and linear integro-dynamic inequalities with several iterated integrals. The material in this chapter is based on some results in [24] and [25].

Let $\mathbb{T}$ be a time scale with forward jump operator and delta differentiation operator σ and Δ, respectively. Suppose that $0 \in \mathbb{T}$.

2.1 Pachpatte's inequalities

We will start with the following useful lemma.

Lemma 2.1.1. *Let $x, x^{\Delta}, a, b \in \mathcal{C}_{rd}(\mathbb{R}_+ \cap \mathbb{T})$, $b \in \mathcal{R}^+$ and*

$$x^{\Delta}(t) \le a(t) + b(t)x(t), \quad t \in \mathbb{R}_+ \cap \mathbb{T}.$$

Then

$$x(t) \le x(0)e_b(t,0) + \int_0^t a(s)e_{\ominus b}(\sigma(s),t)\Delta s, \quad t \in \mathbb{R}_+ \cap \mathbb{T}.$$

Proof. Since $b \in \mathcal{R}^+(\mathbb{R}_+ \cap \mathbb{T})$, we have that

$$e_b(t,0) > 0 \quad \text{and} \quad e_{\ominus b}(t,0) > 0, \quad t \in \mathbb{R}_+ \cap \mathbb{T}.$$

We have

$$\begin{aligned}
(x(t)e_{\ominus b}(t,0))^{\Delta} &= x^{\Delta}(t)e_{\ominus b}(\sigma(t),0) \\
&\quad + x(t)(\ominus b)(t)e_{\ominus b}(t,0) \\
&= x^{\Delta}(t)e_{\ominus b}(\sigma(t),0) \\
&\quad - \frac{x(t)b(t)}{1+\mu(t)b(t)}e_{\ominus b}(t,0) \\
&= x^{\Delta}(t)e_{\ominus b}(\sigma(t),0) \\
&\quad - x(t)b(t)e_{\ominus b}(\sigma(t),0) \\
&= (x^{\Delta}(t) - x(t)b(t))e_{\ominus b}(\sigma(t),0) \\
&\le a(t)e_{\ominus b}(\sigma(t),0), \quad t \in \mathbb{R}_+ \cap \mathbb{T}.
\end{aligned}$$

Hence,

$$x(t)e_{\ominus b}(t,0) \le x(0) + \int_0^t a(s)e_{\ominus b}(\sigma(s),0)\Delta s, \quad t \in \mathbb{R}_+ \cap \mathbb{T},$$

https://doi.org/10.1515/9783110705553-002

or

$$\begin{aligned} x(t) &\leq x(0)e_b(t,0) + \int_0^t a(s)e_{\ominus b}(\sigma(s),0)e_{\ominus b}(0,t)\Delta s \\ &= x(0)e_b(t,0) + \int_0^t a(s)e_{\ominus b}(\sigma(s),t)\Delta s, \quad t \in \mathbb{R}_+ \cap \mathbb{T}. \end{aligned}$$

This completes the proof. □

Theorem 2.1.2 (Pachpatte's Inequality). *Let $x, x^\Delta, a, b, c \in \mathcal{C}_{rd}(\mathbb{R}_+ \cap \mathbb{T})$ be nonnegative functions. If $b(t) \geq 1$, $t \in \mathbb{R}_+ \cap \mathbb{T}$, and*

$$x^\Delta(t) \leq a(t) + b(t)\int_a^t c(s)(x(s) + x^\Delta(s))\Delta s, \quad t \in \mathbb{R}_+ \cap \mathbb{T},$$

then

$$x^\Delta(t) \leq a(t) + b(t)\int_0^t c(s)(A(s) + b(s)B(s))\Delta s, \quad t \in \mathbb{R}_+ \cap \mathbb{T},$$

where

$$\begin{aligned} A(t) &= x(0) + \int_0^t a(s)\Delta s + a(t), \\ f(t) &= b(t)(c(t) + 1), \\ B(t) &= \int_0^t c(s)A(s)e_{\ominus f}(\sigma(s),t)\Delta s, \quad t \in \mathbb{R}_+ \cap \mathbb{T}. \end{aligned}$$

Proof. Let

$$m(t) = \int_0^t c(s)(x(s) + x^\Delta(s))\Delta s, \quad t \in \mathbb{R}_+ \cap \mathbb{T}.$$

Then

$$x^\Delta(t) \leq a(t) + b(t)m(t), \quad t \in \mathbb{R}_+ \cap \mathbb{T}, \tag{2.1}$$

and

$$\begin{aligned} m^\Delta(t) &= c(t)(x(t) + x^\Delta(t)) \\ &\leq c(t)x(t) + c(t)(a(t) + b(t)m(t)) \\ &= a(t)c(t) + c(t)x(t) + b(t)c(t)m(t), \quad t \in \mathbb{R}_+ \cap \mathbb{T}. \end{aligned} \tag{2.2}$$

By (2.1), we get

$$x(t) \le x(0) + \int_0^t (a(s) + b(s))m(s)\Delta s, \quad t \in \mathbb{R}_+ \cap \mathbb{T}.$$

Hence, from (2.2), we obtain

$$\begin{aligned} m^\Delta(t) &\le c(t)x(0) + c(t)\int_0^t a(s)\Delta s \\ &\quad + c(t)\int_0^t b(s)m(s)\Delta s \\ &\quad + c(t)a(t) + b(t)c(t)m(t) \qquad (2.3) \\ &= c(t)A(t) + c(t)\left(b(t)m(t) + \int_0^t b(s)m(s)\Delta s\right) \\ &\le c(t)\left(A(t) + b(t)\left(m(t) + \int_0^t b(s)m(s)\Delta s\right)\right), \end{aligned}$$

$t \in \mathbb{R}_+ \cap \mathbb{T}$. Let

$$r(t) = m(t) + \int_0^t b(s)m(s)\Delta s, \quad t \in \mathbb{R}_+ \cap \mathbb{T}.$$

Then

$$\begin{aligned} r(0) &= m(0) \\ &= 0, \\ m(t) &\le r(t), \quad t \in \mathbb{R}_+ \cap \mathbb{T}, \\ m^\Delta(t) &\le c(t)(A(t) + b(t)r(t)), \quad t \in \mathbb{R}_+ \cap \mathbb{T}, \qquad (2.4) \end{aligned}$$

and

$$\begin{aligned} r^\Delta(t) &= m^\Delta(t) + b(t)m(t) \\ &\le c(t)(A(t) + b(t)r(t)) + b(t)r(t) \\ &= c(t)A(t) + b(t)(c(t) + 1)r(t) \\ &= c(t)A(t) + f(t)r(t), \quad t \in \mathbb{R}_+ \cap \mathbb{T}. \end{aligned}$$

Hence, by Lemma 2.1.1, we get

$$r(t) \le r(0)e_f(t,0) + \int_0^t c(s)A(s)e_{\ominus f}(\sigma(s), t)\Delta s$$

$$\begin{aligned}&= \int_0^t c(s)A(s)e_{\ominus f}(\sigma(s),t)\Delta s\\ &= B(t), \quad t \in \mathbb{R}_+ \cap \mathbb{T}.\end{aligned}$$

Thus, from (2.4), we obtain

$$m^\Delta(t) \le c(t)(A(t) + b(t)B(t)), \quad t \in \mathbb{R}_+ \cap \mathbb{T},$$

whereupon

$$m(t) \le \int_0^t c(s)(A(s) + b(s)B(s))\Delta s, \quad t \in \mathbb{R}_+ \cap \mathbb{T},$$

and, using (2.1), we get

$$\begin{aligned}x^\Delta(t) &\le a(t) + b(t)m(t)\\ &\le a(t) + b(t)\int_0^t c(s)(A(s) + b(s)B(s))\Delta s,\end{aligned}$$

$t \in \mathbb{R}_+ \cap \mathbb{T}$. This completes the proof. □

Theorem 2.1.3. *Let $x, x^\Delta, a, b, c \in \mathcal{C}_{rd}(\mathbb{R}_+ \cap \mathbb{T})$ be nonnegative functions, and*

$$x^\Delta(t) \le a(t) + b(t)\left(x(t) + \int_0^t c(s)(x(s) + x^\Delta(s))\Delta s\right), \quad t \in \mathbb{R}_+ \cap \mathbb{T}.$$

Then

$$\begin{aligned}x^\Delta(t) &\le a(t) + b(t)x(0)e_f(t,0)\\ &\quad + b(t)\int_0^t a(s)(1 + c(s))e_{\ominus f}(\sigma(s),t)\Delta s,\end{aligned}$$

$t \in \mathbb{R}_+ \cap \mathbb{T}$, where

$$f(t) = c(t) + b(t)(1 + c(t)), \quad t \in \mathbb{R}_+ \cap \mathbb{T}.$$

Proof. Let

$$z(t) = x(t) + \int_0^t c(s)(x(s) + x^\Delta(s))\Delta s, \quad t \in \mathbb{R}_+ \cap \mathbb{T}.$$

Then

$$x(0) = z(0),$$

$$\begin{aligned} x(t) &\le z(t), \quad t \in \mathbb{R}_+ \cap \mathbb{T}, \\ x^\Delta(t) &\le a(t) + b(t)z(t), \quad t \in \mathbb{R}_+ \cap \mathbb{T}, \end{aligned} \tag{2.5}$$

and hence,

$$\begin{aligned} z^\Delta(t) &= x^\Delta(t) + c(t)(x(t) + x^\Delta(t)) \\ &= c(t)x(t) + (1 + c(t))x^\Delta(t) \\ &\le c(t)z(t) + (1 + c(t))(a(t) + b(t)z(t)) \\ &= a(t)(1 + c(t)) + (c(t) + b(t)(1 + c(t)))z(t) \\ &= a(t)(1 + c(t)) + f(t)z(t), \quad t \in \mathbb{R}_+ \cap \mathbb{T}. \end{aligned}$$

From Lemma 2.1.1, we now get

$$\begin{aligned} z(t) &\le z(0)e_f(t,0) + \int_0^t a(s)(1 + c(s))e_{\ominus f}(\sigma(s),t)\Delta s \\ &= x(0)e_f(t,0) + \int_0^t a(s)(1 + c(s))e_{\ominus f}(\sigma(s),t)\Delta s, \end{aligned}$$

$t \in \mathbb{R}_+ \cap \mathbb{T}$. From the last inequality and (2.5), we obtain

$$\begin{aligned} x^\Delta(t) \le\; & a(t) + b(t)x(0)e_f(t,0) \\ & + b(t)\int_0^t a(s)(1 + c(s))e_{\ominus f}(\sigma(s),t)\Delta s, \end{aligned}$$

$t \in \mathbb{R}_+ \cap \mathbb{T}$. This completes the proof. □

Theorem 2.1.4 (Pachpatte's Inequality). *Let* $x, x^\Delta, a, b \in C_{rd}(\mathbb{R}_+ \cap \mathbb{T})$ *be nonnegative functions and*

$$\begin{aligned} x^\Delta(t) \le\; & x(0) + \int_0^t a(s)(x(s) + x^\Delta(s))\Delta s \\ & + \int_0^t \left(\int_0^s b(y)x^\Delta(y)\Delta y \right)\Delta s, \quad t \in \mathbb{R}_+ \cap \mathbb{T}. \end{aligned}$$

Then

$$x^\Delta(t) \le x(0)\left(1 + 2\int_0^t a(s)e_c(s,0)\Delta s\right), \quad t \in \mathbb{R}_+ \cap \mathbb{T},$$

where

$$c(t) = 1 + a(t) + b(t), \quad t \in \mathbb{R}_+ \cap \mathbb{T}.$$

Proof. Let

$$m(t) = x(0) + \int_0^t a(s)(x(s) + x^\Delta(s))\Delta s + \int_0^t a(s)\left(\int_0^s b(y)x^\Delta(y)\Delta y\right)\Delta s, \quad t \in \mathbb{R}_+ \cap \mathbb{T}.$$

Then $m(0) = x(0)$ and

$$x^\Delta(t) \le m(t), \quad t \in \mathbb{R}_+ \cap \mathbb{T}. \tag{2.6}$$

We have

$$\begin{aligned} m^\Delta(t) &= a(t)(x(t) + x^\Delta(t)) + a(t)\int_0^t b(y)x^\Delta(y)\Delta y \\ &= a(t)\left(x(t) + x^\Delta(t) + \int_0^t b(y)x^\Delta(y)\Delta y\right) \\ &\le a(t)\left(x(t) + m(t) + \int_0^t b(y)m(y)\Delta y\right), \quad t \in \mathbb{R}_+ \cap \mathbb{T}. \end{aligned} \tag{2.7}$$

By (2.6), we get

$$\begin{aligned} x(t) &\le x(0) + \int_0^t m(y)\Delta y \\ &= m(0) + \int_0^t m(y)\Delta y, \quad t \in \mathbb{R}_+ \cap \mathbb{T}. \end{aligned}$$

Hence, by (2.7), we obtain

$$\begin{aligned} m^\Delta(t) &\le a(t)\left(m(0) + \int_0^t m(y)\Delta y + m(t) + \int_0^t b(y)m(y)\Delta y\right) \\ &= a(t)\left(m(0) + m(t) + \int_0^t (1 + b(y))m(y)\Delta y\right), \quad t \in \mathbb{R}_+ \cap \mathbb{T}. \end{aligned}$$

Let

$$z(t) = m(0) + m(t) + \int_0^t (1 + b(y))m(y)\Delta y, \quad t \in \mathbb{R}_+ \cap \mathbb{T}.$$

Then

$$z(0) = 2m(0)$$

and

$$\begin{aligned} m(t) &\le z(t), \\ m^\Delta(t) &\le a(t)z(t), \quad t \in \mathbb{R}_+ \cap \mathbb{T}. \end{aligned}$$

By the definition of function z, we obtain

$$\begin{aligned} z^\Delta(t) &= m^\Delta(t) + (1 + b(t))m(t) \\ &\le a(t)z(t) + (1 + b(t))z(t) \\ &= (1 + a(t) + b(t))z(t), \quad t \in \mathbb{R}_+ \cap \mathbb{T}. \end{aligned}$$

Hence, using Lemma 2.1.1, we get

$$\begin{aligned} z(t) &\le z(0)e_c(t, 0) \\ &= 2x(0)e_c(t, 0), \\ m^\Delta(t) &\le a(t)z(t) \\ &\le 2x(0)a(t)e_c(t, 0), \quad t \in \mathbb{R}_+ \cap \mathbb{T}. \end{aligned}$$

From here,

$$\begin{aligned} m(t) &\le m(0) + 2x(0) \int_0^t a(s)e_c(s, 0)\Delta s \\ &= x(0) + 2x(0) \int_0^t a(s)e_c(s, 0)\Delta s \\ &= x(0)\left(1 + 2\int_0^t a(s)e_c(s, 0)\Delta s\right), \\ x^\Delta(t) &\le m(t) \\ &\le x(0)\left(1 + 2\int_0^t a(s)e_c(s, 0)\Delta s\right), \quad t \in \mathbb{R}_+ \cap \mathbb{T}. \end{aligned}$$

This completes the proof. □

Theorem 2.1.5 (Pachpatte's Inequality). *Let $x, x^\Delta, a, b \in \mathcal{C}_{rd}(\mathbb{R}_+ \cap \mathbb{T})$ and*

$$\begin{aligned} x^\Delta(t) &\le x(0) + \int_0^t a(s)(x(s) + x^\Delta(s))\Delta s \\ &\quad + \int_0^t a(s)\left(\int_0^s b(y)(x(y) + x^\Delta(y))\Delta y\right)\Delta s, \quad t \in \mathbb{R}_+ \cap \mathbb{T}. \end{aligned}$$

Then

$$x^{\Delta}(t) \le x(0)\left(1 + 2\int_0^t a(s)e_c(s,0)\Delta s + \int_0^t a(s)\left(\int_0^s (1+2b(y))e_{\ominus c}(\sigma(y),s)\Delta y\right)\Delta s\right), \quad t \in \mathbb{R}_+ \cap \mathbb{T},$$

where

$$c(t) = 2 + 2a(t) + b(t), \quad t \in \mathbb{R}_+ \cap \mathbb{T}.$$

Proof. Let

$$m(t) = x(0) + \int_0^t a(s)(x(s) + x^{\Delta}(s))\Delta s + \int_0^t a(s)\left(\int_0^s b(y)(x(y) + x^{\Delta}(y))\Delta y\right)\Delta s, \quad t \in \mathbb{R}_+ \cap \mathbb{T}.$$

Then

$$m(0) = x(0),$$
$$x^{\Delta}(t) \le m(t), \quad t \in \mathbb{R}_+ \cap \mathbb{T}.$$

Hence,

$$x(t) \le x(0) + \int_0^t m(s)\Delta s = m(0) + \int_0^t m(s)\Delta s, \quad t \in \mathbb{R}_+ \cap \mathbb{T},$$

and

$$\begin{aligned} m^{\Delta}(t) &= a(t)(x(t) + x^{\Delta}(t)) + a(t)\int_0^t b(y)(x(y) + x^{\Delta}(y))\Delta y \\ &= a(t)\left(x(t) + x^{\Delta}(t) + \int_0^t b(y)(x(y) + x^{\Delta}(y))\Delta y\right) \\ &\le a(t)\left(m(0) + \int_0^t m(s)\Delta s + m(t)\right. \end{aligned}$$

$$
\begin{aligned}
&+ \int_0^t b(s)\left(x(0) + \int_0^s m(y)\Delta y\right)\Delta s \\
&+ \int_0^t b(y)m(y)\Delta y\Bigg) \\
&= a(t)\left(m(0) + m(0)\int_0^t b(s)\Delta s + \int_0^t (1+b(y))m(y)\Delta y\right. \\
&\left. + m(t) + \int_0^t b(s)\int_0^s m(y)\Delta y\Delta s\right), \quad t \in \mathbb{R}_+ \cap \mathbb{T}.
\end{aligned}
$$

Let

$$
\begin{aligned}
z(t) &= m(0) + m(0)\int_0^t b(s)\Delta s + \int_0^t (1+b(y))m(y)\Delta y \\
&+ m(t) + \int_0^t b(s)\int_0^s m(y)\Delta y\Delta s, \quad t \in \mathbb{R}_+ \cap \mathbb{T}.
\end{aligned}
$$

Then

$$
\begin{aligned}
&m(t) \le z(t), \\
&m^{\Delta}(t) \le a(t)z(t), \\
&z(0) = 2m(0), \quad t \in \mathbb{R}_+ \cap \mathbb{T}.
\end{aligned}
$$

Hence,

$$
m(t) \le m(0) + \int_0^t a(s)z(s)\Delta s, \quad t \in \mathbb{R}_+ \cap \mathbb{T},
$$

and

$$
\begin{aligned}
z^{\Delta}(t) &= m^{\Delta}(t) + m(0)b(t) \\
&\quad + (1+b(t))m(t) + b(t)\int_0^t m(y)\Delta y \\
&\le a(t)z(t) + m(0)b(t) \\
&\quad + (1+b(t))\int_0^t a(s)z(s)\Delta s + (1+b(t))m(0) \\
&\quad + b(t)\int_0^t z(y)\Delta y
\end{aligned}
$$

$$\begin{aligned}&\le m(0)(1+2b(t))+a(t)z(t)+(1+b(t))\int_0^t(1+a(s))z(s)\Delta s\\&\le m(0)(1+2b(t))+(1+a(t)+b(t))\left(z(t)+\int_0^t(1+a(s))z(s)\Delta s\right),\end{aligned}$$

$t\in\mathbb{R}_+\cap\mathbb{T}$. Let

$$v(t)=z(t)+\int_0^t(1+a(s))z(s)\Delta s,\quad t\in\mathbb{R}_+\cap\mathbb{T}.$$

Then

$$\begin{aligned}v(0)&=z(0)\\&=2m(0),\\z^\Delta(t)&\le m(0)(1+2b(t))+(1+a(t)+b(t))v(t),\\z(t)&\le v(t),\\v^\Delta(t)&=z^\Delta(t)+(1+a(t))z(t)\\&\le m(0)(1+2b(t))+(1+a(t)+b(t))v(t)\\&\quad+(1+a(t))v(t)\\&\le m(0)(1+2b(t))+(2+2a(t)+b(t))v(t),\quad t\in\mathbb{R}_+\cap\mathbb{T}.\end{aligned}$$

Hence, employing Lemma 2.1.1, we get

$$v(t)\le 2m(0)e_c(t,0)+m(0)\int_0^t(1+2b(s))e_{\ominus c}(\sigma(s),t)\Delta s,\quad t\in\mathbb{R}_+\cap\mathbb{T}.$$

Therefore

$$\begin{aligned}z(t)&\le x(0)\left(2e_c(t,0)+\int_0^t(1+2b(s))e_{\ominus c}(\sigma(s),t)\Delta\right),\\m^\Delta(t)&\le x(0)a(t)\left(2e_c(t,0)+\int_0^t(1+2b(s))e_{\ominus c}(\sigma(s),t)\Delta s\right),\quad t\in\mathbb{R}_+\cap\mathbb{T}.\end{aligned}$$

Consequently,

$$\begin{aligned}m(t)&\le x(0)+2x(0)\int_0^t a(s)e_c(s,0)\Delta s\\&\quad+x(0)\int_0^t a(s)\left(\int_0^s(1+2b(y))e_{\ominus c}(\sigma(y),t)\Delta y\right)\Delta s,\end{aligned}$$

$t \in \mathbb{R}_+ \cap \mathbb{T}$, and

$$x^\Delta(t) \le x(0)\left(1 + 2\int_0^t a(s)e_c(s,0)\Delta s + \int_0^t a(s)\left(\int_0^s (1+2b(y))e_{\ominus c}(\sigma(y),t)\Delta y\right)\Delta s\right),$$

$t \in \mathbb{R}_+ \cap \mathbb{T}$. This completes the proof. □

2.2 Modifications of Pachpatte's inequalities

We will start with the following useful lemma.

Lemma 2.2.1. *Let $f \in \mathcal{C}_{rd}(\mathbb{R}_+ \cap \mathbb{T})$. Then*

$$\int_0^t \int_0^s f(y)\Delta y\Delta s = \int_0^t (t-\sigma(y))f(y)\Delta y, \quad t \in \mathbb{R}_+ \cap \mathbb{T}.$$

Proof. Let

$$g(t) = \int_0^t \int_0^s f(y)\Delta y\Delta s - \int_0^t (t-\sigma(y))f(y)\Delta y, \quad t \in \mathbb{R}_+ \cap \mathbb{T}.$$

Then $g(0) = 0$ and

$$\begin{aligned} g^\Delta(t) &= \int_0^t f(y)\Delta y - \int_0^t f(y)\Delta y - (\sigma(t)-\sigma(t))f(t) \\ &= 0, \quad t \in \mathbb{R}_+ \cap \mathbb{T}. \end{aligned}$$

Therefore

$$g(t) = 0, \quad t \in \mathbb{R}_+ \cap \mathbb{T}.$$

This completes the proof. □

Theorem 2.2.2. *Let $x, x^\Delta, x^{\Delta^2}, f, g, h \in \mathcal{C}_{rd}(\mathbb{R}_+ \cap \mathbb{T})$ be nonnegative functions and*

$$x^{\Delta^2}(t) \le f(t) + h(t)\int_0^t g(s)(x(s) + x^\Delta(s))\Delta s, \quad t \in \mathbb{R}_+ \cap \mathbb{T}.$$

Then

$$x^{\Delta^2}(t) \le f(t) + h(t)\int_0^t p(s)e_{\ominus q}(\sigma(s), t)\Delta s, \quad t \in \mathbb{R}_+ \cap \mathbb{T},$$

where

$$p(t) = g(t)\left(x(0) + (t+1)x^{\Delta}(0) + (t+1)\int_0^t f(s)\Delta s \right),$$

$$q(t) = (t+1)g(t)\int_0^t h(s)\Delta s, \quad t \in \mathbb{R}_+ \cap \mathbb{T}.$$

Proof. Let

$$z(t) = \int_0^t g(s)(x(s) + x^{\Delta}(s))\Delta s, \quad t \in \mathbb{R}_+ \cap \mathbb{T}.$$

Then

$$x^{\Delta^2}(t) \le f(t) + h(t)z(t), \quad t \in \mathbb{R}_+ \cap \mathbb{T},$$

and

$$z(0) = 0.$$

Hence, using Lemma 2.2.1, we get

$$\begin{aligned} x^{\Delta}(t) &\le x^{\Delta}(0) + \int_0^t (f(s) + h(s)z(s))\Delta s, \\ x(t) &\le x(0) + tx^{\Delta}(0) + \int_0^t \int_0^s (f(y) + h(y)z(y))\Delta y \Delta s \\ &= x(0) + tx^{\Delta}(0) + \int_0^t (t - \sigma(s))(f(s) + h(s)z(s))\Delta s \\ &\le x(0) + tx^{\Delta}(0) + \int_0^t (t - s)(f(s) + h(s)z(s))\Delta s, \quad t \in \mathbb{R}_+ \cap \mathbb{T}. \end{aligned}$$

Next, $z(t)$ is nondecreasing on $\mathbb{R}_+ \cap \mathbb{T}$ and

$$z^{\Delta}(t) = g(t)(x(t) + x^{\Delta}(t))$$

$$\begin{aligned}
&\leq g(t)\left(x(0) + tx^{\Delta}(0) + \int_0^t (t-s)(f(s) + h(s)z(s))\Delta s\right.\\
&\quad \left. + x^{\Delta}(0) + \int_0^t (f(s) + h(s)z(s))\Delta s\right)\\
&\leq g(t)\left(x(0) + tx^{\Delta}(0) + t\int_0^t f(s)\Delta s + t\int_0^t h(s)z(s)\Delta s\right.\\
&\quad \left. + x^{\Delta}(0) + \int_0^t f(s)\Delta s + \int_0^t h(s)z(s)\Delta s\right)\\
&\leq g(t)\left(x(0) + tx^{\Delta}(0) + t\int_0^t f(s)\Delta s + \left(t\int_0^t h(s)\Delta s\right)z(t)\right.\\
&\quad \left. + x^{\Delta}(0) + \int_0^t f(s)\Delta s + \left(\int_0^t h(s)\Delta s\right)z(t)\right)\\
&= p(t) + q(t)z(t), \quad t \in \mathbb{R}_+ \cap \mathbb{T}.
\end{aligned}$$

Hence, using Lemma 2.1.1, we obtain

$$\begin{aligned}
z(t) &\leq z(0)e_q(t,0) + \int_0^t p(s)e_{\ominus q}(\sigma(s),t)\Delta s\\
&= \int_0^t p(s)e_{\ominus q}(\sigma(s),t)\Delta s, \quad t \in \mathbb{R}_+ \cap \mathbb{T},
\end{aligned}$$

and

$$\begin{aligned}
x^{\Delta^2}(t) &\leq f(t) + h(t)z(t)\\
&\leq f(t) + h(t)\int_0^t p(s)e_{\ominus q}(\sigma(s),t)\Delta s, \quad t \in \mathbb{R}_+ \cap \mathbb{T}.
\end{aligned}$$

This completes the proof. □

Theorem 2.2.3. *Let $x, x^{\Delta}, x^{\Delta^2}, f, g, h \in \mathcal{C}_{rd}(\mathbb{R}_+ \cap \mathbb{T})$ be nonnegative functions and*

$$x^{\Delta^2}(t) \leq f(t) + h(t)\left(x^{\Delta}(t) + \int_0^t g(s)(x(s) + x^{\Delta}(s))\Delta s\right), \quad t \in \mathbb{R}_+ \cap \mathbb{T}.$$

Then

$$\begin{aligned}
x^{\Delta^2}(t) &\leq f(t) + x^{\Delta}(0)h(t)e_q(t,0)\\
&\quad + h(t)\int_0^t p(s)e_{\ominus q}(\sigma(s),t)\Delta s, \quad t \in \mathbb{R}_+ \cap \mathbb{T},
\end{aligned}$$

where

$$p(t) = f(t) + g(t)x(0) + (1+t)g(t)x^{\Delta}(0) + (1+t)g(t)\int_0^t f(s)\Delta s,$$

$$q(t) = h(t) + (1+t)g(t)\int_0^t h(s)\Delta s, \quad t \in \mathbb{R}_+ \cap \mathbb{T}.$$

Proof. Let

$$z(t) = x^{\Delta}(t) + \int_0^t g(s)(x(s) + x^{\Delta}(s))\Delta s, \quad t \in \mathbb{R}_+ \cap \mathbb{T}.$$

Then

$$x^{\Delta^2}(t) \le f(t) + h(t)z(t), \quad t \in \mathbb{R}_+ \cap \mathbb{T}.$$

Applying Lemma 2.2.1, we get

$$x^{\Delta}(t) \le x^{\Delta}(0) + \int_0^t (f(s) + h(s)z(s))\Delta s,$$

$$\begin{aligned}
x(t) &\le x(0) + tx^{\Delta}(0) + \int_0^t \int_0^s (f(y) + h(y)z(y))\Delta y\Delta s \\
&= x(0) + tx^{\Delta}(0) + \int_0^t (t - \sigma(s))(f(s) + h(s)z(s))\Delta s \\
&\le x(0) + tx^{\Delta}(0) + \int_0^t (t - s)(f(s) + h(s)z(s))\Delta s, \quad t \in \mathbb{R}_+ \cap \mathbb{T}.
\end{aligned}$$

Therefore, since $z(t)$ is a nondecreasing function on $\mathbb{R}_+ \cap \mathbb{T}$, we obtain

$$\begin{aligned}
z^{\Delta}(t) &= x^{\Delta^2}(t) + g(t)(x(t) + x^{\Delta}(t)) \\
&\le f(t) + h(t)z(t) \\
&\quad + g(t)x(0) + tg(t)x^{\Delta}(0) + tg(t)\int_0^t f(s)\Delta s \\
&\quad + tg(t)\left(\int_0^t h(s)\Delta s\right)z(t) + g(t)x^{\Delta}(0) \\
&\quad + g(t)\int_0^t f(s)\Delta s + g(t)\left(\int_0^t h(s)\Delta s\right)z(t) \\
&= p(t) + q(t)z(t), \quad t \in \mathbb{R}_+ \cap \mathbb{T}.
\end{aligned}$$

Hence, by Lemma 2.1.1, we obtain

$$\begin{aligned} z(t) &\le z(0)e_q(t,0) + \int_0^t p(s)e_{\ominus q}(\sigma(s),t)\Delta s \\ &= x^{\Delta}(0)e_q(t,0) + \int_0^t p(s)e_{\ominus q}(\sigma(s),t)\Delta s, \\ x^{\Delta^2}(t) &\le f(t) + x^{\Delta}(0)h(t)e_q(t,0) \\ &\quad + h(t)\int_0^t p(s)e_{\ominus q}(\sigma(s),t)\Delta s, \quad t \in \mathbb{R}_+ \cap \mathbb{T}. \end{aligned}$$

This completes the proof. □

2.3 Inequalities with several iterated integrals

Theorem 2.3.1. *Let $x, h \in \mathcal{C}_{rd}(\mathbb{R}_+ \cap \mathbb{T})$ be nonnegative functions, $r \in \mathcal{C}_{rd}(\mathbb{R}_+ \cap \mathbb{T})$ be a positive function, x_0 be a positive constant, and*

$$x(t) \le x_0 + \int_0^t r(t_1)\int_0^{t_1} h(t_2)x(t_2)\Delta t_2\Delta t_1, \quad t \in \mathbb{R}_+ \cap \mathbb{T}.$$

Then

$$x(t) \le x_0 e_g(t,0), \quad t \in \mathbb{R}_+ \cap \mathbb{T},$$

where

$$g(t) = r(t)\int_0^t h(t_1)\Delta t_1, \quad t \in \mathbb{R}_+ \cap \mathbb{T}.$$

Proof. Let

$$v(t) = x_0 + \int_0^t r(t_1)\int_0^{t_1} h(t_2)x(t_2)\Delta t_2\Delta t_1, \quad t \in \mathbb{R}_+ \cap \mathbb{T}.$$

Then

$$\begin{aligned} v(0) &= x_0, \\ x(t) &\le v(t), \quad t \in \mathbb{R}_+ \cap \mathbb{T}, \end{aligned}$$

and

$$v^\Delta(t) = r(t)\int_0^t h(t_2)x(t_2)\Delta t_2, \quad t \in \mathbb{R}_+ \cap \mathbb{T}.$$

Hence,

$$\begin{aligned} v^\Delta(0) &= 0, \\ \frac{v^\Delta(t)}{r(t)} &= \int_0^t h(t_2)x(t_2)\Delta t_2 \\ &\le \int_0^t h(t_2)v(t_2)\Delta t_2, \quad t \in \mathbb{R}_+ \cap \mathbb{T}. \end{aligned}$$

Note that v is a nondecreasing function on $\mathbb{R}_+ \cap \mathbb{T}$. Therefore

$$\left(\frac{v^\Delta(t)}{r(t)}\right)^\Delta \le h(t)v(t),$$

or

$$\begin{aligned} \frac{(\frac{v^\Delta(t)}{r(t)})^\Delta v(t)}{v(t)v(\sigma(t))} &\le h(t) \\ &\le h(t) + \frac{\frac{(v^\Delta(t))^2}{r(t)}}{v(t)v(\sigma(t))}, \quad t \in \mathbb{R}_+ \cap \mathbb{T}. \end{aligned}$$

From here,

$$\frac{(\frac{v^\Delta(t)}{r(t)})^\Delta v(t) - \frac{(v^\Delta(t))^2}{r(t)}}{v(t)v(\sigma(t))} \le h(t),$$

or

$$\left(\frac{v^\Delta(t)}{r(t)v(t)}\right)^\Delta \le h(t), \quad t \in \mathbb{R}_+ \cap \mathbb{T}.$$

Consequently,

$$\frac{v^\Delta(t)}{r(t)v(t)} \le \int_0^t h(t_1)\Delta t_1,$$

or

$$\begin{aligned} v^\Delta(t) &\le v(t)r(t)\int_0^t h(t_1)\Delta t_1 \\ &= v(t)g(t), \quad t \in \mathbb{R}_+ \cap \mathbb{T}. \end{aligned}$$

Hence, by Lemma 2.1.1, we obtain

$$\begin{aligned} v(t) &\le v(0)e_g(t,0) \\ &= x_0 e_g(t,0) \end{aligned}$$

and

$$\begin{aligned} x(t) &\le v(t) \\ &\le x_0 e_g(t,0), \quad t \in \mathbb{R}_+ \cap \mathbb{T}. \end{aligned}$$

This completes the proof. □

Theorem 2.3.2. *Let $x, h \in \mathcal{C}_{rd}(\mathbb{R}_+ \cap \mathbb{T})$ be nonnegative functions, $g \in \mathcal{C}_{rd}(\mathbb{R}_+ \cap \mathbb{T})$ be a positive function, x_0 be a positive constant, and*

$$x(t) \le x_0 + \int_0^t \frac{1}{g(t_1)} \int_0^{t_1} h(t_2)x(t_2)\Delta t_2 \Delta t_1, \quad t \in \mathbb{R}_+ \cap \mathbb{T}.$$

Then

$$x(t) \le x_0 e_p(t,0), \quad t \in \mathbb{R}_+ \cap \mathbb{T},$$

where

$$p(t) = \frac{\int_0^t h(s)\Delta s}{g(t)}, \quad t \in \mathbb{R}_+ \cap \mathbb{T}.$$

Proof. Let

$$z(t) = x_0 + \int_0^t \frac{1}{g(t_1)} \int_0^{t_1} h(t_2)x(t_2)\Delta t_2 \Delta t_1, \quad t \in \mathbb{R}_+ \cap \mathbb{T}.$$

Then

$$z(0) = x_0$$

and

$$\begin{aligned} x(t) &\le z(t), \\ z^\Delta(t) &= \frac{1}{g(t)} \int_0^t h(t_2)x(t_2)\Delta t_2, \\ g(t)z^\Delta(t) &= \int_0^t h(t_2)x(t_2)\Delta t_2, \quad t \in \mathbb{R}_+ \cap \mathbb{T}. \end{aligned}$$

Hence,

$$\begin{aligned}(g(t)z^{\Delta}(t))^{\Delta} &= h(t)x(t)\\ &\le h(t)z(t),\end{aligned}$$

or

$$\frac{(g(t)z^{\Delta}(t))^{\Delta}}{z(t)} \le h(t), \quad t \in \mathbb{R}_+ \cap \mathbb{T}.$$

Note that $z(t)$ is a positive nondecreasing function on $\mathbb{R}_+ \cap \mathbb{T}$. Therefore

$$\begin{aligned}\frac{(g(t)z^{\Delta}(t))^{\Delta}z(t)}{z(t)z(\sigma(t))} &\le \frac{(g(t)z^{\Delta}(t))^{\Delta}}{z(t)}\\ &\le h(t) + \frac{g(t)(z^{\Delta}(t))^2}{z(t)z(\sigma(t))}, \quad t \in \mathbb{R}_+ \cap \mathbb{T}.\end{aligned}$$

Consequently,

$$\left(\frac{g(t)z^{\Delta}(t)}{z(t)}\right)^{\Delta} \le h(t), \quad t \in \mathbb{R}_+ \cap \mathbb{T}.$$

From here,

$$\frac{g(t)z^{\Delta}(t)}{z(t)} \le \int_0^t h(s)\Delta s,$$

or

$$\begin{aligned}z^{\Delta}(t) &\le \frac{\int_0^t h(s)\Delta s}{g(t)} z(t)\\ &= p(t)z(t), \quad t \in \mathbb{R}_+ \cap \mathbb{T}.\end{aligned}$$

Hence, employing Lemma 2.1.1, we obtain

$$\begin{aligned}z(t) &\le z(0)e_p(t,0)\\ &= x_0 e_p(t,0), \quad t \in \mathbb{R}_+ \cap \mathbb{T}.\end{aligned}$$

Consequently,

$$\begin{aligned}x(t) &\le z(t)\\ &\le x_0 e_p(t,0), \quad t \in \mathbb{R}_+ \cap \mathbb{T}.\end{aligned}$$

This completes the proof. □

Theorem 2.3.3. *Let $x, h \in \mathcal{C}_{rd}(\mathbb{R}_+ \cap \mathbb{T})$ be nonnegative functions, $g \in \mathcal{C}_{rd}(\mathbb{R}_+ \cap \mathbb{T})$ be a positive function, x_0 be a nonnegative constant, and*

$$x(t) \le x_0 + \int_0^t \frac{1}{g(t_1)} \int_0^{t_1} \int_0^{\tau_1} h(t_2)x(t_2)\Delta t_2 \Delta t_1, \quad t \in \mathbb{R}_+ \cap \mathbb{T}.$$

Then

$$x(t) \le x_0\left(1 + \int_0^t \frac{t_1}{g(t_1)} \int_0^{t_1} h(s)\left(1 + \frac{s}{g(s)} \int_0^s h(y)e_{\ominus f}(\sigma(y), s)\Delta y\right)\Delta s_1 \Delta t_1 + 1\right),$$

$t \in \mathbb{R}_+ \cap \mathbb{T}$, where

$$f(t) = \frac{t}{g(t)}(h(t) + 1), \quad t \in \mathbb{R}_+ \cap \mathbb{T}.$$

Proof. Let

$$z(t) = x_0 + \int_0^t \frac{1}{g(t_1)} \int_0^{t_1} \int_0^{\tau_1} h(t_2)x(t_2)\Delta t_2 \Delta t_1, \quad t \in \mathbb{R}_+ \cap \mathbb{T}.$$

Then

$$x(t) \le z(t),$$
$$z(0) = x_0,$$

and

$$\begin{aligned} z^\Delta(t) &= \frac{1}{g(t)} \int_0^t \int_0^{t_1} h(t_2)x(t_2)\Delta t_2 \Delta t_1 \\ &\le \frac{1}{g(t)} \int_0^t \int_0^{t_1} h(t_2)z(t_2)\Delta t_2 \Delta t_1 \\ &= \frac{1}{g(t)} \int_0^t (t - \sigma(t_1))h(t_1)z(t_1)\Delta t_1 \\ &\le \frac{t}{g(t)} \int_0^t h(t_1)z(t_1)\Delta t_1 \\ &\le \frac{t}{g(t)} \int_0^t h(t_1)(z^\Delta(t_1) + z(t_1))\Delta t_1, \quad t \in \mathbb{R}_+ \cap \mathbb{T}. \end{aligned}$$

Hence, by Theorem 2.1.2, we get

$$\begin{aligned}
z^{\Delta}(t) &\le \frac{t}{g(t)}\int_0^t h(s)\left(z(0)+\frac{s}{g(s)}z(0)\int_0^s h(y)e_{\ominus f}(\sigma(y),s)\Delta y\right)\Delta s\\
&= z(0)\frac{t}{g(t)}\int_0^t h(s)\left(1+\frac{s}{g(s)}\int_0^s h(y)e_{\ominus f}(\sigma(y),s)\Delta y\right)\Delta s\\
&= x_0\frac{t}{g(t)}\int_0^t h(s)\left(1+\frac{s}{g(s)}\int_0^s h(y)e_{\ominus f}(\sigma(y),s)\Delta y\right)\Delta s, \quad t\in\mathbb{R}_+\cap\mathbb{T}.
\end{aligned}$$

Then

$$z(t)\le x_0\left(1+\int_0^t \frac{t_1}{g(t_1)}\int_0^{t_1} h(s)\left(1+\frac{s}{g(s)}\int_0^s h(y)e_{\ominus f}(\sigma(y),s)\Delta y\right)\Delta s\Delta t_1\right)$$

and

$$\begin{aligned}
x(t)&\le z(t)\\
&\le x_0\left(1+\int_0^t \frac{t_1}{g(t_1)}\int_0^{t_1} h(s)\left(1+\frac{s}{g(s)}\int_0^s h(y)e_{\ominus f}(\sigma(y),s)\Delta y\right)\Delta s\Delta t_1\right),
\end{aligned}$$

$t\in\mathbb{R}_+\cap\mathbb{T}$. This completes the proof. □

Theorem 2.3.4. *Let $x,p,q\in\mathcal{C}_{rd}(\mathbb{R}_+\cap\mathbb{T})$ be nonnegative functions, $r\in\mathcal{C}_{rd}(\mathbb{R}_+\cap\mathbb{T})$ be a positive function, x_0 be a nonnegative constant, and*

$$x(t)\le x_0+\int_0^t r(t_1)\int_0^{t_1} p(t_2)\left(x(t_2)+\int_0^{t_2} r(t_3)\int_0^{t_3} q(t_4)x(t_4)\Delta t_4\Delta t_3\right)\Delta t_2\Delta t_1,$$

$t\in\mathbb{R}_+\cap\mathbb{T}$. Then

$$x(t)\le x_0\left(1+\int_0^t r(t_1)\int_0^{t_1} p(t_2)\left(1+\int_0^{t_2} g(t_3)\Delta t_3\right)\Delta t_2\Delta t_1\right),\quad t\in\mathbb{R}_+\cap\mathbb{T},$$

where

$$\begin{aligned}
f(t)&=r(t)(p(t)+q(t)+1),\\
g(t)&=1+r(t)\int_0^t (p(s)+q(s))\left(1+r(s)\int_0^s (p(\tau)+q(\tau))e_{\ominus f}(\sigma(\tau),s)\Delta\tau\right)\Delta s,
\end{aligned}$$

$t\in\mathbb{R}_+\cap\mathbb{T}$.

Proof. Let

$$z(t) = x_0 + \int_0^t r(t_1) \int_0^{t_1} p(t_2)\left(x(t_2) + \int_0^{t_2} r(t_3) \int_0^{t_3} q(t_4)x(t_4)\Delta t_4 \Delta t_3 \right)\Delta t_2 \Delta t_1, \quad t \in \mathbb{R}_+ \cap \mathbb{T}.$$

Then

$$z(0) = x_0,$$
$$x(t) \le z(t), \quad t \in \mathbb{R}_+ \cap \mathbb{T},$$

and

$$z^\Delta(t) = r(t) \int_0^t p(t_1)\left(x(t_1) + \int_0^{t_1} r(t_2) \int_0^{t_2} q(t_3)x(t_3)\Delta t_3 \Delta t_2 \right)\Delta t_1, \tag{2.8}$$

and

$$\begin{aligned} \frac{z^\Delta(t)}{r(t)} &= \int_0^t p(t_1)\left(x(t_1) + \int_0^{t_1} r(t_2) \int_0^{t_2} q(t_3)x(t_3)\Delta t_3 \Delta t_2 \right)\Delta t_1, \\ \left(\frac{z^\Delta(t)}{r(t)}\right)^\Delta &= p(t)\left(x(t) + \int_0^t r(t_1) \int_0^{t_1} q(t_2)x(t_2)\Delta t_2 \Delta t_1 \right) \\ &\le p(t)\left(z(t) + \int_0^t r(t_1) \int_0^{t_1} q(t_2)z(t_2)\Delta t_2 \Delta t_1 \right), \quad t \in \mathbb{R}_+ \cap \mathbb{T}. \end{aligned}$$

Let

$$v(t) = z(t) + \int_0^t r(t_1) \int_0^{t_1} q(t_2)z(t_2)\Delta t_2 \Delta t_1, \quad t \in \mathbb{R}_+ \cap \mathbb{T}.$$

Hence,

$$\begin{aligned} v(0) &= z(0) \\ &= x_0, \\ z(t) &\le v(t), \quad t \in \mathbb{R}_+ \cap \mathbb{T}, \end{aligned}$$

and

$$v^\Delta(t) = z^\Delta(t) + r(t) \int_0^t q(t_1)z(t_1)\Delta t_1, \quad t \in \mathbb{R}_+ \cap \mathbb{T}. \tag{2.9}$$

By (2.8), we get

$$z^{\Delta}(t) \le r(t)\int_0^t p(t_1)\left(z(t_1)+\int_0^{t_1} r(t_2)\int_0^{t_2} q(t_3)z(t_3)\Delta t_3\Delta t_2\right)\Delta t_1$$
$$= r(t)\int_0^t p(t_1)v(t_1)\Delta t_1, \quad t \in \mathbb{R}_+ \cap \mathbb{T}.$$

Hence, from (2.9), we obtain

$$v^{\Delta}(t) \le r(t)\int_0^t p(t_1)v(t_1)\Delta t_1 + r(t)\int_0^t q(t_1)v(t_1)\Delta t_1$$
$$= r(t)\int_0^t (p(t_1)+q(t_1))v(t_1)\Delta t_1$$
$$\le r(t)\int_0^t (p(t_1)+q(t_1))(v(t_1)+v^{\Delta}(t_1))\Delta t_1, \quad t \in \mathbb{R}_+ \cap \mathbb{T}.$$

Hence Theorem 2.1.2 now yields

$$v^{\Delta}(t) \le v(0) + r(t)\int_0^t (p(s)+q(s))\left(v(0)+r(s)v(0)\int_0^s (p(\tau)+q(\tau))e_{\ominus f}(\sigma(\tau),s)\Delta\tau\right)$$
$$= v(0)\left(1+r(t)\int_0^t (p(s)+q(s))\left(1+r(s)\int_0^s (p(\tau)+q(\tau))e_{\ominus f}(\sigma(\tau),s)\Delta\tau\right)\Delta s\right)$$
$$= x_0\left(1+r(t)\int_0^t (p(s)+q(s))\left(1+r(s)\int_0^s (p(\tau)+q(\tau))e_{\ominus f}(\sigma(\tau),s)\Delta\tau\right)\Delta s\right)$$
$$= x_0 g(t), \quad t \in \mathbb{R}_+ \cap \mathbb{T}.$$

Hence,

$$v(t) \le v(0) + x_0\int_0^t g(s)\Delta s$$
$$= x_0\left(1+\int_0^t g(s)\Delta s\right),$$
$$\left(\frac{z^{\Delta}(t)}{r(t)}\right)^{\Delta} \le x_0 p(t)\left(1+\int_0^t g(s)\Delta s\right),$$

$$\frac{z^{\Delta}(t)}{r(t)} \le x_0 \int_0^t p(t_1)\left(1 + \int_0^{t_1} g(t_2)\Delta t_2\right)\Delta t_1,$$

$$z^{\Delta}(t) \le x_0 r(t) \int_0^t p(t_1)\left(1 + \int_0^{t_1} g(t_2)\Delta t_2\right)\Delta t_1,$$

$$\begin{aligned} z(t) &\le x_0 + x_0 \int_0^t r(t_1) \int_0^{t_1} p(t_2)\left(1 + \int_0^{t_2} g(t_3)\Delta t_3\right)\Delta t_2 \Delta t_1 \\ &= x_0\left(1 + \int_0^t r(t_1) \int_0^{t_1} p(t_2)\left(1 + \int_0^{t_2} g(t_3)\Delta t_3\right)\Delta t_2 \Delta t_1\right), \end{aligned}$$

$$\begin{aligned} x(t) &\le z(t) \\ &\le x_0\left(1 + \int_0^t r(t_1) \int_0^{t_1} p(t_2)\left(1 + \int_0^{t_2} g(t_3)\Delta t_3\right)\Delta t_2 \Delta t_1\right) \quad t \in \mathbb{R}_+ \cap \mathbb{T}. \end{aligned}$$

This completes the proof. □

Theorem 2.3.5. *Let $x, q \in \mathcal{C}_{rd}(\mathbb{R}_+ \cap \mathbb{T})$ be nonnegative functions, $p, r \in \mathcal{C}_{rd}(\mathbb{R}_+ \cap \mathbb{T})$ be positive functions, x_0 be a nonnegative constant, and*

$$\begin{aligned} x(t) \le x_0 + \int_0^t r(t_1) \int_0^{t_1}\int_0^{t_2} p(t_3)\Bigg(x(t_3) + \\ + \int_0^{t_3} r(t_4) \int_0^{t_4}\int_0^{t_5} q(t_6)x(t_6)\Delta t_6 \Delta t_5 \Delta t_4\Bigg)\Delta t_3 \Delta t_2 \Delta t_1, \quad t \in \mathbb{R}_+ \cap \mathbb{T}. \end{aligned}$$

Then

$$x(t) \le x_0\left(1 + \int_0^t r(t_1) \int_0^{t_1} (t_1 - t_2)p(t_2)\left(1 + \int_0^{t_2} g(t_3)\Delta t_3\right)\Delta t_2 \Delta t_1\right),$$

$t \in \mathbb{R}_+ \cap \mathbb{T}$, where

$$f(t) = tr(t)(p(t) + q(t) + 1),$$

$$g(t) = tr(t) \int_0^t (p(t_1) + q(t_1))\left(1 + t_1 r(t_1) \int_0^{t_1} (p(t_2) + q(t_2)) e_{\ominus f}(\sigma(t_2), t_1)\Delta t_2\right)\Delta t_1,$$

$t \in \mathbb{R}_+ \cap \mathbb{T}$.

Proof. Let

$$z(t) = x_0 + \int_0^t r(t_1) \int_0^{t_1} \int_0^{t_2} p(t_3) \Bigg(x(t_3) + \\ + \int_0^{t_3} r(t_4) \int_0^{t_4} \int_0^{t_5} q(t_6)x(t_6)\Delta t_6 \Delta t_5 \Delta t_4 \Bigg) \Delta t_3 \Delta t_2 \Delta t_1, \quad t \in \mathbb{R}_+ \cap \mathbb{T}.$$

Then

$$z(0) = x_0$$

and

$$z^\Delta(t) = r(t) \int_0^t \int_0^{t_1} p(t_2) \Bigg(x(t_2) + \int_0^{t_2} r(t_3) \int_0^{t_3} \int_0^{t_4} q(t_5)x(t_5)\Delta t_5 \Delta t_4 \Delta t_3 \Bigg) \Delta t_2 \Delta t_1, \tag{2.10}$$

$t \in \mathbb{R}_+ \cap \mathbb{T}$. Also,

$$\frac{z^\Delta(t)}{r(t)} = \int_0^t \int_0^{t_1} p(t_2) \Bigg(x(t_2) + \int_0^{t_2} r(t_3) \int_0^{t_3} \int_0^{t_4} q(t_5)x(t_5)\Delta t_5 \Delta t_4 \Delta t_3 \Bigg) \Delta t_2 \Delta t_1,$$

$$\left(\frac{z^\Delta(t)}{r(t)} \right)^{\Delta^2} = p(t) \Bigg(x(t) + \int_0^t r(t_1) \int_0^{t_1} \int_0^{t_2} q(t_3)x(t_3)\Delta t_3 \Delta t_2 \Delta t_1 \Bigg) \Delta t,$$

$$\begin{aligned} \frac{1}{p(t)} \left(\frac{z^\Delta(t)}{r(t)} \right)^{\Delta^2} &= x(t) + \int_0^t r(t_1) \int_0^{t_1} \int_0^{t_2} q(t_3)x(t_3)\Delta t_3 \Delta t_2 \Delta t_1 \\ &\le z(t) + \int_0^t r(t_1) \int_0^{t_1} \int_0^{t_2} q(t_3)z(t_3)\Delta t_3 \Delta t_2 \Delta t_1, \quad t \in \mathbb{R}_+ \cap \mathbb{T}. \end{aligned}$$

Let

$$v(t) = z(t) + \int_0^t r(t_1) \int_0^{t_1} \int_0^{t_2} q(t_3)z(t_3)\Delta t_3 \Delta t_2 \Delta t_1, \quad t \in \mathbb{R}_+ \cap \mathbb{T}.$$

Then

$$v^\Delta(t) = z^\Delta(t) + r(t) \int_0^t \int_0^{t_1} q(t_2)z(t_2)\Delta t_2 \Delta t_1, \quad t \in \mathbb{R}_+ \cap \mathbb{T}. \tag{2.11}$$

By (2.10), we get

$$z^\Delta(t) \le r(t) \int_0^t \int_0^{t_1} p(t_2) \Bigg(z(t_2) + \int_0^{t_2} r(t_3) \int_0^{t_3} \int_0^{t_4} q(t_5)z(t_5)\Delta t_5 \Delta t_4 \Delta t_3 \Bigg) \Delta t_2 \Delta t_1$$

$$= r(t) \int_0^t \int_0^{t_1} p(t_2) v(t_2) \Delta t_2 \Delta t_1, \quad t \in \mathbb{R}_+ \cap \mathbb{T}.$$

Hence, from (2.11), we obtain

$$\begin{aligned} v^\Delta(t) &\le r(t) \int_0^t \int_0^{t_1} p(t_2) v(t_2) \Delta t_2 \Delta t_1 \\ &\quad + r(t) \int_0^t \int_0^{t_1} q(t_2) v(t_2) \Delta t_2 \Delta t_1 \\ &= r(t) \int_0^t \int_0^{t_1} (p(t_2) + q(t_2)) v(t_2) \Delta t_2 \Delta t_1 \\ &= r(t) \int_0^t (t - \sigma(t_1))(p(t_1) + q(t_1)) v(t_1) \Delta t_1 \\ &\le tr(t) \int_0^t (p(t_1) + q(t_1)) v(t_1) \Delta t_1 \\ &\le tr(t) \int_0^t (p(t_1) + q(t_1))(v(t_1) + v^\Delta(t_1)) \Delta t_1, \quad t \in \mathbb{R}_+ \cap \mathbb{T}. \end{aligned}$$

Theorem 2.1.2 now implies

$$\begin{aligned} v^\Delta(t) &\le tr(t) \int_0^t (p(t_1) + q(t_1)) \left(v(0) + t_1 r(t_1) v(0) \int_0^{t_1} (p(t_2) + q(t_2)) e_{\ominus f}(\sigma(t_2), t_1) \Delta t_2 \right) \Delta t_1 \\ &= x_0 tr(t) \int_0^t (p(t_1) + q(t_1)) \left(1 + t_1 r(t_1) \int_0^{t_1} (p(t_2) + q(t_2)) e_{\ominus f}(\sigma(t_2), t_1) \Delta t_2 \right) \Delta t_1 \\ &= x_0 g(t), \quad t \in \mathbb{R}_+ \cap \mathbb{T}. \end{aligned}$$

Therefore

$$\begin{aligned} v(t) &\le x_0 + x_0 \int_0^t g(t_1) \Delta t_1 \\ &= x_0 \left(1 + \int_0^t g(t_1) \Delta t_1 \right), \end{aligned}$$

$$\frac{1}{p(t)} \left(\frac{z^\Delta(t)}{r(t)} \right)^{\Delta^2} \le v(t)$$

$$\begin{aligned}
&\leq x_0\left(1+\int_0^t g(t_1)\Delta t_1\right),\\
\left(\frac{z^{\Delta}(t)}{r(t)}\right)^{\Delta^2} &\leq x_0 p(t)\left(1+\int_0^t g(t_1)\Delta t_1\right),\\
\frac{z^{\Delta}(t)}{r(t)} &\leq x_0\int_0^t\int_0^{t_1} p(t_2)\left(1+\int_0^{t_2} g(t_3)\Delta t_3\right)\Delta t_2\Delta t_1\\
&= x_0\int_0^t (t-\sigma(t_1))p(t_1)\left(1+\int_0^{t_1} g(t_2)\Delta t_2\right)\Delta t_1,\\
z^{\Delta}(t) &\leq x_0 r(t)\int_0^t (t-t_1)p(t_1)\left(1+\int_0^{t_1} g(t_2)\Delta t_2\right)\Delta t_1,\\
z(t) &\leq x_0 + x_0\int_0^t r(t_1)\int_0^{t_1}(t_1-t_2)p(t_2)\left(1+\int_0^{t_2} g(t_3)\Delta t_3\right)\Delta t_2\Delta t_1,
\end{aligned}$$

$t \in \mathbb{R}_+ \cap \mathbb{T}$. Then

$$\begin{aligned}
x(t) &\leq z(t)\\
&\leq x_0\left(1+\int_0^t r(t_1)\int_0^{t_1}(t_1-t_2)p(t_2)\left(1+\int_0^{t_2} g(t_3)\Delta t_3\right)\Delta t_2\Delta t_1\right),
\end{aligned}$$

$t \in \mathbb{R}_+ \cap \mathbb{T}$. This completes the proof. □

Theorem 2.3.6. *Let $x, f, g, h \in \mathcal{C}_{rd}(\mathbb{R}_+ \cap \mathbb{T})$ be nonnegative functions, $r \in \mathcal{C}_{rd}(\mathbb{R}_+ \cap \mathbb{T})$ be a positive function such that $r(t) \geq 1$, $t \in \mathbb{R}_+ \cap \mathbb{T}$, and*

$$x(t) \leq f(t) + g(t)\int_0^t r(t_1)\int_0^{t_1} h(t_2)x(t_2)\Delta t_2\Delta t_1, \quad t \in \mathbb{R}_+ \cap \mathbb{T}.$$

Then

$$x(t) \leq f(t) + g(t)\int_0^t\left(h(t_1) + r(t_1)\int_0^{t_1} h_4(t_2)\Delta t_2\right)\Delta t_1, \quad t \in \mathbb{R}_+ \cap \mathbb{T},$$

where

$$\begin{aligned}
h_1(t) &= r(t)\int_0^t h(t_1)f(t_1)\Delta t_1,\\
h_2(t) &= h(t)g(t),
\end{aligned}$$

$$h_3(t) = r(t)(h_2(t) + 1),$$

$$h_4(t) = h_2(t)\Bigg(\int_0^t h_1(s)\Delta s + h_1(t) + r(t)\int_0^t h_2(s)\Bigg(\int_0^s h_1(y)\Delta y + h_1(s)\Bigg)e_{\ominus h_3}(\sigma(s), t)\Delta s\Bigg), \quad t \in \mathbb{R}_+ \cap \mathbb{T}.$$

Proof. Let

$$z(t) = \int_0^t r(t_1)\int_0^{t_1} h(t_2)x(t_2)\Delta t_2\Delta t_1, \quad t \in \mathbb{R}_+ \cap \mathbb{T}.$$

Then

$$z(0) = 0$$

and

$$x(t) \le f(t) + g(t)z(t), \quad t \in \mathbb{R}_+ \cap \mathbb{T}.$$

Next we can estimate

$$\begin{aligned} z^\Delta(t) &= r(t)\int_0^t h(t_1)x(t_1)\Delta t_1 \\ &\le r(t)\int_0^t h(t_1)(f(t_1) + g(t_1)z(t_1))\Delta t_1 \\ &= r(t)\int_0^t h(t_1)f(t_1)\Delta t_1 + r(t)\int_0^t h(t_1)g(t_1)z(t_1)\Delta t_1 \\ &\le h_1(t) + r(t)\int_0^t h_2(t_1)(z(t_1) + z^\Delta(t_1))\Delta t_1, \quad t \in \mathbb{R}_+ \cap \mathbb{T}. \end{aligned}$$

Hence, using Theorem 2.1.2, we obtain

$$\begin{aligned} z^\Delta(t) \le h_1(t) + r(t)\int_0^t h_2(t_1)\Bigg(z(0) + \int_0^{t_1} h_1(s)\Delta s + h_1(t_1) + r(t_1)\int_0^{t_1} h_2(s)\Bigg(z(0) \\ + \int_0^s h_1(y)\Delta y + h_1(s)\Bigg)e_{\ominus h_3}(\sigma(s), t_1)\Delta s\Bigg)\Delta t_1 \\ = h_1(t) + r(t)\int_0^t h_2(t_1)\Bigg(\int_0^{t_1} h_1(s)\Delta s + h_1(t_1) \end{aligned}$$

$$+ r(t_1)\int_0^{t_1} h_2(s)\left(\int_0^s h_1(y)\Delta y + h_1(s)\right)e_{\ominus h_3}(\sigma(s),t_1)\Delta s\Bigg)\Delta t_1$$

$$= h_1(t) + r(t)\int_0^t h_4(t_1)\Delta t_1,$$

$$z(t) \le \int_0^t \left(h_1(t_1) + r(t_1)\int_0^{t_1} h_4(t_2)\Delta t_2\right)\Delta t_1,$$

$$x(t) \le f(t) + g(t)\int_0^t \left(h(t_1) + r(t_1)\int_0^{t_1} h_4(t_2)\Delta t_2\right)\Delta t_1, \quad t \in \mathbb{R}_+ \cap \mathbb{T}.$$

This completes the proof. □

Theorem 2.3.7. *Let $f, q, p, x \in \mathcal{C}_{rd}(\mathbb{R}_+ \cap \mathbb{T})$ be nonnegative functions, $g \in \mathcal{C}_{rd}(\mathbb{R}_+ \cap \mathbb{T})$ be a positive function such that $tg(t) \ge 1$, $t \in \mathbb{R}_+ \cap \mathbb{T}$, and*

$$x(t) \le f(t) + q(t)\int_0^t g(t_1)\int_0^{t_1}\int_0^{\tau_1} p(\tau_2)x(\tau_2)\Delta\tau_2\Delta\tau_1\Delta t_1, \quad t \in \mathbb{R}_+ \cap \mathbb{T}.$$

Then

$$x(t) \le f(t) + g(t)\int_0^t h_4(s)\Delta s, \quad t \in \mathbb{R}_+ \cap \mathbb{T},$$

where

$$h_1(t) = g(t)\int_0^t\int_0^{t_1} p(t_2)f(t_2)\Delta t_2\Delta t_1,$$

$$h_2(t) = p(t)g(t),$$

$$h_3(t) = tg(t)(h_2(t) + 1),$$

$$h_4(t) = h_1(t) + tg(t)\int_0^t h_2(t_1)\left(\int_0^{t_1} h_1(s)\Delta s + h_1(t_1)\right.$$

$$\left. + t_1 g(t_1)\int_0^{t_1} h_2(s)\left(\int_0^s h_1(y)\Delta y + h_1(s)\right)e_{\ominus h_3}(\sigma(s),t)\Delta s\right)\Delta t_1, \quad t \in \mathbb{R}_+ \cap \mathbb{T}.$$

Proof. Let

$$z(t) = \int_0^t g(t_1)\int_0^{t_1}\int_0^{\tau_1} p(\tau_2)x(\tau_2)\Delta\tau_2\Delta\tau_1\Delta t_1, \quad t \in \mathbb{R}_+ \cap \mathbb{T}.$$

Then

$$z(0) = 0$$

and

$$x(t) \le f(t) + g(t)z(t), \quad t \in \mathbb{R}_+ \cap \mathbb{T}.$$

Next we can write

$$\begin{aligned}
z^\Delta(t) &= g(t)\int_0^t\int_0^{t_1} p(t_2)x(t_2)\Delta t_2\Delta t_1\\
&\le g(t)\int_0^t\int_0^{t_1} p(t_2)f(t_2)\Delta t_2\Delta t_1\\
&\quad + g(t)\int_0^t\int_0^{t_1} p(t_2)g(t_2)z(t_2)\Delta t_2\Delta t_1\\
&= h_1(t) + g(t)\int_0^t (t-\sigma(t_1))p(t_1)g(t_1)z(t_1)\Delta t_1\\
&\le h_1(t) + tg(t)\int_0^t p(t_1)g(t_1)z(t_1)\Delta t_1\\
&= h_1(t) + tg(t)\int_0^t h_2(t_1)z(t_1)\Delta t_1\\
&\le h_1(t) + tg(t)\int_0^t h_2(t_1)(z(t_1)+z^\Delta(t_1))\Delta t_1, \quad t \in \mathbb{R}_+ \cap \mathbb{T}.
\end{aligned}$$

Hence, by Theorem 2.1.2, we get

$$\begin{aligned}
z^\Delta(t) &\le h_1(t) + tg(t)\int_0^t h_2(t_1)\Bigg(\int_0^{t_1} h_1(s)\Delta s + h_1(t_1)\\
&\quad + t_1g(t_1)\int_0^{t_1}\Bigg(\int_0^s h_1(y)\Delta y + h_1(s)\Bigg)e_{\ominus h_3}(\sigma(s),t)\Delta s\Bigg)\Delta t_1\\
&= h_4(t), \quad t \in \mathbb{R}_+ \cap \mathbb{T}.
\end{aligned}$$

Therefore

$$z(t) \le \int_0^t h_4(s)\Delta s,$$

$$\begin{aligned} x(t) &\le f(t) + g(t)z(t) \\ &\le f(t) + g(t)\int_0^t h_4(s)\Delta s, \quad t \in \mathbb{R}_+ \cap \mathbb{T}. \end{aligned}$$

This completes the proof. □

Theorem 2.3.8. *Let $n \ge 2$, $x, f_k \in \mathcal{C}_{rd}(\mathbb{R}_+ \cap \mathbb{T})$, $k \in \{1, \ldots, n\}$, be nonnegative functions, $p_l \in \mathcal{C}_{rd}(\mathbb{R}_+ \cap \mathbb{T})$, $l \in \{1, \ldots, n\}$, be positive functions, u_0 be a positive constant, and*

$$\begin{aligned} x(t) \le u_0 + \sum_{k=1}^{n}\Bigg(&\int_0^t p_1(s_1)\int_0^{s_1} p_2(s_2)\cdots\int_0^{s_{k-2}} p_{k-1}(s_{k-1}) \\ &\times \int_0^{s_{k-1}} p_k(s_k) f_k(s_k) x(s_k)\Delta s_k \Delta s_{k-1}\cdots \Delta s_2 \Delta s_1\Bigg), \end{aligned}$$

$t \in \mathbb{R}_+ \cap \mathbb{T}$. Then

$$x(t) \le u_0 e_{q_1}(t, 0), \quad t \in \mathbb{R}_+ \cap \mathbb{T},$$

where

$$\begin{aligned} q(t) &= \sum_{k=2}^{n}\Bigg(\int_0^t p_2(s_2)\int_0^{s_2}\cdots\int_0^{s_{k-2}} p_{k-1}(s_{k-1})\int_0^{s_{k-1}} p_k(s_k) f_k(s_k)\Delta s_k \Delta s_{k-1}\cdots \Delta s_3 \Delta s_2\Bigg), \\ q_1(t) &= f_1(t)p_1(t) + p_1(t)q(t), \quad t \in \mathbb{R}_+ \cap \mathbb{T}. \end{aligned}$$

Proof. Let

$$\begin{aligned} z(t) = u_0 + \sum_{k=1}^{n}\Bigg(&\int_0^t p_1(s_1)\int_0^{s_1} p_2(s_2)\cdots\int_0^{s_{k-2}} p_{k-1}(s_{k-1}) \\ &\times \int_0^{s_{k-1}} p_k(s_k) f_k(s_k) x(s_k)\Delta s_k \Delta s_{k-1}\cdots \Delta s_2 \Delta s_1\Bigg), \end{aligned}$$

$t \in \mathbb{R}_+ \cap \mathbb{T}$. Then

$$z(0) = u_0 > 0,$$

and

$$\begin{aligned} x(t) &\le z(t), \\ z^{\Delta}(t) &= p_1(t) f_1(t) x(t) \\ &\quad + p_1(t)\sum_{k=2}^{n}\Bigg(\int_0^t p_2(s_2)\cdots\int_0^{s_{k-2}} p_{k-1}(s_{k-1}) \end{aligned}$$

$$\times \int_0^{s_{k-1}} p_k(s_k) f_k(s_k) x(s_k) \Delta s_k \Delta s_{k-1} \cdots \Delta s_2 \Bigg),$$

$t \in \mathbb{R}_+ \cap \mathbb{T}$. Let

$$z_1(t) = \sum_{k=2}^{n} \Bigg(\int_0^t p_2(s_2) \cdots \int_0^{s_{k-2}} p_{k-1}(s_{k-1}) \int_0^{s_{k-1}} p_k(s_k) f_k(s_k) x(s_k) \Delta s_k \Delta s_{k-1} \cdots \Delta s_2 \Bigg),$$

$t \in \mathbb{R}_+ \cap \mathbb{T}$. Hence,

$$z^\Delta(t) = p_1(t) f_1(t) x(t) + p_1(t) z_1(t), \quad t \in \mathbb{R}_+ \cap \mathbb{T},$$

or

$$\frac{z^\Delta(t)}{p_1(t)} - f_1(t) x(t) = z_1(t), \quad t \in \mathbb{R}_+ \cap \mathbb{T}.$$

Similarly,

$$\frac{z_1^\Delta(t)}{p_2(t)} - f_2(t) x(t) = z_2(t), \quad t \in \mathbb{R}_+ \cap \mathbb{T},$$

where

$$z_2(t) = \sum_{k=3}^{n} \Bigg(\int_0^t p_3(s_3) \cdots \int_0^{s_{k-1}} p_k(s_k) f_k(s_k) x(s_k) \Delta s_k \Delta s_{k-1} \cdots \Delta s_3 \Bigg), \quad t \in \mathbb{R}_+ \cap \mathbb{T}.$$

And so on,

$$\frac{z_{n-2}^\Delta(t)}{p_{n-1}(t)} - f_{n-1}(t) x(t) = z_{n-1}(t), \quad t \in \mathbb{R}_+ \cap \mathbb{T},$$

where

$$z_{n-1}(t) = \int_0^t p_n(s_n) f_n(s_n) x(s_n) \Delta s_n, \quad t \in \mathbb{R}_+ \cap \mathbb{T}.$$

Therefore

$$z_{n-1}^\Delta(t) = p_n(t) f_n(t) x(t), \quad t \in \mathbb{R}_+ \cap \mathbb{T}.$$

Hence,

$$\begin{aligned} \frac{z_{n-1}^\Delta(t)}{z(t)} &= \frac{p_n(t) f_n(t) x(t)}{z(t)} \\ &\leq p_n(t) f_n(t), \quad t \in \mathbb{R}_+ \cap \mathbb{T}. \end{aligned}$$

Now, using that

$$\begin{aligned}\int_0^t \frac{z_{n-1}^{\Delta}(s)}{z(s)}\Delta s &= \int_0^t \left(\left(\frac{z_{n-1}(s)}{z(s)}\right)^{\Delta} + \frac{z_{n-1}(s)z^{\Delta}(s)}{z(s)z(\sigma(s))}\right)\Delta s\\ &\geq \int_0^t \left(\frac{z_{n-1}(s)}{z(s)}\right)^{\Delta}\Delta s\\ &= \frac{z_{n-1}(t)}{z(t)}, \quad t \in \mathbb{R}_+ \cap \mathbb{T},\end{aligned}$$

we get

$$\begin{aligned}\frac{z_{n-1}(t)}{z(t)} &\leq \int_0^t \frac{z_{n-1}^{\Delta}(s)}{z(s)}\Delta s\\ &\leq \int_0^t p_n(s)f_n(s)\Delta s, \quad t \in \mathbb{R}_+ \cap \mathbb{T},\end{aligned}$$

and

$$\begin{aligned}\frac{z_{n-2}(t)}{z(t)} &\leq \int_0^t \frac{z_{n-2}^{\Delta}(s_{n-1})}{z(s_{n-1})}\Delta s_{n-1}\\ &= \int_0^t \frac{z_{n-1}(s_{n-1})p_{n-1}(s_{n-1}) + f_{n-1}(s_{n-1})p_{n-1}(s_{n-1})x(s_{n-1})}{z(s_{n-1})}\Delta s_{n-1}\\ &\leq \int_0^t f_{n-1}(s_{n-1})p_{n-1}(s_{n-1})\Delta s_{n-1}\\ &\quad + \int_0^t p_{n-1}(s_{n-1})\frac{z_{n-1}(s_{n-1})}{z(s_{n-1})}\Delta s_{n-1}\\ &\leq \int_0^t f_{n-1}(s_{n-1})p_{n-1}(s_{n-1})\Delta s_{n-1}\\ &\quad + \int_0^t p_{n-1}(s_{n-1})\int_0^{s_{n-1}} p_n(s_n)f_n(s_n)\Delta s_n\Delta s_{n-1}, \quad t \in \mathbb{R}_+ \cap \mathbb{T}.\end{aligned}$$

Similarly,

$$\frac{z^{\Delta}(t)}{p_1(t)} - f_1(t)x(t) \leq z(t)\sum_{k=2}^{n}\left(\int_0^t p_2(s_2)\int_0^{s_2} p_3(s_3)\cdots\int_0^{s_{k-2}} p_{k-1}(s_{k-1})\right.$$

$$\begin{aligned} &\times \int_0^{s_{k-1}} p_k(s_k) f_k(s_k) \Delta s_k \Delta s_{k-1} \cdots \Delta s_3 \Delta s_2 \Bigg) \\ &= z(t) q(t), \quad t \in \mathbb{R}_+ \cap \mathbb{T}, \end{aligned}$$

i. e.,

$$\begin{aligned} \frac{z^\Delta(t)}{z(t)} &\leq f_1(t) p_1(t) \frac{x(t)}{z(t)} + p_1(t) q(t) \\ &\leq f_1(t) p_1(t) + p_1(t) q(t) \\ &= q_1(t), \quad t \in \mathbb{R}_+ \cap \mathbb{T}, \end{aligned}$$

or

$$z^\Delta(t) \leq q_1(t) z(t), \quad t \in \mathbb{R}_+ \cap \mathbb{T}.$$

Hence, due to Lemma 2.1.1, we get

$$\begin{aligned} z(t) &\leq z(0) e_{q_1}(t, 0) \\ &= u_0 e_{q_1}(t, 0), \quad t \in \mathbb{R}_+ \cap \mathbb{T}. \end{aligned}$$

Consequently,

$$\begin{aligned} x(t) &\leq z(t) \\ &\leq u_0 e_{q_1}(t, 0), \quad t \in \mathbb{R}_+ \cap \mathbb{T}. \end{aligned}$$

This completes the proof. □

3 Nonlinear integral inequalities

This chapter is concerned with Dragomir type and Pachpatte type nonlinear integral inequalities. The results contained in this chapter can be found in [3, 9, 11, 17] and [18].

Let $\mathbb{T}$ be a time scale with forward jump operator and delta differentiation operator σ and Δ, respectively.

3.1 Dragomir's inequalities

Suppose that $a, b \in \mathbb{T}$, $a < b$, $J = [a, b]$ is a time scale interval.

Theorem 3.1.1 (Dragomir's Inequality). *Let $x, f, g \in \mathcal{C}(J)$ be nonnegative functions, $L \in \mathcal{C}(J \times \mathbb{R}_+)$ be a nonnegative function such that*

$$0 \le L(t,x) - L(t,y) \le M(t,y)(x-y)$$

for any $t \in J$ and for any $x \ge y \ge 0$, where $M \in \mathcal{C}(J \times \mathbb{R})$ is a nonnegative function. If

$$x(t) \le f(t) + g(t) \int_a^t L(s, x(s))\Delta s, \quad t \in J,$$

then

$$x(t) \le f(t) + g(t) \int_a^t h_1(s) e_{\ominus h_2}(\sigma(s), t)\Delta s, \quad t \in J,$$

where

$$\begin{aligned} h_1(t) &= L(t, f(t)), \\ h_2(t) &= g(t) M(t, f(t)), \quad t \in J. \end{aligned}$$

Proof. Let

$$z(t) = \int_a^t L(s, x(s))\Delta s, \quad t \in J.$$

Then

$$x(t) \le f(t) + g(t) z(t), \quad t \in J, \tag{3.1}$$

and

$$z(0) = 0.$$

https://doi.org/10.1515/9783110705553-003

We have

$$
\begin{aligned}
z^{\Delta}(t) &= L(t, x(t)) \\
&\le L(t, f(t) + g(t)z(t)) \\
&= L(t, f(t) + g(t)z(t)) - L(t, f(t)) + L(t, f(t)) \\
&\le M(t, f(t))(f(t) + g(t)z(t) - f(t)) + L(t, f(t)) \\
&= L(t, f(t)) + g(t)M(t, f(t))z(t) \\
&\le h_1(t) + h_2(t)z(t), \quad t \in J.
\end{aligned}
$$

Hence, by Lemma 2.1.1, we get

$$
z(t) \le \int_a^t h_1(s)e_{\ominus h_2}(\sigma(s), t)\Delta s, \quad t \in J.
$$

From here and (3.1), we obtain

$$
x(t) \le f(t) + g(t)\int_a^t h_1(s)e_{\ominus h_2}(\sigma(s), t)\Delta s, \quad t \in J.
$$

This completes the proof. □

Theorem 3.1.2 (Dragomir's Inequality). *Let $x, f, g \in C(J)$ be nonnegative functions, $L \in C(J \times \mathbb{R}_+)$ be a nonnegative function, $\psi \in C(\mathbb{R}_+)$ be a nonnegative strictly increasing function with $\psi(0) = 0$ and*

$$
0 \le L(t, x) - L(t, y) \le M(t, y)\psi^{-1}(x - y)
$$

for any $t \in J$ and for any $x \ge y \ge 0$, where $M \in C(J \times \mathbb{R}_+)$ is a nonnegative function and ψ^{-1} is the inverse of ψ. If

$$
x(t) \le f(t) + \psi\left(g(t)\int_a^t L(s, x(s))\Delta s\right), \quad t \in J,
$$

then

$$
x(t) \le f(t) + \psi\left(g(t)\int_a^t h_1(s)e_{\ominus h_2}(\sigma(s), t)\Delta s\right), \quad t \in J,
$$

where h_1 and h_2 are defined as in Theorem 3.1.1.

Proof. Let

$$
z(t) = \int_a^t L(s, x(s))\Delta s, \quad t \in J.
$$

Then

$$x(t) \leq f(t) + \psi(g(t)z(t)), \quad t \in J, \tag{3.2}$$

and

$$z(0) = 0.$$

We have

$$\begin{aligned} z^{\Delta}(t) &= L(t, x(t)) \\ &\leq L(t, f(t) + \psi(g(t)z(t))) \\ &= L(t, f(t) + \psi(g(t)z(t))) - L(t, f(t)) + L(t, f(t)) \\ &\leq M(t, f(t))\psi^{-1}(f(t) + \psi(g(t))z(t) - f(t)) + L(t, f(t)) \\ &= M(t, f(t))\psi^{-1}(\psi(g(t)z(t))) + h_1(t) \\ &= h_1(t) + M(t, f(t))g(t)z(t) \\ &= h_1(t) + h_2(t)z(t), \quad t \in J. \end{aligned}$$

Hence, by Lemma 2.1.1, we get

$$z(t) \leq \int_a^t h_1(s)e_{\ominus h_2}(\sigma(s), t)\Delta s, \quad t \in J.$$

From here and from (3.2), we obtain

$$x(t) \leq f(t) + \psi\left(g(t)\int_a^t h_1(s)e_{\ominus h_2}(\sigma(s), t)\Delta s\right), \quad t \in J.$$

This completes the proof. □

3.2 Pachpatte's inequalities

In this section we suppose that $0 \in \mathbb{T}$.

Theorem 3.2.1 (Pachpatte's Inequality). *Let $x, f, g \in \mathcal{C}(\mathbb{R}_+ \cap \mathbb{T})$ be nonnegative functions, $F \in \mathcal{C}(\mathbb{R}_+)$ be nonnegative, strictly increasing, convex and submultiplicative function on $(0, \infty)$, and*

$$\lim_{u \to \infty} F(u) = \infty,$$

let F^{-1} denote the inverse function of F, and $\alpha, \beta \in \mathcal{C}(\mathbb{R}_+ \cap \mathbb{T})$ be positive functions such that

$$\alpha(t) + \beta(t) = 1, \quad t \in \mathbb{R}_+ \cap \mathbb{T}.$$

Let also $L \in \mathcal{C}((\mathbb{R}_+ \cap \mathbb{T}) \times \mathbb{R}_+)$ be a nonnegative function such that

$$0 \le L(t,x) - L(t,y) \le M(t,y)(x-y), \quad t \in \mathbb{R}_+ \cap \mathbb{T},$$

for any $x \ge y \ge 0$, where $M \in \mathcal{C}((\mathbb{R}_+ \cap \mathbb{T}) \times \mathbb{R})$ is a nonnegative function. If

$$x(t) \le f(t) + g(t)F^{-1}\left(\int_0^t L(s, F(x(s)))\Delta s\right), \quad t \in \mathbb{R}_+ \cap \mathbb{T}, \tag{3.3}$$

then

$$x(t) \le f(t) + g(t)F^{-1}\left(\int_0^t L(s, h_5(s))\Delta s\right), \quad t \in \mathbb{R}_+ \cap \mathbb{T},$$

where

$$\begin{aligned}
h_1(t) &= \alpha(t)F(f(t)(\alpha(t))^{-1}),\\
h_2(t) &= \beta(t)F(g(t)(\beta(t))^{-1}),\\
h_3(t) &= L(t, h_1(t)),\\
h_4(t) &= h_2(t)M(t, h_1(t)),\\
h_5(t) &= h_1(t) + h_2(t)\int_0^t h_3(s)e_{\ominus h_4}(\sigma(s), t)\Delta s, \quad t \in \mathbb{R}_+ \cap \mathbb{T}.
\end{aligned}$$

Proof. We rewrite inequality (3.3) in the following way:

$$x(t) \le \alpha(t)(f(t)(\alpha(t))^{-1}) + \beta(t)(g(t)(\beta(t))^{-1})F^{-1}\left(\int_0^t L(s, F(x(s)))\Delta s\right),$$

$t \in \mathbb{R}_+ \cap \mathbb{T}$. Hence, since F is strictly increasing, convex and submultiplicative on $(0, \infty)$, we get

$$\begin{aligned}
F(x(t)) &\le F\left(\alpha(t)(f(t)(\alpha(t))^{-1}) + \beta(t)(g(t)(\beta(t))^{-1})F^{-1}\left(\int_0^t L(s, F(x(s)))\Delta s\right)\right)\\
&\le \alpha(t)F(f(t)(\alpha(t))^{-1}) + \beta(t)F(y(t)(\beta(t))^{-1})F^{-1}\left(\int_0^t L(s, F(x(s)))\Delta s\right)\\
&\le \alpha(t)F(f(t)(\alpha(t))^{-1})\\
&\quad + \beta(t)F(g(t)(\beta(t))^{-1})F\left(F^{-1}\left(\int_0^t L(s, F(x(s)))\Delta s\right)\right)
\end{aligned}$$

$$= h_1(t) + h_2(t) \int_0^t L(s, F(x(s)))\Delta s, \quad t \in \mathbb{R}_+ \cap \mathbb{T}.$$

Now we apply Theorem 3.1.1 for $F(x(t))$, $t \in \mathbb{R}_+ \cap \mathbb{T}$, and get

$$\begin{aligned} F(x(t)) &\le h_1(t) + h_2(t) \int_0^t h_3(s) e_{\ominus h_4}(\sigma(s), t)\Delta s \\ &= h_5(t), \quad t \in \mathbb{R}_+ \cap \mathbb{T}. \end{aligned}$$

Hence,

$$x(t) \le f(t) + g(t) F^{-1}\left(\int_0^t L(s, h_5(s))\Delta s \right), \quad t \in \mathbb{R}_+ \cap \mathbb{T}.$$

This completes the proof. □

Theorem 3.2.2 (Pachpatte's Inequality). *Let x, f, g, L and M be as in Theorem 3.2.1. Let also $H \in \mathcal{C}(\mathbb{R}_+)$ be a positive, strictly increasing, subadditive and submultiplicative function on $(0, \infty)$ with $\lim_{y\to\infty} H(y) = \infty$. If*

$$x(t) \le f(t) + g(t) H^{-1}\left(\int_0^t L(s, H(x(s)))\Delta s \right), \quad t \in \mathbb{R}_+ \cap \mathbb{T}, \tag{3.4}$$

then

$$x(t) \le H^{-1}\left(h_1(t) + h_2(t) \int_0^t h_3(s) e_{\ominus h_4}(\sigma(s), t)\Delta s \right), \quad t \in \mathbb{R}_+ \cap \mathbb{T},$$

where

$$\begin{aligned} h_1(t) &= H(f(t)), \\ h_2(t) &= H(g(t)), \\ h_3(t) &= L(t, h_1(t)), \\ h_4(t) &= h_2(t) M(t, h_1(t)), \quad t \in \mathbb{R}_+ \cap \mathbb{T}. \end{aligned}$$

Proof. Since H is monotonic, subadditive and submultiplicative on $(0, \infty)$, by inequality (3.4), we get

$$H(x(t)) \le H\left(f(t) + g(t) H^{-1}\left(\int_0^t L(s, H(x(s)))\Delta s \right)\right)$$

$$
\begin{aligned}
&\le H(f(t)) + H\left(g(t)H^{-1}\left(\int_0^t L(s, H(x(s)))\Delta s\right)\right) \\
&\le H(f(t)) + H(g(t))H\left(H^{-1}\left(\int_0^t L(s, H(x(s)))\Delta s\right)\right) \\
&= H(f(t)) + H(g(t))\int_0^t L(s, H(x(s)))\Delta s \\
&= h_1(t) + h_2(t)\int_0^t L(s, H(x(s)))\Delta s, \quad t \in \mathbb{R}_+ \cap \mathbb{T}.
\end{aligned}
$$

Now we apply Theorem 3.1.1 for $H(x(t))$, $t \in \mathbb{R}_+ \cap \mathbb{T}$, and get

$$
H(x(t)) \le h_1(t) + h_2(t)\int_0^t h_3(s)e_{\ominus h_4}(\sigma(s), t)\Delta s, \quad t \in \mathbb{R}_+ \cap \mathbb{T}.
$$

Since H^{-1} is increasing, we get

$$
x(t) \le H^{-1}\left(h_1(t) + h_2(t)\int_0^t h_3(s)e_{\ominus h_4}(\sigma(s), t)\Delta s\right), \quad t \in \mathbb{R}_+ \cap \mathbb{T}.
$$

This completes the proof. □

Theorem 3.2.3 (Pachpatte's Inequality). *Let x, f, g, L, and M be as in Theorem 3.2.1. Let also $k \in C((\mathbb{R}_+ \cap \mathbb{T}) \times (\mathbb{R}_+ \cap \mathbb{T}))$ and suppose there exists $k_t^\Delta \in C((\mathbb{R}_+ \cap \mathbb{T}) \times (\mathbb{R}_+ \cap \mathbb{T}))$, such that $k(t, s)$ and $k_t^\Delta(t, s)$ are nonnegative functions for $0 \le s \le t < \infty$. If*

$$
x(t) \le f(t) + g(t)\int_0^t k(t, s)(x(s) + L(s, x(s)))\Delta s, \quad t \in \mathbb{R}_+ \cap \mathbb{T},
$$

then

$$
x(t) \le f(t) + g(t)\int_0^t A(s)e_{\ominus B}(\sigma(s), t)\Delta s, \quad t \in \mathbb{R}_+ \cap \mathbb{T},
$$

where

$$
\begin{aligned}
A(t) &= k(\sigma(t), t)(f(t) + L(t, f(t))) \\
&\quad + \int_0^t k_t^\Delta(t, s)(f(s) + L(s, f(s)))\Delta s,
\end{aligned}
$$

$$B(t) = k(\sigma(t),t)(M(t,f(t)) + 1)g(t) + \int_0^t k_t^\Delta(t,s)g(s)(M(s,f(s)) + 1)\Delta s, \quad t \in \mathbb{R}_+ \cap \mathbb{T}.$$

Proof. Let

$$z(t) = \int_0^t k(t,s)(x(s) + L(s,x(s)))\Delta s, \quad t \in \mathbb{R}_+ \cap \mathbb{T}.$$

Then $z(t)$ is a nondecreasing function on $\mathbb{R}_+ \cap \mathbb{T}$,

$$x(t) \le f(t) + g(t)z(t), \quad t \in \mathbb{R}_+ \cap \mathbb{T},$$

and

$$z(0) = 0.$$

We have

$$\begin{aligned}
z^\Delta(t) &= k(\sigma(t),t)(x(t) + L(t,x(t))) \\
&\quad + \int_0^t k_t^\Delta(t,s)(x(s) + L(s,x(s)))\Delta s \\
&\le k(\sigma(t),t)(f(t) + g(t)z(t) + L(t,f(t) + g(t)z(t)) \\
&\quad - L(t,f(t)) + L(t,f(t))) \\
&\quad + \int_0^t k_t^\Delta(t,s)(f(s) + g(s)z(s) + L(s,f(s) + g(s)z(s)) \\
&\quad - L(s,f(s)z(s)) + L(s,f(s)z(s)))\Delta s \\
&\le k(\sigma(t),t)(f(t) + g(t)z(t) + M(t,f(t))g(t)z(t) + L(t,f(t))) \\
&\quad + \int_0^t k_t^\Delta(t,s)(f(s) + g(s)z(s) + M(s,f(s))g(s)z(s) + L(s,f(s)))\Delta s \\
&= k(\sigma(t),t)(f(t) + L(t,f(t))) \\
&\quad + \int_0^t k_t^\Delta(t,s)(f(s) + L(s,f(s)))\Delta s \\
&\quad + k(\sigma(t),t)(M(t,f(t)) + 1)g(t)z(t) \\
&\quad + \int_0^t k_t^\Delta(t,s)g(s)(M(s,f(s)) + 1)z(s)\Delta s
\end{aligned}$$

$$\begin{aligned}
&\le A(t) + k(\sigma(t), t)(M(t, f(t)) + 1)g(t)z(t) \\
&\quad + \left(\int_0^t k_t^{\Delta}(t,s)g(s)(M(s,f(s)) + 1)\Delta s\right)z(t) \\
&= A(t) + B(t)z(t), \quad t \in \mathbb{R}_+ \cap \mathbb{T}.
\end{aligned}$$

Hence, by Lemma 2.1.1, we get

$$z(t) \le \int_0^t A(s)e_{\ominus B}(\sigma(s), t)\Delta s, \quad t \in \mathbb{R}_+ \cap \mathbb{T},$$

and

$$x(t) \le f(t) + g(t)\int_0^t A(s)e_{\ominus B}(\sigma(s), t)\Delta s, \quad t \in \mathbb{R}_+ \cap \mathbb{T}.$$

This completes the proof. □

Theorem 3.2.4. *Let $p \in (1, \infty)$, $f \in \mathcal{C}(\mathbb{R}_+ \cap \mathbb{T})$ be a nonnegative function, $x \in \mathcal{C}(\mathbb{R}_+ \cap \mathbb{T})$ be a positive function such that*

$$\begin{aligned}
&x(t) \ge 1, \\
&(x(t))^p \le c_1 + c_2\int_0^t f(y)x(y)\Delta y, \quad t \in \mathbb{R}_+ \cap \mathbb{T},
\end{aligned}$$

for some positive constants c_1 and c_2. Then

$$x(t) \le \left(c_1 e_{c_2 f}(t, 0)\right)^{\frac{1}{p}}, \quad t \in \mathbb{R}_+ \cap \mathbb{T}.$$

Proof. Since $x(t) \ge 1$, $t \in \mathbb{R}_+ \cap \mathbb{T}$, we get

$$x(t) \le (x(t))^p, \quad t \in \mathbb{R}_+ \cap \mathbb{T}.$$

Let

$$z(t) = c_1 + c_2\int_0^t f(y)x(y)\Delta y, \quad t \in \mathbb{R}_+ \cap \mathbb{T}.$$

Then

$$\begin{aligned}
&z(0) = c_1, \\
&(x(t))^p \le z(t), \quad t \in \mathbb{R}_+ \cap \mathbb{T},
\end{aligned}$$

and

$$\begin{aligned} z^{\Delta}(t) &= c_2 f(t)x(t) \\ &\le c_2 f(t)(x(t))^p \\ &\le c_2 f(t)z(t), \quad t \in \mathbb{R}_+ \cap \mathbb{T}. \end{aligned}$$

Hence, by Lemma 2.1.1, we get

$$\begin{aligned} z(t) &\le z(0)e_{c_2 f}(t,0) \\ &= c_1 e_{c_2 f}(t,0), \quad t \in \mathbb{R}_+ \cap \mathbb{T}. \end{aligned}$$

Therefore

$$(x(t))^p \le c_1 e_{c_2 f}(t,0), \quad t \in \mathbb{R}_+ \cap \mathbb{T},$$

whereupon

$$x(t) \le (c_1 e_{c_2 f}(t,0))^{\frac{1}{p}}, \quad t \in \mathbb{R}_+ \cap \mathbb{T}.$$

This completes the proof. □

Theorem 3.2.5. *Let $p \in (1,\infty)$, $f \in \mathcal{C}(\mathbb{R}_+ \cap \mathbb{T})$ be a positive function, $g, h_1, h_2 \in \mathcal{C}(\mathbb{R}_+ \cap \mathbb{T})$ be nonnegative functions, $x \in \mathcal{C}(\mathbb{R}_+ \cap \mathbb{T})$ be a positive function such that*

$$\begin{aligned} & x(t) \ge 1, \\ & (x(t))^p \le f(t) + g(t)\int_0^t (h_1(s)(x(s))^p + h_2(s)x(s))\Delta s, \quad t \in \mathbb{R}_+ \cap \mathbb{T}. \end{aligned}$$

Then

$$x(t) \le \left(f(t) + g(t)\int_0^t h_3(s)e_{\ominus h_4}(\sigma(s),t)\Delta s \right)^{\frac{1}{p}}, \quad t \in \mathbb{R}_+ \cap \mathbb{T},$$

where

$$\begin{aligned} h_3(t) &= f(t)(h_1(t) + h_2(t)), \\ h_4(t) &= g(t)(h_1(t) + h_2(t)), \quad t \in \mathbb{R}_+ \cap \mathbb{T}. \end{aligned}$$

Proof. Since $x(t) \ge 1$, $t \in \mathbb{R}_+ \cap \mathbb{T}$, we get

$$x(t) \le (x(t))^p, \quad t \in \mathbb{R}_+ \cap \mathbb{T}.$$

Let

$$z(t) = \int_0^t (h_1(s)(x(s))^p + h_2(s)x(s))\Delta s, \quad t \in \mathbb{R}_+ \cap \mathbb{T}.$$

Then

$$(x(t))^p \le f(t) + g(t)z(t), \quad t \in \mathbb{R}_+ \cap \mathbb{T},$$

and

$$z(0) = 0.$$

We have

$$\begin{aligned} z^\Delta(t) &= h_1(t)(x(t))^p + h_2(t)x(t) \\ &\le (h_1(t) + h_2(t))(x(t))^p \\ &\le (h_1(t) + h_2(t))(f(t) + g(t)z(t)) \\ &= f(t)(h_1(t) + h_2(t)) + g(t)(h_1(t) + h_2(t))z(t) \\ &= h_3(t) + h_4(t)z(t), \quad t \in \mathbb{R}_+ \cap \mathbb{T}. \end{aligned}$$

Hence, from Lemma 2.1.1, we get

$$z(t) \le \int_0^t h_3(s)e_{\ominus h_4}(\sigma(s), t)\Delta s, \quad t \in \mathbb{R}_+ \cap \mathbb{T}.$$

Then

$$(x(t))^p \le f(t) + g(t)\int_0^t h_3(s)e_{\ominus h_4}(\sigma(s), t)\Delta s, \quad t \in \mathbb{R}_+ \cap \mathbb{T},$$

whereupon

$$x(t) \le \left(f(t) + g(t)\int_0^t h_3(s)e_{\ominus h_4}(\sigma(s), t)\Delta s\right)^{\frac{1}{p}}, \quad t \in \mathbb{R}_+ \cap \mathbb{T}.$$

This completes the proof. □

Theorem 3.2.6. *Let $p \in (1, \infty)$, $x, f \in \mathcal{C}(\mathbb{R}_+ \cap \mathbb{T})$ be nonnegative functions, c_1 and c_2 be nonnegative constants. If*

$$\begin{aligned} x(t) &\le 1, \\ (x(t))^{\frac{1}{p}} &\le c_1 + c_2\int_0^t f(s)x(s)\Delta s, \quad t \in \mathbb{R}_+ \cap \mathbb{T}, \end{aligned}$$

then

$$x(t) \le (c_1 e_{c_2 f}(t, 0))^p, \quad t \in \mathbb{R}_+ \cap \mathbb{T}.$$

Proof. Since

$$0 \le x(t) \le 1, \quad t \in \mathbb{R}_+ \cap \mathbb{T},$$

we have

$$x(t) \le (x(t))^{\frac{1}{p}}, \quad t \in \mathbb{R}_+ \cap \mathbb{T}.$$

Let

$$z(t) = c_1 + c_2 \int_0^t f(s)x(s)\Delta s, \quad t \in \mathbb{R}_+ \cap \mathbb{T}.$$

Then

$$\begin{aligned} z(0) &= c_1, \\ (x(t))^{\frac{1}{p}} &\le z(t), \\ z^{\Delta}(t) &= c_2 f(t)x(t) \\ &\le c_2 f(t)(x(t))^{\frac{1}{p}} \\ &\le c_2 f(t)z(t), \quad t \in \mathbb{R}_+ \cap \mathbb{T}. \end{aligned}$$

Hence, using Lemma 2.1.1, we get

$$z(t) \le c_1 e_{c_2 f}(t, 0), \quad t \in \mathbb{R}_+ \cap \mathbb{T}.$$

Therefore

$$(x(t))^{\frac{1}{p}} \le c_1 e_{c_2 f}(t, 0), \quad t \in \mathbb{R}_+ \cap \mathbb{T},$$

whereupon

$$x(t) \le (c_1 e_{c_2 f}(t, 0))^p, \quad t \in \mathbb{R}_+ \cap \mathbb{T}.$$

This completes the proof. □

Theorem 3.2.7. *Let $p \in (1, \infty)$, $x, f, g, h_1, h_2 \in \mathcal{C}(\mathbb{R}_+ \cap \mathbb{T})$ be nonnegative functions. If*

$$\begin{aligned} x(t) &\le 1, \\ (x(t))^{\frac{1}{p}} &\le f(t) + g(t) \int_0^t (h_1(s)(x(s))^{\frac{1}{p}} + h_2(s)x(s))\Delta s, \quad t \in \mathbb{R}_+ \cap \mathbb{T}, \end{aligned}$$

then

$$x(t) \le \left(f(t) + g(t) \int_0^t h_3(s) e_{\ominus h_4}(\sigma(s), t)\Delta s \right)^p, \quad t \in \mathbb{R}_+ \cap \mathbb{T},$$

where

$$\begin{aligned} h_3(t) &= f(t)(h_1(t) + h_2(t)), \\ h_4(t) &= g(t)(h_1(t) + h_2(t)), \quad t \in \mathbb{R}_+ \cap \mathbb{T}. \end{aligned}$$

Proof. Since

$$0 \le x(t) \le 1, \quad t \in \mathbb{R}_+ \cap \mathbb{T},$$

we have

$$x(t) \le (x(t))^{\frac{1}{p}}, \quad t \in \mathbb{R}_+ \cap \mathbb{T}.$$

Let

$$z(t) = \int_0^t (h_1(s)(x(s))^{\frac{1}{p}} + h_2(s)x(s))\Delta s, \quad t \in \mathbb{R}_+ \cap \mathbb{T}.$$

Then

$$z(0) = 0$$

and

$$\begin{aligned} (x(t))^{\frac{1}{p}} &\le f(t) + g(t)z(t), \\ z^{\Delta}(t) &= h_1(t)(x(t))^{\frac{1}{p}} + h_2(t)x(t) \\ &\le (h_1(t) + h_2(t))(x(t))^{\frac{1}{p}} \\ &\le (h_1(t) + h_2(t))(f(t) + g(t)z(t)) \\ &= (h_1(t) + h_2(t))f(t) + (h_1(t) + h_2(t))g(t)z(t) \\ &= h_3(t) + h_4(t)z(t), \quad t \in \mathbb{R}_+ \cap \mathbb{T}. \end{aligned}$$

Hence, by Lemma 2.1.1, we conclude that

$$z(t) \le \int_0^t h_3(s)e_{\ominus h_4}(\sigma(s), t)\Delta s, \quad t \in \mathbb{R}_+ \cap \mathbb{T},$$

and

$$\begin{aligned} (x(t))^{\frac{1}{p}} &\le f(t) + g(t)z(t) \\ &\le f(t) + g(t)\int_0^t h_3(s)e_{\ominus h_4}(\sigma(s), t)\Delta s, \quad t \in \mathbb{R}_+ \cap \mathbb{T}, \end{aligned}$$

whereupon

$$x(t) \le \left(f(t) + g(t) \int_0^t h_3(s) e_{\ominus h_4}(\sigma(s), t) \Delta s \right)^p, \quad t \in \mathbb{R}_+ \cap \mathbb{T}.$$

This completes the proof. □

Theorem 3.2.8. *Let $p \in (1, \infty)$, c_1 and c_2 be positive constants, $x \in \mathcal{C}(\mathbb{R}_+ \cap \mathbb{T})$ be a positive function, $f, g \in \mathcal{C}(\mathbb{R}_+ \cap \mathbb{T})$ be nonnegative functions, $L \in \mathcal{C}((\mathbb{R}_+ \cap \mathbb{T}) \times \mathbb{R}_+)$ be a function that satisfies*

$$0 \le L(t, x) - L(t, y) \le k(t, y)(x - y)$$

for $t \in \mathbb{R}_+ \cap \mathbb{T}$, $x \ge y \ge 0$, where $k \in \mathcal{C}((\mathbb{R}_+ \cap \mathbb{T}) \times \mathbb{R}_+)$ is a nonnegative function. If

$$x(t) \ge 1,$$

$$(x(t))^p \le c_1 + c_2 \int_0^t (f(s) L(s, x(s)) + g(s) x(s)) \Delta s, \quad t \in \mathbb{R}_+ \cap \mathbb{T},$$

then

$$x(t) \le \left(c_1 e_{h_1}(t, 0) + \int_0^t h_2(s) e_{\ominus h_1}(\sigma(s), t) \Delta s \right)^{\frac{1}{p}}, \quad t \in \mathbb{R}_+ \cap \mathbb{T},$$

where

$$h_1(t) = c_2(f(t) k(t, g(t)) + g(t)),$$
$$h_2(t) = c_2 f(t) L(t, g(t)), \quad t \in \mathbb{R}_+ \cap \mathbb{T}.$$

Proof. Since

$$x(t) \ge 1, \quad t \in \mathbb{R}_+ \cap \mathbb{T},$$

then

$$x(t) \le (x(t))^p, \quad t \in \mathbb{R}_+ \cap \mathbb{T}.$$

Let

$$z(t) = c_1 + c_2 \int_0^t (f(s) L(s, x(s)) + g(s) x(s)) \Delta s, \quad t \in \mathbb{R}_+ \cap \mathbb{T}.$$

Then

$$z(0) = c_1$$

and

$$(x(t))^p \leq z(t), \quad t \in \mathbb{R}_+ \cap \mathbb{T}.$$

We have

$$\begin{aligned}
z^\Delta(t) &= c_2(f(t)L(t,x(t)) + g(t)x(t)) \\
&\leq c_2(f(t)L(t,(x(t))^p) + g(t)(x(t))^p) \\
&\leq c_2(f(t)L(t,z(t)) + g(t)z(t)) \\
&\leq c_2(f(t)L(t,z(t)+g(t)) + g(t)z(t)) \\
&= c_2(f(t)L(t,z(t)+g(t)) - f(t)L(t,g(t)) \\
&\quad + f(t)L(t,g(t)) + g(t)z(t)) \\
&\leq c_2(f(t)k(t,g(t))z(t) + f(t)L(t,g(t)) + g(t)z(t)) \\
&= c_2((f(t)k(t,g(t)) + g(t))z(t) + f(t)L(t,g(t))) \\
&= h_1(t)z(t) + h_2(t), \quad t \in \mathbb{R}_+ \cap \mathbb{T}.
\end{aligned}$$

Hence, by Lemma 2.1.1, we get

$$z(t) \leq c_1 e_{h_1}(t,0) + \int_0^t h_2(s) e_{\ominus h_1}(\sigma(s),t)\Delta s, \quad t \in \mathbb{R}_+ \cap \mathbb{T}.$$

Therefore

$$(x(t))^p \leq c_1 e_{h_1}(t,0) + \int_0^t h_2(s) e_{\ominus h_1}(\sigma(s),t)\Delta s, \quad t \in \mathbb{R}_+ \cap \mathbb{T},$$

whereupon

$$x(t) \leq \left(c_1 e_{h_1}(t,0) + \int_0^t h_2(s) e_{\ominus h_1}(\sigma(s),t)\Delta s \right)^{\frac{1}{p}}, \quad t \in \mathbb{R}_+ \cap \mathbb{T}.$$

This completes the proof. □

Theorem 3.2.9. *Let $x, y, g_i, h_i \in \mathcal{C}(\mathbb{R}_+ \cap \mathbb{T})$, $i \in \{1,2,3,4\}$, be nonnegative functions, c_1 and c_2 be nonnegative constants, and*

$$\begin{aligned}
(x(t))^2 &\leq c_1 + \int_0^t (g_1(s)(x(s))^2 + h_1(s)x(s))\Delta s \\
&\quad + \int_0^t (g_2(s)(y(s))^2 + h_2(s)y(s))\Delta s,
\end{aligned}$$

$$(y(t))^2 \leq c_2 + \int_0^t (g_3(s)(x(s))^2 + h_3(s)x(s))\Delta s$$
$$+ \int_0^t (g_4(s)(y(s))^2 + h_4(s)y(s))\Delta s, \quad t \in \mathbb{R}_+ \cap \mathbb{T}.$$

Then

$$x(t) + y(t) \leq \left(c_3 + c_3 \int_0^t p(s)e_{\ominus p}(\sigma(s), t)\Delta s \right)^{\frac{1}{2}}, \quad t \in \mathbb{R}_+ \cap \mathbb{T},$$

if $x(t) + y(t) \geq 1$, $t \in \mathbb{R}_+ \cap \mathbb{T}$, *and*

$$x(t) + y(t) \leq \left(c_3 + \int_0^t p(s)\Delta s \right)^{\frac{1}{2}}, \quad t \in \mathbb{R}_+ \cap \mathbb{T},$$

if $x(t) + y(t) \leq 1$, $t \in \mathbb{R}_+ \cap \mathbb{T}$. *Here*

$$G(t) = \max\{g_1(t) + g_3(t), g_2(t) + g_4(t)\},$$
$$H(t) = \max\{h_1(t) + h_3(t), h_2(t) + h_4(t)\},$$
$$p(t) = 2(G(t) + H(t)), \quad t \in \mathbb{R}_+ \cap \mathbb{T},$$
$$c_3 = 2(c_1 + c_2).$$

Proof. Let

$$z(t) = x(t) + y(t), \quad t \in \mathbb{R}_+ \cap \mathbb{T}.$$

Then

$$(x(t))^2 + (y(t))^2 \leq c_1 + c_2 + \int_0^t ((g_1(s) + g_3(s))(x(s))^2 + (h_1(s) + h_3(s))x(s))\Delta s$$
$$+ \int_0^t ((g_2(s) + g_4(s))(y(s))^2 + (h_2(s) + h_4(s))y(s))\Delta s$$
$$\leq c_1 + c_2 + \int_0^t (G(s)(x(s))^2 + H(s)x(s))\Delta s$$
$$+ \int_0^t (G(s)(y(s))^2 + H(s)y(s))\Delta s$$
$$= c_1 + c_2 + \int_0^t G(s)((x(s))^2 + (y(s))^2)\Delta s$$

$$+\int_0^t H(s)(x(s)+y(s))\Delta s, \quad t\in\mathbb{R}_+\cap\mathbb{T}.$$

Hence, using that

$$a^2+b^2\geq\frac{1}{2}(a+b)^2,$$
$$a^2+b^2\leq(a+b)^2, \quad a\geq 0, \quad b\geq 0,$$

we get

$$\begin{aligned}\frac{1}{2}(x(t)+y(t))^2 &\leq (x(t))^2+(y(t))^2\\ &\leq c_1+c_2+\int_0^t G(s)(x(s)+y(s))^2\Delta s\\ &\quad+\int_0^t H(s)(x(s)+y(s))\Delta s, \quad t\in\mathbb{R}_+\cap\mathbb{T},\end{aligned}$$

or

$$\frac{1}{2}(z(t))^2\leq c_1+c_2+\int_0^t G(s)(z(s))^2\Delta s+\int_0^t H(s)z(s)\Delta s, \quad t\in\mathbb{R}_+\cap\mathbb{T},$$

or

$$\begin{aligned}(z(t))^2 &\leq 2(c_1+c_2)+2\int_0^t G(s)(z(s))^2\Delta s+2\int_0^t H(s)z(s)\Delta s\\ &= c_3+\int_0^t (2G(s)(z(s))^2+2H(s)z(s))\Delta s, \quad t\in\mathbb{R}_+\cap\mathbb{T}.\end{aligned}$$

1. Let

   $$x(t)+y(t)\geq 1, \quad t\in\mathbb{R}_+\cap\mathbb{T}.$$

   Then

   $$z(t)\geq 1, \quad t\in\mathbb{R}_+\cap\mathbb{T},$$

   and

   $$z(t)\leq(z(t))^2, \quad t\in\mathbb{R}_+\cap\mathbb{T},$$

and

$$\begin{aligned}(z(t))^2 &\le c_3 + \int_0^t (2G(s) + 2H(s))(z(s))^2 \Delta s\\ &= c_3 + \int_0^t p(s)(z(s))^2 \Delta s, \quad t \in \mathbb{R}_+ \cap \mathbb{T}.\end{aligned}$$

Hence, from Theorem 1.1.10, we obtain

$$(z(t))^2 \le c_3 + c_3 \int_0^t p(s) e_{\ominus p}(\sigma(s), t) \Delta s, \quad t \in \mathbb{R}_+ \cap \mathbb{T}.$$

Therefore

$$x(t) + y(t) \le \left(c_3 + c_3 \int_0^t p(s) e_{\ominus p}(\sigma(s), t) \Delta s \right)^{\frac{1}{2}}, \quad t \in \mathbb{R}_+ \cap \mathbb{T}.$$

2. Let

$$x(t) + y(t) \le 1, \quad t \in \mathbb{R}_+ \cap \mathbb{T}.$$

Then

$$z(t) \le 1, \quad t \in \mathbb{R}_+ \cap \mathbb{T}.$$

Therefore

$$\begin{aligned}(z(t))^2 &\le c_3 + \int_0^t (2G(s) + 2H(s)) z(s) \Delta s\\ &= c_3 + \int_0^t p(s) z(s) \Delta s\\ &\le c_3 + \int_0^t p(s) \Delta s, \quad t \in \mathbb{R}_+ \cap \mathbb{T}.\end{aligned}$$

From here,

$$z(t) \le \left(c_3 + \int_0^t p(s) \Delta s \right)^{\frac{1}{2}}, \quad t \in \mathbb{R}_+ \cap \mathbb{T},$$

or

$$x(t) + y(t) \le \left(c_3 + \int_0^t p(s)\Delta s \right)^{\frac{1}{2}}, \quad t \in \mathbb{R}_+ \cap \mathbb{T}.$$

This completes the proof. □

Theorem 3.2.10. *Let $p \in (1, \infty)$, $g, h_1, h_2 \in \mathcal{C}(\mathbb{R}_+ \cap \mathbb{T})$ be nonnegative functions, $x, f \in \mathcal{C}(\mathbb{R}_+ \cap \mathbb{T})$ be positive functions, and*

$$\begin{aligned} & x(t) \ge 1, \\ & (x(t))^p \le f(t) + g(t) \int_0^t h_1(s) \left(\int_0^s h_2(y) x(y) \Delta y \right) \Delta s, \quad t \in \mathbb{R}_+ \cap \mathbb{T}. \end{aligned}$$

Then

$$x(t) \le \left(f(t) + g(t) \int_0^t \left(h_3(s) + h_1(s) \int_0^s h_4(y)(A(y) + h_1(y)B(y))\Delta y \right) \Delta s \right)^{\frac{1}{p}}, \quad t \in \mathbb{R}_+ \cap \mathbb{T},$$

where

$$\begin{aligned} h_3(t) &= h_1(t) \int_0^t h_2(s) f(s) \Delta s, \\ h_4(t) &= h_2(t) g(t), \\ A(t) &= \int_0^t h_3(s)\Delta s + h_3(t), \\ h_5(t) &= h_1(t)(h_4(t) + 1), \\ B(t) &= \int_0^t h_4(s) A(s) e_{\ominus h_5}(\sigma(s), t)\Delta s, \quad t \in \mathbb{R}_+ \cap \mathbb{T}. \end{aligned}$$

Proof. Since

$$x(t) \ge 1, \quad t \in \mathbb{R}_+ \cap \mathbb{T},$$

we have that

$$x(t) \le (x(t))^p, \quad t \in \mathbb{R}_+ \cap \mathbb{T}.$$

Let

$$z(t) = \int_0^t h_1(s) \left(\int_0^s h_2(y) x(y) \Delta y \right) \Delta s, \quad t \in \mathbb{R}_+ \cap \mathbb{T}.$$

Then

$$z(0) = 0,$$
$$(x(t))^p \le f(t) + g(t)z(t), \quad t \in \mathbb{R}_+ \cap \mathbb{T},$$

and

$$\begin{aligned}
z^\Delta(t) &= h_1(t)\int_0^t h_2(s)x(s)\Delta s \\
&\le h_1(t)\int_0^t h_2(s)(x(s))^p\Delta s \\
&\le h_1(t)\int_0^t h_2(s)(f(s) + g(s)z(s))\Delta s \\
&= h_1(t)\int_0^t h_2(s)f(s)\Delta s + h_1(t)\int_0^t h_2(s)g(s)z(s)\Delta s \\
&= h_3(t) + h_1(t)\int_0^t h_4(s)z(s)\Delta s \\
&\le h_3(t) + h_1(t)\int_0^t h_4(s)(z(s) + z^\Delta(s))\Delta s, \quad t \in \mathbb{R}_+ \cap \mathbb{T}.
\end{aligned}$$

Hence, by Theorem 2.1.2, we get

$$z^\Delta(t) \le h_3(t) + h_1(t)\int_0^t h_4(s)(A(s) + h_1(s)B(s))\Delta s,$$
$$z(t) \le \int_0^t \left(h_3(s) + h_1(s)\int_0^s h_4(y)(A(y) + h_1(y)B(y))\Delta y \right)\Delta s, \quad t \in \mathbb{R}_+ \cap \mathbb{T}.$$

Therefore

$$(x(t))^p \le f(t) + g(t)\int_0^t \left(h_3(s) + h_1(s)\int_0^s h_4(y)(A(y) + h_1(y)B(y))\Delta y \right)\Delta s, \quad t \in \mathbb{R}_+ \cap \mathbb{T},$$

whereupon

$$x(t) \le \left(f(t) + g(t)\int_0^t \left(h_3(s) + h_1(s)\int_0^s h_4(y)(A(y) + h_1(y)B(y))\Delta y \right)\Delta s \right)^{\frac{1}{p}}, \quad t \in \mathbb{R}_+ \cap \mathbb{T}.$$

This completes the proof. □

4 Diamond-α integral inequalities

In this chapter are investigated Diamond-α integral inequalities such as Steffensen, Jensen, Radon and Schlömilch inequalities. This material in this chapter cna be found in [4, 12, 23, 24] and [25].

Suppose that $\mathbb{T}$ is a time scale with forward jump operator, backward jump operator, delta differentiation operator, and nabla differentiation operator σ, ρ, Δ, and ∇, respectively.

4.1 Diamond-α calculus on time scales

Definition 4.1.1. Let $\alpha \in [0,1]$ and $f : \mathbb{T} \to \mathbb{R}$ be Δ and ∇ differentiable at $t \in \mathbb{T}$. Define the diamond-α dynamic derivative $f^{\diamond_\alpha}$ of f at t as follows:

$$f^{\diamond_\alpha}(t) = \alpha f^{\Delta}(t) + (1-\alpha) f^{\nabla}(t).$$

Thus, f is diamond-α differentiable at $t \in \mathbb{T}$ if and only if f is Δ and ∇ differentiable at t. When $\alpha = 1$, we have

$$f^{\diamond_\alpha}(t) = f^{\Delta}(t)$$

and for $\alpha = 0$, we get

$$f^{\diamond_\alpha}(t) = f^{\nabla}(t).$$

Example 4.1.2. Let $\mathbb{T} = 2^{\mathbb{N}_0}$, $f : \mathbb{T} \to \mathbb{R}$ be defined as follows:

$$f(t) = t^3 - 2t^2 + t + 1, \quad t \in \mathbb{T}.$$

Then

$$\begin{aligned}
\sigma(t) &= 2t, \\
f^{\Delta}(t) &= (\sigma(t))^2 + t\sigma(t) + t^2 - 2(\sigma(t) + t) + 1 \\
&= 4t^2 + 2t^2 + t^2 - 6t + 1 \\
&= 7t^2 - 6t + 1, \quad t \in \mathbb{T},
\end{aligned}$$

and

$$\begin{aligned}
\rho(t) &= \frac{1}{2}t, \\
f^{\nabla}(t) &= (\rho(t))^2 + t\rho(t) + t^2 - 2(\rho(t) + t) + 1 \\
&= \frac{1}{4}t^2 + \frac{1}{2}t^2 + t^2 - 3t + 1
\end{aligned}$$

https://doi.org/10.1515/9783110705553-004

$$= \frac{7}{4}t^2 - 3t + 1, \quad t \in \mathbb{T}.$$

Hence,

$$\begin{aligned} f^{\diamond_{\frac{1}{2}}}(t) &= \frac{1}{2}f^{\Delta}(t) + \frac{1}{2}f^{\nabla}(t) \\ &= \frac{1}{2}(7t^2 - 6t + 1) + \frac{1}{2}\left(\frac{7}{4}t^2 - 3t + 1\right) \\ &= \frac{7}{2}t^2 - 3t + \frac{1}{2} + \frac{7}{8}t^2 - \frac{3}{2}t + \frac{1}{2} \\ &= \frac{35}{8}t^2 - \frac{7}{2}t + 1, \end{aligned}$$

and

$$\begin{aligned} f^{\diamond_{\frac{1}{3}}}(t) &= \frac{1}{3}f^{\Delta}(t) + \frac{2}{3}f^{\nabla}(t) \\ &= \frac{1}{3}(7t^2 - 6t + 1) + \frac{2}{3}\left(\frac{7}{4}t^2 - 3t + 1\right) \\ &= \frac{7}{3}t^2 - 2t + \frac{1}{3} + \frac{7}{6}t^2 - 2t + \frac{2}{3} \\ &= \frac{7}{2}t^2 - 4t + 1, \quad t \in \mathbb{T}. \end{aligned}$$

Below we will list some of the properties of the diamond-α derivative. Let $f, g : \mathbb{T} \to \mathbb{R}$ be diamond-α differentiable at $t \in \mathbb{T}$.

Theorem 4.1.3. *$f + g$ is diamond-α differentiable at t and*

$$(f + g)^{\diamond_\alpha}(t) = f^{\diamond_\alpha}(t) + g^{\diamond_\alpha}(t).$$

Proof. We have

$$\begin{aligned} (f + g)^{\diamond_\alpha}(t) &= \alpha(f + g)^{\Delta}(t) + (1 - \alpha)(f + g)^{\nabla}(t) \\ &= \alpha(f^{\Delta}(t) + g^{\Delta}(t)) + (1 - \alpha)(f^{\nabla}(t) + g^{\nabla}(t)) \\ &= \alpha f^{\Delta}(t) + (1 - \alpha)f^{\nabla}(t) \\ &\quad + \alpha g^{\Delta}(t) + (1 - \alpha)g^{\nabla}(t) \\ &= f^{\diamond_\alpha}(t) + g^{\diamond_\alpha}(t). \end{aligned}$$

This completes the proof. □

Theorem 4.1.4. *For any $c \in \mathbb{C}$, we have that cf is diamond-α differentiable at t and*

$$(cf)^{\diamond_\alpha}(t) = cf^{\diamond_\alpha}(t).$$

Proof. We have

$$(cf)^{\diamond_\alpha}(t) = \alpha(cf)^{\Delta}(t) + (1 - \alpha)(cf)^{\nabla}(t)$$

$$
\begin{aligned}
&= \alpha(cf^{\Delta}(t)) + (1-\alpha)(cf^{\nabla}(t)) \\
&= c(\alpha f^{\Delta}(t) + (1-\alpha)f^{\nabla}(t)) \\
&= cf^{\diamond_\alpha}(t).
\end{aligned}
$$

This completes the proof. □

Theorem 4.1.5. *fg is diamond-α differentiable at t and*

$$
\begin{aligned}
(fg)^{\diamond_\alpha}(t) &= f^{\diamond_\alpha}(t)g(t) + \alpha f^{\sigma}(t)g^{\Delta}(t) + (1-\alpha)f^{\rho}(t)g^{\nabla}(t) \\
&= f(t)g^{\diamond_\alpha}(t) + \alpha f^{\Delta}(t)g^{\sigma}(t) + (1-\alpha)f^{\nabla}(t)g^{\rho}(t).
\end{aligned}
$$

Proof. We have

$$
\begin{aligned}
(fg)^{\diamond_\alpha}(t) &= \alpha(fg)^{\Delta}(t) + (1-\alpha)(fg)^{\nabla}(t) \\
&= \alpha(f^{\Delta}(t)g(t) + f^{\sigma}(t)g^{\Delta}(t)) + (1-\alpha)(f^{\nabla}(t)g(t) + f^{\rho}(t)g^{\nabla}(t)) \\
&= (\alpha f^{\Delta}(t) + (1-\alpha)f^{\nabla}(t))g(t) + \alpha f^{\sigma}(t)g^{\Delta}(t) + (1-\alpha)f^{\rho}(t)g^{\nabla}(t) \\
&= f^{\diamond_\alpha}(t)g(t) + \alpha f^{\sigma}(t)g^{\Delta}(t) + (1-\alpha)f^{\rho}(t)g^{\nabla}(t) \\
&= \alpha(f^{\Delta}(t)g^{\sigma}(t) + f(t)g^{\Delta}(t)) + (1-\alpha)(f^{\nabla}(t)g^{\rho}(t) + f(t)g^{\nabla}(t)) \\
&= (\alpha g^{\Delta}(t) + (1-\alpha)g^{\nabla}(t))f(t) + \alpha f^{\Delta}(t)g^{\sigma}(t) + (1-\alpha)f^{\nabla}(t)g^{\rho}(t) \\
&= f(t)g^{\diamond_\alpha} + \alpha f^{\Delta}(t)g^{\sigma}(t) + (1-\alpha)f^{\nabla}(t)g^{\rho}(t).
\end{aligned}
$$

This completes the proof. □

Definition 4.1.6. Let $\alpha \in [0,1]$, $a, b \in \mathbb{T}$, $a < b$, $t \in [a,b]$, and suppose f is Δ and ∇ integrable over $[a,b]$. Define diamond-α integral from a to t of f as follows:

$$
\int_a^t f(s)\diamond_\alpha s = \alpha \int_a^t f(s)\Delta s + (1-\alpha)\int_a^t f(s)\nabla s.
$$

Example 4.1.7. Let $\mathbb{T} = 3^{\mathbb{N}_0} \cup \{\frac{1}{3}\}$,

$$
f(t) = t^2 + t, \quad t \in \mathbb{T}.
$$

We will compute

$$
\int_1^9 f(s)\diamond_{\frac{1}{3}} s.
$$

Let

$$
\begin{aligned}
h_1(t) &= t^3, \\
h_2(t) &= t^2, \quad t \in \mathbb{T}.
\end{aligned}
$$

We have

$$\begin{aligned}
\sigma(t) &= 3t,\\
h_1^{\Delta}(t) &= (\sigma(t))^2 + t\sigma(t) + t^2\\
&= 9t^2 + 3t^2 + t^2\\
&= 13t^2,\\
h_2^{\Delta}(t) &= \sigma(t) + t\\
&= 3t + t\\
&= 4t, \quad t \in \mathbb{T},\\
\rho(t) &= \frac{1}{3}t,\\
h_1^{\nabla}(t) &= (\rho(t))^2 + t\rho(t) + t^2\\
&= \frac{1}{9}t^2 + \frac{1}{3}t^2 + t^2\\
&= \frac{13}{9}t^2,\\
h_2^{\nabla}(t) &= \rho(t) + t\\
&= \frac{1}{3}t + t\\
&= \frac{4}{3}t, \quad t \in \mathbb{T}, \quad t \geq 1.
\end{aligned}$$

Hence,

$$\begin{aligned}
\int_1^9 f(s)\diamond_{\frac{1}{3}} s &= \frac{1}{3}\int_1^9 f(s)\Delta s + \frac{2}{3}\int_1^9 f(s)\nabla s\\
&= \frac{1}{3}\int_1^9 (s^2 + s)\Delta s + \frac{2}{3}\int_1^9 (s^2 + s)\nabla s\\
&= \frac{1}{3}\int_1^9 \left(\frac{1}{13}h_1^{\Delta}(s) + \frac{1}{4}h_2^{\Delta}(s)\right)\Delta s\\
&\quad + \frac{2}{3}\int_1^9 \left(\frac{9}{13}h_1^{\nabla}(s) + \frac{3}{4}h_2^{\nabla}(s)\right)\nabla s\\
&= \frac{1}{39}(h_1(9) - h_1(1)) + \frac{1}{12}(h_2(9) - h_2(1))\\
&\quad + \frac{18}{39}(h_1(9) - h_1(1)) + \frac{1}{2}(h_2(9) - h_2(1))\\
&= \frac{19}{39}(729 - 1) + \frac{7}{12}(81 - 1)
\end{aligned}$$

$$= \frac{1064}{3} + \frac{140}{3}$$
$$= \frac{1204}{3}.$$

Remark 4.1.8. Note that

$$\begin{aligned}\left(\int_a^t f(s)\diamond_\alpha s\right)^{\diamond_\alpha} &= \alpha\left(\int_a^t f(s)\diamond_\alpha s\right)^{\Delta} + (1-\alpha)\left(\int_a^t f(s)\diamond_\alpha s\right)^{\nabla} \\ &= \alpha\left(\alpha\int_a^t f(s)\Delta s + (1-\alpha)\int_a^t f(s)\nabla s\right)^{\Delta} \\ &\quad + (1-\alpha)\left(\alpha\int_a^t f(s)\Delta s + (1-\alpha)\int_a^t f(s)\nabla s\right)^{\nabla} \\ &= \alpha^2 f(t) + \alpha(1-\alpha)f(\sigma(t)) + \alpha(1-\alpha)f(\rho(t)) + (1-\alpha)^2 f(t) \\ &= (2\alpha^2 - 2\alpha + 1)f(t) + \alpha(1-\alpha)(f(\sigma(t)) + f(\rho(t))), \quad t \in [a,b].\end{aligned}$$

Thus, in the general case we do not have

$$\left(\int_a^t f(s)\diamond_\alpha s\right)^{\diamond_\alpha} = f(t).$$

Below we suppose that $f, g : \mathbb{T} \to \mathbb{R}$ are diamond-α integrable over $[a,b]$.

Theorem 4.1.9. *For any $c \in \mathbb{C}$, the function cf is diamond-α integrable over $[a,b]$ and*

$$\int_a^b (cf)(s)\diamond_\alpha s = c\int_a^b f(s)\diamond_\alpha s.$$

Proof. We have

$$\begin{aligned}\int_a^b (cf)(s)\diamond_\alpha s &= \alpha\int_a^b (cf)(s)\Delta s + (1-\alpha)\int_a^b (cf)(s)\nabla s \\ &= c\left(\alpha\int_a^b f(s)\Delta s + (1-\alpha)\int_a^b f(s)\nabla s\right) \\ &= c\int_a^b f(s)\diamond_\alpha s.\end{aligned}$$

This completes the proof. □

Theorem 4.1.10. *$f+g$ is diamond-α integrable over $[a,b]$ and*

$$\int_a^b (f+g)(s)\diamond_\alpha s = \int_a^b f(s)\diamond_\alpha s + \int_a^b g(s)\diamond_\alpha s.$$

Proof. We have

$$\begin{aligned}\int_a^b (f+g)(s)\diamond_\alpha s &= \alpha \int_a^b (f+g)(s)\Delta s + (1-\alpha)\int_a^b (f+g)(s)\nabla s\\ &= \alpha \int_a^b f(s)\Delta s + (1-\alpha)\int_a^b f(s)\nabla s\\ &\quad + \alpha \int_a^b g(s)\Delta s + (1-\alpha)\int_a^b g(s)\nabla s\\ &= \int_a^b f(s)\diamond_\alpha s + \int_a^b g(s)\diamond_\alpha s.\end{aligned}$$

This completes the proof. □

Theorem 4.1.11. *We have*

$$\int_a^b f(s)\diamond_\alpha s = \int_a^t f(s)\diamond_\alpha s + \int_t^b f(s)\diamond_\alpha s$$

for any $t \in [a,b]$.

Proof. We have

$$\begin{aligned}\int_a^b f(s)\diamond_\alpha s &= \alpha \int_a^b f(s)\Delta s + (1-\alpha)\int_a^b f(s)\nabla s\\ &= \alpha\left(\int_a^t f(s)\Delta s + \int_t^b f(s)\Delta s\right)\\ &\quad + (1-\alpha)\left(\int_a^t f(s)\nabla s + \int_t^b f(s)\nabla s\right)\\ &= \int_a^t f(s)\diamond_\alpha s + \int_t^b f(s)\diamond_\alpha s, \quad t\in[a,b].\end{aligned}$$

This completes the proof. □

Theorem 4.1.12. *Let $a, b \in \mathbb{T}_\kappa^\kappa$ with $a < b$ and suppose $f, g, h : [a, b] \to \mathbb{R}$ are $\diamond_\alpha$-integrable functions. Suppose that $l, y \in [a, b]$ are such that*

$$\int_a^y h(t)\diamond_\alpha t = \int_a^b g(t)\diamond_\alpha t = \int_l^b h(t)\diamond_\alpha t.$$

Then

$$\int_a^b f(t)g(t)\diamond_\alpha t = \int_a^y (f(t)h(t) - (f(t) - f(y))(h(t) - g(t)))\diamond_\alpha t + \int_y^b (f(t) - f(y))g(t)\diamond_\alpha t,$$

$$\int_a^b f(t)g(t)\diamond_\alpha t = \int_a^l (f(t) - f(l))g(t)\diamond_\alpha t + \int_l^b (f(t)h(t) - (f(t) - f(l))(h(t) - g(t)))\diamond_\alpha t.$$

Proof. We have

$$\begin{aligned}
&\int_a^y (f(t)h(t) - (f(t) - f(y))(h(t) - g(t)))\diamond_\alpha t - \int_a^b f(t)g(t)\diamond_\alpha t \\
&= \int_a^y (f(t)h(t) - f(t)g(t) - (f(t) - f(y))(h(t) - g(t)))\diamond_\alpha t \\
&\quad + \int_a^y f(t)g(t)\diamond_\alpha t - \int_a^b f(t)g(t)\diamond_\alpha t \\
&= \int_a^y f(y)(h(t) - g(t))\diamond_\alpha t - \int_y^b f(t)g(t)\diamond_\alpha t \\
&= f(y)\left(\int_a^y h(t)\diamond_\alpha t - \int_a^y g(t)\diamond_\alpha t\right) - \int_y^b f(t)g(t)\diamond_\alpha t \\
&= f(y)\left(\int_a^b g(t)\diamond_\alpha t - \int_a^y g(t)\diamond_\alpha t\right) - \int_y^b f(t)g(t)\diamond_\alpha t \\
&= f(y)\int_y^b g(t)\diamond_\alpha t - \int_y^b f(t)g(t)\diamond_\alpha t
\end{aligned}$$

$$= \int_{y}^{b} (f(y) - f(t))g(t)\diamond_\alpha t.$$

Next we get

$$\int_{l}^{b} (f(t)h(t) - (f(t) - f(l))(h(t) - g(t)))\diamond_\alpha t - \int_{a}^{b} f(t)g(t)\diamond_\alpha t$$
$$= \int_{l}^{b} (f(t)h(t) - f(t)g(t) - (f(t) - f(l))(h(t) - g(t)))\diamond_\alpha t$$
$$+ \int_{l}^{b} f(t)g(t)\diamond_\alpha t - \int_{a}^{b} f(t)g(t)\diamond_\alpha t$$
$$= f(l) \int_{l}^{b} (h(t) - g(t))\diamond_\alpha t - \int_{a}^{l} f(t)g(t)\diamond_\alpha t$$
$$= f(l) \int_{l}^{b} h(t)\diamond_\alpha t - f(l) \int_{l}^{b} g(t)\diamond_\alpha t - \int_{a}^{l} f(t)g(t)\diamond_\alpha t$$
$$= f(l) \int_{a}^{b} g(t)\diamond_\alpha t - f(l) \int_{l}^{b} g(t)\diamond_\alpha t - \int_{a}^{l} f(t)g(t)\diamond_\alpha t$$
$$= f(l)\left(\int_{a}^{b} g(t)\diamond_\alpha t - \int_{l}^{b} g(t)\diamond_\alpha t \right) - \int_{a}^{l} f(t)g(t)\diamond_\alpha t$$
$$= f(l) \int_{a}^{l} g(t)\diamond_\alpha t - \int_{a}^{l} f(t)g(t)\diamond_\alpha t$$
$$= \int_{a}^{l} (f(l) - f(t))g(t)\diamond_\alpha t.$$

This completes the proof. □

Theorem 4.1.13. *Assume*

(A1) *k is a positive $\diamond_\alpha$-integrable function on $[a, b]$,*

(A2) *$\phi, \psi, h : [a, b] \to \mathbb{R}$ are $\diamond_\alpha$-integrable functions on $[a, b]$,*

(A3) *$[c, d] \subseteq [a, b]$ with*

$$\int_{c}^{d} h(t)k(t)\diamond_\alpha t = \int_{a}^{b} \psi(t)k(t)\diamond_\alpha t,$$

(A4) *$z \in [a, b]$.*

Then

$$\int_c^d \phi(t)h(t)\diamond_\alpha t - \int_a^b \phi(t)\psi(t)\diamond_\alpha t = \int_a^c \left(\frac{\phi(z)}{k(z)} - \frac{\phi(t)}{k(t)}\right)\psi(t)k(t)\diamond_\alpha t$$
$$+ \int_c^d \left(\frac{\phi(t)}{k(t)} - \frac{\phi(z)}{k(z)}\right)k(t)(h(t) - \psi(t))\diamond_\alpha t$$
$$+ \int_d^b \left(\frac{\phi(z)}{k(z)} - \frac{\phi(t)}{k(t)}\right)\psi(t)k(t)\diamond_\alpha t.$$

Proof. We have

$$\int_c^d \phi(t)h(t)\diamond_\alpha t - \int_a^b \phi(t)\psi(t)\diamond_\alpha t$$
$$= \int_c^d k(t)h(t)\frac{\phi(t)}{k(t)}\diamond_\alpha t - \int_a^b \frac{\phi(t)}{k(t)}k(t)\psi(t)\diamond_\alpha t$$
$$= \int_c^d k(t)(h(t) - \psi(t))\frac{\phi(t)}{k(t)}\diamond_\alpha t + \int_c^d k(t)\psi(t)\frac{\phi(t)}{k(t)}\diamond_\alpha t$$
$$- \int_a^b \frac{\phi(t)}{k(t)}k(t)\psi(t)\diamond_\alpha t$$
$$= \int_c^d k(t)(h(t) - \psi(t))\frac{\phi(t)}{k(t)}\diamond_\alpha t$$
$$- \left(\int_d^c k(t)\psi(t)\frac{\phi(t)}{k(t)}\diamond_\alpha t + \int_a^b \frac{\phi(t)}{k(t)}k(t)\psi(t)\diamond_\alpha t\right)$$
$$= \int_c^d k(t)(h(t) - \psi(t))\frac{\phi(t)}{k(t)}\diamond_\alpha t$$
$$- \left(\int_a^c k(t)\psi(t)\frac{\phi(t)}{k(t)}\diamond_\alpha t + \int_d^b \frac{\phi(t)}{k(t)}k(t)\psi(t)\diamond_\alpha t\right)$$
$$= \int_a^c \left(\frac{\phi(z)}{k(z)} - \frac{\phi(t)}{k(t)}\right)k(t)\psi(t)\diamond_\alpha t$$
$$+ \int_c^d \left(\frac{\phi(t)}{k(t)} - \frac{\phi(z)}{k(z)}\right)k(t)(h(t) - \psi(t))\diamond_\alpha t$$

$$
\begin{aligned}
&+\int_d^b\left(\frac{\phi(z)}{k(z)}-\frac{\phi(t)}{k(t)}\right)k(t)\psi(t)\diamond_\alpha t\\
&+\frac{\phi(z)}{k(z)}\left(\int_c^d k(t)(h(t)-\psi(t))\diamond_\alpha t-\int_a^c k(t)\psi(T)\diamond_\alpha t\right.\\
&\left.-\int_d^b k(t)\psi(t)\diamond_\alpha t\right)\\
=&\int_a^c\left(\frac{\phi(z)}{k(z)}-\frac{\phi(t)}{k(t)}\right)k(t)\psi(t)\diamond_\alpha t\\
&+\int_c^d\left(\frac{\phi(t)}{k(t)}-\frac{\phi(z)}{k(z)}\right)k(t)(h(t)-\psi(t))\diamond_\alpha t\\
&+\int_d^b\left(\frac{\phi(z)}{k(z)}-\frac{\phi(t)}{k(t)}\right)k(t)\psi(t)\diamond_\alpha t\\
&+\frac{\phi(z)}{k(z)}\left(\int_c^d k(t)h(t)\diamond_\alpha t-\int_a^b k(t)\psi(t)\diamond_\alpha t\right)\\
=&\int_a^c\left(\frac{\phi(z)}{k(z)}-\frac{\phi(t)}{k(t)}\right)k(t)\psi(t)\diamond_\alpha t\\
&+\int_c^d\left(\frac{\phi(t)}{k(t)}-\frac{\phi(z)}{k(z)}\right)k(t)(h(t)-\psi(t))\diamond_\alpha t\\
&+\int_d^b\left(\frac{\phi(z)}{k(z)}-\frac{\phi(t)}{k(t)}\right)k(t)\psi(t)\diamond_\alpha t.
\end{aligned}
$$

This completes the proof. □

4.2 Steffensen's inequalities

Theorem 4.2.1. *Let $a,b\in\mathbb{T}^\kappa_\kappa$ with $a<b$ and $f,g,h:[a,b]\to\mathbb{R}$ be $\diamond_\alpha$-integrable with f of one sign and decreasing, and*

$$0\le g(t)\le h(t),\quad t\in[a,b].$$

Assume that $l,\gamma\in[a,b]$ and

$$\int_l^b h(t)\diamond_\alpha t\le\int_a^b g(t)\diamond_\alpha t\le\int_a^\gamma h(t)\diamond_\alpha t\quad\text{if}\quad f\ge 0,$$

$$\int_a^y h(t)\diamond_\alpha t \le \int_a^b g(t)\diamond_\alpha t \le \int_l^b h(t)\diamond_\alpha t \quad \textit{if} \quad f \le 0.$$

Then

$$\int_l^b f(t)h(t)\diamond_\alpha t \le \int_a^b f(t)g(t)\diamond_\alpha t \le \int_a^b f(t)h(t)\diamond_\alpha t.$$

Proof. Assume that $f \ge 0$ on $[a,b]$. Then

$$\begin{aligned}
\int_a^b f(t)g(t)\diamond_\alpha t - \int_l^b f(t)h(t)\diamond_\alpha t &= \int_a^l f(t)g(t)\diamond_\alpha t + \int_l^b f(t)g(t)\diamond_\alpha t \\
&\quad - \int_l^b f(t)h(t)\diamond_\alpha t \\
&= \int_a^l f(t)g(t)\diamond_\alpha t - \int_l^b f(t)(h(t)-g(t))\diamond_\alpha t \\
&\ge \int_a^l f(t)g(t)\diamond_\alpha t - f(l)\int_l^b (h(t)-g(t))\diamond_\alpha t \\
&= \int_a^l f(t)g(t)\diamond_\alpha t - f(l)\int_l^b h(t)\diamond_\alpha t \\
&\quad + f(l)\int_l^b g(t)\diamond_\alpha t \\
&\ge \int_a^l f(t)g(t)\diamond_\alpha t - f(l)\int_a^b g(t)\diamond_\alpha t \\
&\quad + f(l)\int_l^b g(t)\diamond_\alpha t \\
&= \int_a^l f(t)g(t)\diamond_\alpha t \\
&\quad - f(l)\left(\int_a^b g(t)\diamond_\alpha t - \int_l^b g(t)\diamond_\alpha t\right) \\
&= \int_a^l f(t)g(t)\diamond_\alpha t - f(l)\int_a^l g(t)\diamond_\alpha t
\end{aligned}$$

$$= \int_a^l (f(t) - f(l))g(t)\diamond_\alpha t$$
$$\geq 0,$$

i. e.,

$$\int_a^b f(t)g(t)\diamond_\alpha t \geq \int_l^b f(t)h(t)\diamond_\alpha t.$$

Next we have

$$\begin{aligned}
\int_a^y f(t)h(t)\diamond_\alpha t - \int_a^b f(t)g(t)\diamond_\alpha t &= \int_a^y f(t)h(t)\diamond_\alpha t - \int_a^y f(t)g(t)\diamond_\alpha t \\
&\quad - \int_y^b f(t)g(t)\diamond_\alpha t \\
&= \int_a^y f(t)(h(t) - g(t))\diamond_\alpha t - \int_y^b f(t)g(t)\diamond_\alpha t \\
&\geq f(y)\int_a^y (h(t) - g(t))\diamond_\alpha t - \int_y^b f(t)g(t)\diamond_\alpha t \\
&= f(y)\int_a^y h(t)\diamond_\alpha t - f(y)\int_a^y g(t)\diamond_\alpha t \\
&\quad - \int_y^b f(t)g(t)\diamond_\alpha t \\
&\geq f(y)\int_a^y h(t)\diamond_\alpha t - f(y)\int_a^y g(t)\diamond_\alpha t \\
&\quad - f(y)\int_y^b g(t)\diamond_\alpha t \\
&= f(y)\int_a^y h(t)\diamond_\alpha t \\
&\quad - f(y)\left(\int_a^y g(t)\diamond_\alpha t + \int_y^b g(t)\diamond_\alpha t\right) \\
&= f(y)\int_a^y h(t)\diamond_\alpha t - f(y)\int_a^b g(t)\diamond_\alpha t
\end{aligned}$$

$$\geq f(y)\int_a^b g(t)\diamond_\alpha t - f(y)\int_a^b g(t)\diamond_\alpha t$$
$$= 0,$$

i. e.,

$$\int_a^y f(t)h(t)\diamond_\alpha t \geq \int_a^b f(t)g(t)\diamond_\alpha t.$$

The case when $f \leq 0$ we leave to the reader as an exercise. This completes the proof. □

Theorem 4.2.2. *Let $a, b \in \mathbb{T}_\kappa^\kappa$ with $a < b$ and $f, g, h : [a, b] \to \mathbb{R}$ be $\diamond_\alpha$-integrable functions, f of one sign and decreasing, and*

$$0 \leq g(t) \leq h(t), \quad t \in [a, b].$$

Assume that $l, y \in [a, b]$ are such that

$$\int_a^y h(t)\diamond_\alpha t = \int_a^b g(t)\diamond_\alpha t = \int_l^b h(t)\diamond_\alpha t.$$

Then

$$\begin{aligned}\int_l^b f(t)h(t)\diamond_\alpha t &\leq \int_l^b (f(t)h(t) - (f(t) - f(l))(h(t) - g(t)))\diamond_\alpha t\\ &\leq \int_a^b f(t)g(t)\diamond_\alpha t\\ &\leq \int_a^y (f(t)h(t) - (f(t) - f(y))(h(t) - g(t)))\diamond_\alpha t\\ &\leq \int_a^y f(t)h(t)\diamond_\alpha t.\end{aligned}$$

Proof. From the assumptions on the functions f, g, h, we get

$$\int_a^l (f(t) - f(l))g(t)\diamond_\alpha t \geq 0,$$

$$\int_l^b (f(l) - f(t))(h(t) - g(t))\diamond_\alpha t \geq 0.$$

By the second integral identity of Theorem 4.1.12, we have

$$\int_a^b f(t)g(t)\diamond_\alpha t = \int_a^l (f(t)-f(l))g(t)\diamond_\alpha t$$
$$+ \int_l^b (f(t)h(t) - (f(t)-f(l))(h(t)-g(t)))\diamond_\alpha t,$$

whereupon

$$\int_l^b f(t)h(t)\diamond_\alpha t \le \int_l^b (f(t)h(t) - (f(t)-f(l))(h(t)-g(t)))\diamond_\alpha t$$
$$\le \int_a^b f(t)g(t)\diamond_\alpha t.$$

By the first integral identity of Theorem 4.1.12, we have

$$\int_a^b f(t)g(t)\diamond_\alpha t = \int_a^y (f(t)h(t) - (f(t)-f(y))(h(t)-g(t)))\diamond_\alpha t$$
$$+ \int_y^b (f(t)-f(y))g(t)\diamond_\alpha t$$
$$\le \int_a^y (f(t)h(t) - (f(t)-f(y))(h(t)-g(t)))\diamond_\alpha t$$
$$\le \int_a^y f(t)h(t)\diamond_\alpha t.$$

This completes the proof. □

Theorem 4.2.3. *Let $a, b \in \mathbb{T}^\kappa_\kappa$ with $a < b$ and $f, g, h, \phi : [a, b] \to \mathbb{R}$ be $\diamond_\alpha$-integrable functions, f of one sign and decreasing and*

$$0 \le \phi(t) \le g(t) \le h(t) - \phi(t), \quad t \in [a, b].$$

Assume $l, y \in [a, b]$ are such that

$$\int_a^y h(t)\diamond_\alpha t = \int_a^b g(t)\diamond_\alpha t = \int_l^b h(t)\diamond_\alpha t.$$

Then

$$\begin{aligned}\int_l^b f(t)h(t)\diamond_\alpha t + \int_a^b |f(t) - f(l)|\phi(t)\diamond_\alpha t \\ \leq \int_a^b f(t)g(t)\diamond_\alpha t \\ \leq \int_a^y f(t)h(t)\diamond_\alpha t - \int_a^b |f(t) - f(y)|\phi(t)\diamond_\alpha t.\end{aligned}$$

Proof. From the assumptions on the functions f, g, h, and ϕ, we get

$$\begin{aligned}&\int_a^y (f(t) - f(y))(h(t) - g(t))\diamond_\alpha t + \int_y^b (f(y) - f(t))g(t)\diamond_\alpha t \\ &\geq \int_a^y (f(t) - f(y))\phi(t)\diamond_\alpha t + \int_y^b (f(y) - f(t))\phi(t)\diamond_\alpha t \\ &= \int_a^b |f(t) - f(y)|\phi(t)\diamond_\alpha t\end{aligned}$$

and

$$\begin{aligned}&\int_a^l (f(t) - f(l))g(t)\diamond_\alpha t + \int_l^b (f(l) - f(t))(h(t) - g(t))\diamond_\alpha t \\ &\geq \int_a^l (f(t) - f(l))\phi(t)\diamond_\alpha t + \int_l^b (f(l) - f(t))\phi(t)\diamond_\alpha t \\ &= \int_a^b |f(t) - f(l)|\phi(t)\diamond_\alpha t.\end{aligned}$$

Hence, due to the identities of Theorem 4.1.12, we find

$$\begin{aligned}\int_a^b f(t)g(t)\diamond_\alpha t &= \int_a^y (f(t)h(t) - (f(t) - f(y))(h(t) - g(t)))\diamond_\alpha t \\ &\quad + \int_y^b (f(t) - f(y))g(t)\diamond_\alpha t \\ &= \int_a^y f(t)h(t)\diamond_\alpha t - \int_a^y (f(t) - f(y))(h(t) - g(t))\diamond_\alpha t\end{aligned}$$

$$-\int_y^b (f(y) - f(t))g(t) \diamond_\alpha t$$

$$\leq \int_a^y f(t)h(t)\diamond_\alpha t - \int_a^b |f(t) - f(y)|\phi(t)\diamond_\alpha t$$

and

$$\int_a^b f(t)g(t)\diamond_\alpha t = \int_a^l (f(t) - f(l))g(t)\diamond_\alpha t$$

$$+ \int_l^b (f(t)h(t) - (f(t) - f(l))(h(t) - g(t)))\diamond_\alpha t$$

$$= \int_l^b f(t)h(t)\diamond_\alpha t - \int_l^b (f(t) - f(l))(h(t) - g(t))\diamond_\alpha t$$

$$+ \int_a^l (f(t) - f(l))g(t) \diamond_\alpha t$$

$$= \int_l^b f(t)h(t) \diamond_\alpha t + \int_l^b (f(l) - f(t))(h(t) - g(t)) \diamond_\alpha t$$

$$+ \int_a^l (f(t) - f(l))g(t) \diamond_\alpha t$$

$$\geq \int_l^b f(t)h(t)\diamond_\alpha t + \int_a^b |f(t) - f(l)|\phi(t)\diamond_\alpha t.$$

This completes the proof. $\qquad\qquad\qquad\qquad\qquad\qquad\qquad\qquad\qquad\qquad\qquad\square$

Theorem 4.2.4. *Suppose* (A1)–(A3) *and*
(A5) $\frac{\phi}{k}$ *is nonincreasing on* $[a, b]$,
(A6) $0 \leq \psi(t) \leq h(t), t \in [a, b]$.

Then

$$\int_a^b \phi(t)\psi(t)\diamond_\alpha t \leq \int_c^d \phi(t)\psi(t)\diamond_\alpha t + \int_a^c \left(\frac{\phi(t)}{k(t)} - \frac{\phi(d)}{k(d)} \right)\psi(t)k(t)\diamond_\alpha t.$$

Proof. Since $\frac{\phi}{k}$ is nonincreasing, k is positive, and $0 \leq \psi \leq h$ on $[a, b]$, we have

$$\int_c^d \left(\frac{\phi(t)}{k(t)} - \frac{\phi(d)}{k(d)} \right)k(t)(h(t) - \psi(t))\diamond_\alpha t \geq 0,$$

$$\int_d^b \left(\frac{\phi(d)}{k(d)} - \frac{\phi(t)}{k(t)}\right)\psi(t)k(t)\diamond_\alpha t \geq 0.$$

Now, we apply Theorem 4.1.13 for $z = d$ and get

$$\begin{aligned}
&\int_c^d \phi(t)h(t)\diamond_\alpha t - \int_a^b \phi(t)\psi(t)\diamond_\alpha t - \int_a^c \left(\frac{\phi(d)}{k(d)} - \frac{\phi(t)}{k(t)}\right)\psi(t)k(t)\diamond_\alpha t \\
&\quad = \int_c^d \left(\frac{\phi(t)}{k(t)} - \frac{\phi(d)}{k(d)}\right)k(t)(h(t) - \psi(t))\diamond_\alpha t + \int_d^b \left(\frac{\phi(d)}{k(d)} - \frac{\phi(t)}{k(t)}\right)\psi(t)k(t)\diamond_\alpha t \\
&\quad \geq 0.
\end{aligned}$$

This completes the proof. □

Theorem 4.2.5. *Suppose* (A1)–(A3) *and* (A5). *Then*

$$\int_c^d \phi(t)\psi(t)\diamond_\alpha t - \int_d^b \left(\frac{\phi(c)}{k(c)} - \frac{\phi(t)}{k(t)}\right)\psi(t)k(t)\diamond_\alpha t \leq \int_a^b \phi(t)\psi(t)\diamond_\alpha t.$$

Proof. Since $\frac{\phi}{k}$ is nonincreasing and $0 \leq \psi \leq h$ on $[a, b]$, we have

$$\int_c^d \left(\frac{\phi(c)}{k(c)} - \frac{\phi(t)}{k(t)}\right)k(t)(h(t) - \psi(t))\diamond_\alpha t \geq 0,$$

$$\int_a^c \left(\frac{\phi(t)}{k(t)} - \frac{\phi(c)}{k(c)}\right)\psi(t)k(t)\diamond_\alpha t \geq 0.$$

Hence, by Theorem 4.1.13 for $z = c$, we find

$$\begin{aligned}
&\int_c^d \phi(t)h(t)\diamond_\alpha t - \int_d^b \left(\frac{\phi(c)}{k(c)} - \frac{\phi(t)}{k(t)}\right)\psi(t)k(t)\diamond_\alpha t \\
&\quad = \int_a^b \phi(t)\psi(t)\diamond_\alpha t \\
&\qquad + \int_c^d \left(\frac{\phi(t)}{k(t)} - \frac{\phi(c)}{k(c)}\right)k(t)(h(t) - \psi(t))\diamond_\alpha t \\
&\qquad + \int_a^c \left(\frac{\phi(c)}{k(c)} - \frac{\phi(t)}{k(t)}\right)\psi(t)k(t)\diamond_\alpha t \\
&\quad \leq \int_a^b \phi(t)\psi(t)\diamond_\alpha t.
\end{aligned}$$

This completes the proof. □

4.3 The Jensen inequality

Theorem 4.3.1 (Jensen's Inequality). *Let $g \in C([a,b])$ and let $\Phi : \mathbb{R} \to \mathbb{R}$ be convex. Then*

$$\Phi\left(\frac{1}{b-a}\int_a^b g(t)\diamond_\alpha t\right) \le \frac{1}{b-a}\int_a^b \Phi(g(t))\diamond_\alpha t.$$

Proof. Since Φ is convex, we have

$$\begin{aligned}\Phi\left(\frac{1}{b-a}\int_a^b g(t)\diamond_\alpha t\right) &= \Phi\left(\frac{\alpha}{b-a}\int_a^b g(t)\Delta t + \frac{1-\alpha}{b-a}\int_a^b g(t)\nabla t\right)\\ &\le \alpha\Phi\left(\frac{1}{b-a}\int_a^b g(t)\Delta t\right) + (1-\alpha)\Phi\left(\frac{1}{b-a}\int_a^b g(t)\nabla t\right).\end{aligned} \tag{4.1}$$

Let

$$\begin{aligned}x_0 &= \frac{\int_a^b g(t)\Delta t}{b-a},\\ x &= g(t).\end{aligned}$$

Since Φ is convex, there exists a $\beta \in \mathbb{R}$ such that

$$\Phi(x) - \Phi(x_0) \ge \beta(x - x_0), \tag{4.2}$$

or

$$\Phi(g(t)) - \Phi\left(\frac{1}{b-a}\int_a^b g(t)\Delta t\right) \ge \beta\left(g(t) - \frac{1}{b-a}\int_a^b g(t)\Delta t\right).$$

We integrate the last inequality with respect to t from a to b and find

$$\begin{aligned}\int_a^b \Phi(g(t))\Delta t - (b-a)\Phi\left(\frac{1}{b-a}\int_a^b g(t)\Delta t\right) &\ge \beta\left(\int_a^b g(t)\Delta t - \int_a^b g(t)\Delta t\right)\\ &= 0,\end{aligned}$$

i. e.,

$$\Phi\left(\frac{1}{b-a}\int_a^b g(t)\Delta t\right) \le \frac{1}{b-a}\int_a^b \Phi(g(t))\Delta t. \tag{4.3}$$

Similarly,

$$\Phi\left(\frac{1}{b-a}\int_a^b g(t)\nabla t\right) \le \frac{1}{b-a}\int_a^b \Phi(g(t))\nabla t.$$

By the last inequality and (4.1), (4.3), we find

$$\Phi\left(\frac{1}{b-a}\int_a^b g(t)\diamond_\alpha t\right) \le \frac{\alpha}{b-a}\int_a^b \Phi(g(t))\Delta t + \frac{1-\alpha}{b-a}\int_a^b \Phi(g(t))\nabla t$$
$$= \frac{1}{b-a}\int_a^b \Phi(g(t))\diamond_\alpha t.$$

This completes the proof. □

Theorem 4.3.2 (The Generalized Jensen Inequality). *Let $g, h \in \mathcal{C}([a,b])$ with*

$$\int_a^b |h(t)|\diamond_\alpha t > 0,$$

and let $\Phi : \mathbb{R} \to \mathbb{R}$ be convex. Then

$$\Phi\left(\frac{\int_a^b |h(t)|g(t)\diamond_\alpha t}{\int_a^b |h(t)|\diamond_\alpha t}\right) \le \frac{\int_a^b |h(t)|\Phi(g(t))\diamond_\alpha t}{\int_a^b |h(t)|\diamond_\alpha t}.$$

Proof. Take

$$x_0 = \frac{\int_a^b |h(t)|g(t)\diamond_\alpha t}{\int_a^b |h(t)|\diamond_\alpha t},$$
$$x = g(t)$$

in (4.2) and find

$$\int_a^b |h(t)|\Phi(g(t))\diamond_\alpha t - \left(\int_a^b |h(t)|\diamond_\alpha t\right)\Phi\left(\frac{\int_a^b |h(t)|g(t)\diamond_\alpha t}{\int_a^b |h(t)|\diamond_\alpha t}\right)$$
$$= \int_a^b |h(t)|\Phi(g(t))\diamond_\alpha t - \left(\int_a^b |h(t)|\diamond_\alpha t\right)\Phi(x_0)$$
$$= \int_a^b |h(t)|(\Phi(g(t)) - \Phi(x_0))\diamond_\alpha t$$
$$\ge \beta\int_a^b |h(t)|(g(t) - x_0)\diamond_\alpha t$$
$$= \beta\left(\int_a^b |h(t)|g(t)\diamond_\alpha t - \int_a^b |h(t)|x_0\diamond_\alpha t\right)$$

$$= \beta\left(\int_a^b |h(t)|g(t)\diamond_\alpha t - \int_a^b |h(t)|g(t)\diamond_\alpha t\right)$$
$$= 0.$$

This completes the proof. □

4.4 The Radon inequality

Theorem 4.4.1. *Let $w, f, g \in \mathcal{C}([a,b])$ be positive and $\diamond_\alpha$-integrable functions and c_1, c_2, c_3, c_4 be positive constants. If $\beta \geq 1$, $\gamma, \zeta, \eta, \lambda \in [0, \infty)$, and*

$$c_3\left(\int_a^b |w(t)|g(t)\diamond_\alpha t\right)^\lambda \geq c_4 \sup_{t\in[a,b]} (g(t))^\lambda,$$

then

$$\begin{aligned}&\frac{(c_1(\int_a^b |w(t)|\diamond_\alpha t)^\eta(\int_a^b |w(t)|f(t)\diamond_\alpha t)^{\beta+\zeta} + c_2(\int_a^b |w(t)|f(t)\diamond_\alpha t)^{\beta+\eta})^{\gamma+1}}{(c_3(\int_a^b |w(t)|\diamond_\alpha t)^\lambda - c_4)(\int_a^b |w(t)|g(t)\diamond_\alpha t)^{\gamma(\lambda+1)}}\\ &\quad\times \frac{1}{(\int_a^b |w(t)|\diamond_\alpha t)^{(\gamma+1)(\beta+\eta-1)-\gamma\lambda}}\\ &\leq \int_a^b |w(t)|\frac{(c_1(\int_a^b |w(t)|f(t)\diamond_\alpha t)^\zeta + c_2|f(t)|^\eta)^{\gamma+1}|f(t)|^{\beta(\gamma+1)}}{(c_3(\int_a^b |w(t)|g(t)\diamond_\alpha t)^\lambda - c_4|g(t)|^\lambda)^\gamma(g(t))^\gamma}\diamond_\alpha t.\end{aligned} \tag{4.4}$$

Proof. Set

$$\Lambda = \int_a^b |w(t)|\diamond_\alpha t,$$
$$Y = \int_a^b |w(t)|f(t)\diamond_\alpha t,$$
$$\Omega = \int_a^b |w(t)|g(t)\diamond_\alpha t,$$
$$J(t) = (c_1 Y^\zeta + c_2|f(t)|^\eta)|f(t)|^\beta,$$
$$k(t) = (c_3\Omega^\lambda - c_4(g(t))^\lambda)g(t).$$

Then the right-hand side of inequality (4.4) can be rewritten in the form

$$\int_a^b |w(t)|\frac{(J(t))^{\gamma+1}}{(k(t))^\gamma}\diamond_\alpha t$$

$$= \int_a^b |w(t)|k(t)\left(\frac{J(t)}{k(t)}\right)^{\gamma+1} \diamond_\alpha t$$

$$= \left(\int_a^b |w(t)|k(t)\diamond_\alpha t\right)\int_a^b \frac{|w(t)|k(t)}{\int_a^b |w(t)|k(t)\diamond_\alpha t}\left(\frac{J(t)}{k(t)}\right)^{\gamma+1} \diamond_\alpha t.$$

Let

$$\Phi(x) = x^{\gamma+1}, \quad x \in (0,\infty).$$

Then Φ is convex and by the Jensen inequality, we get

$$\left(\int_a^b \frac{|w(t)|k(t)}{\int_a^b |w(t)|k(t)\diamond_\alpha t}\left(\frac{J(t)}{k(t)}\right)\diamond_\alpha t\right)^{\gamma+1}$$

$$\leq \int_a^b \frac{|w(t)|k(t)}{\int_a^b |w(t)|k(t)\diamond_\alpha t}\left(\frac{J(t)}{k(t)}\right)^{\gamma+1} \diamond_\alpha t,$$

whereupon

$$\frac{\left(\int_a^b |w(t)|J(t)\diamond_\alpha t\right)^{\gamma+1}}{\left(\int_a^b |w(t)|k(t)\diamond_\alpha t\right)^{\gamma}} \leq \int_a^b |w(t)|\frac{(J(t))^{\gamma+1}}{(k(t))^{\gamma}}\diamond_\alpha t.$$

Now we put the values of J and k in the right-hand side of the last inequality and find

$$\int_a^b |w(t)|\frac{((c_1 Y^{\zeta} + c_2|f(t)|^{\eta})|f(t)|^{\beta})^{\gamma+1}}{((c_3\Omega^{\lambda} - c_4|g(t)|^{\lambda})|g(t)|)^{\gamma}}\diamond_\alpha t$$

$$\geq \frac{(\int_a^b |w(t)|(c_1 Y^{\zeta} + c_2|f(t)|^{\eta})|f(t)|^{\beta}\diamond_\alpha t)^{\gamma+1}}{(\int_a^b |w(t)|(c_3\Omega^{\lambda} - c_4|g(t)|^{\lambda})|g(t)|\diamond_\alpha t)^{\gamma}}$$

$$= \frac{(c_1 Y^{\zeta}\int_a^b |w(t)||f(t)|^{\beta}\diamond_\alpha t + c_2\int_a^b |w(t)||f(t)|^{\beta+\eta}\diamond_\alpha t)^{\gamma+1}}{(c_3\Omega^{\lambda+1} - c_4\int_a^b |w(t)||g(t)|^{\lambda+1}\diamond_\alpha t)^{\gamma}}$$

$$\geq \frac{(c_1 Y^{\beta+\zeta}(\frac{1}{\Lambda})^{\beta-1} + c_2 Y^{\beta+\eta}(\frac{1}{\Lambda})^{\beta+\eta-1})^{\gamma+1}}{(c_3\Omega^{\lambda+1} - c_4\Omega^{\lambda+1}(\frac{1}{\Lambda})^{\lambda})^{\gamma}}$$

$$= \frac{(c_1 Y^{\beta+\zeta}\Lambda^{\eta} + c_2 Y^{\beta+\eta})^{\gamma+1}}{(c_3\Lambda^{\lambda} - c_4)^{\gamma}\Omega^{\gamma(\lambda+1)}\Lambda^{(\gamma+1)(\beta+\eta-1)-\gamma\lambda}}.$$

This completes the proof. □

4.5 The Schlömilch Inequality

Theorem 4.5.1 (The Schlömilch Inequality). *Let* $w, f \in C([a,b])$ *and*

$$\int_a^b |w(t)| \diamond_\alpha t = 1.$$

If $\eta_2 \geq \eta_1 > 0$, *then*

$$\left(\int_a^b |w(t)||f(t)|^{\eta_1} \diamond_\alpha t \right)^{\frac{1}{\eta_1}} \leq \left(\int_a^b |w(t)||f(t)|^{\eta_2} \diamond_\alpha t \right)^{\frac{1}{\eta_2}}. \tag{4.5}$$

Proof. Let

$$\begin{aligned} &\beta = 1, \quad \zeta = \eta = \lambda = 0, \\ &\gamma > 0, \quad \gamma + 1 = \frac{\eta_2}{\eta_1}, \quad g(t) = 1, \\ &c_1 = 0, \quad c_2 = c_3 = 1, \quad c_4 = 0. \end{aligned}$$

Then the Radon inequality takes the form

$$\left(\int_a^b |w(t)||f(t)| \diamond_\alpha t \right)^{\frac{\eta_2}{\eta_1}} \leq \int_a^b |w(t)||f(t)|^{\frac{\eta_2}{\eta_1}} \diamond_\alpha t.$$

Replacing $|f(t)|$ by $|f(t)|^{\eta_1}$ and taking power $\frac{1}{\eta_2}$, we get (4.5). This completes the proof. □

5 Fractional integral inequalities

This chapter is devoted on fractional integral inequalities on time scales. They are considered the Poincaré type inequality, opial type inequalities, Sobolev type inequality and Ostrowski type inequalities. Some of the results in this chapter can be found in [4, 27] and [28].

Suppose that $\mathbb{T}$ is an unbounded time scale with forward jump operator and delta differentiation operator σ and Δ, respectively. Let also $0 \in \mathbb{T}$. For $\alpha \geq 0$, by $h_\alpha : \mathbb{T}\times\mathbb{T} \to \mathbb{R}$ we will denote the generalized polynomials on time scales defined as follows:

$$h_0(t,s) = 1,$$
$$h_\alpha(t,s) = \int_s^t h_{\alpha-1}(t,\sigma(\tau))\Delta\tau, \quad t,s \in \mathbb{T}.$$

For $\alpha \geq 1$ and $f \in C_{rd}(\mathbb{T})$, by D_0^α we denote the delta-Riemann–Liouville operator defined by

$$D_0^\alpha f(t) = \int_0^t h_{\alpha-1}(t,\sigma(\tau))f(\tau)\Delta\tau,$$
$$D_0^0 f(t) = f(t), \quad t \in \mathbb{T}.$$

5.1 Integral inequalities for the Chebyshev functional

Definition 5.1.1. Let $t \in \mathbb{T}$, $t > 0$, and f, g be two integrable functions defined on $[0,t]$. If for any $x, y \in [0,t]$,

$$(f(x) - f(y))(g(x) - g(y)) \geq 0,$$

then the functions f and g are called synchronous functions on $[0,t]$.

Theorem 5.1.2. *Let f and g be two synchronous functions on $[0,\infty)$. Then for any $t \geq 0$, $\alpha \geq 1$, we have*

$$D_0^\alpha(fg)(t) \geq (h_\alpha(t,0))^{-1} D_0^\alpha f(t) D_0^\alpha g(t).$$

Proof. Take $t > 0$ and $\alpha \geq 1$ arbitrarily. Since f and g are two synchronous functions on $[0,\infty)$, for any $\phi, \tau \geq 0$, we have

$$(f(\tau) - f(\phi))(g(\tau) - g(\phi)) \geq 0,$$

or

$$f(\tau)g(\tau) - f(\tau)g(\phi) - f(\phi)g(\tau) + f(\phi)g(\phi) \geq 0,$$

https://doi.org/10.1515/9783110705553-005

whereupon

$$f(\tau)g(\tau)+f(\phi)g(\phi)\geq f(\tau)g(\phi)+f(\phi)g(\tau).$$

Hence,

$$\begin{aligned}&h_{\alpha-1}(t,\sigma(\tau))f(\tau)g(\tau)+h_{\alpha-1}(t,\sigma(\tau))f(\phi)g(\phi)\\&\quad\geq h_{\alpha-1}(t,\sigma(\tau))f(\tau)g(\phi)+h_{\alpha-1}(t,\sigma(\tau))f(\phi)g(\tau)\end{aligned}$$

for any $\tau,\phi\geq 0$. We integrate the last inequality with respect to τ from 0 to t and, using that τ and ϕ are independent, find

$$\begin{aligned}&\int_0^t h_{\alpha-1}(t,\sigma(\tau))f(\tau)g(\tau)\Delta\tau+\left(\int_0^t h_{\alpha-1}(t,\sigma(\tau))\Delta\tau\right)f(\phi)g(\phi)\\&\quad\geq\left(\int_0^t h_{\alpha-1}(t,\sigma(\tau))f(\tau)\Delta\tau\right)g(\phi)+\left(\int_0^t h_{\alpha-1}(t,\sigma(\tau))g(\tau)\Delta\tau\right)f(\phi),\end{aligned}$$

or

$$\begin{aligned}&D_0^\alpha(fg)(t)+f(\phi)g(\phi)h_\alpha(t,0)\\&\quad\geq g(\phi)D_0^\alpha f(t)+f(\phi)D_0^\alpha g(t)\end{aligned}$$

for any $\phi\geq 0$. Now we multiply both sides of the last inequality with $h_{\alpha-1}(t,\sigma(\phi))$ and get

$$\begin{aligned}&D_0^\alpha(fg)(t)h_{\alpha-1}(t,\sigma(\phi))+f(\phi)g(\phi)h_{\alpha-1}(t,\sigma(\phi))h_\alpha(t,0)\\&\quad\geq h_{\alpha-1}(t,\sigma(\phi))g(\phi)D_0^\alpha f(t)+h_{\alpha-1}(t,\sigma(\phi))f(\phi)D_0^\alpha g(t),\end{aligned}$$

which we integrate with respect to ϕ from 0 to t and find

$$\begin{aligned}&D_0^\alpha(fg)(t)\int_0^t h_{\alpha-1}(t,\sigma(\phi))\Delta\phi+\left(\int_0^t f(\phi)g(\phi)h_{\alpha-1}(t,\sigma(\phi))\Delta\phi\right)h_\alpha(t,0)\\&\quad\geq\left(\int_0^t h_{\alpha-1}(t,\sigma(\phi))g(\phi)\Delta\phi\right)D_0^\alpha f(t)+\left(\int_0^t h_{\alpha-1}(t,\sigma(\phi))f(\phi)\Delta\phi\right)D_0^\alpha g(t)\end{aligned}$$

and

$$D_0^\alpha(fg)(t)h_\alpha(t,0)+h_\alpha(t,0)D_0^\alpha(fg)(t)\geq D_0^\alpha f(t)D_0^\alpha g(t)+D_0^\alpha f(t)D_0^\alpha g(t),$$

or

$$h_\alpha(t,0)D_0^\alpha(fg)(t)\geq D_0^\alpha f(t)D_0^\alpha g(t).$$

This completes the proof. □

Theorem 5.1.3. *Let f and g be two synchronous functions on $[0,\infty)$. Then for any $t>0$, $\alpha,\beta \geq 1$, we have*

$$h_\alpha(t,0)D_0^\beta(fg)(t)+h_\beta(t,0)D_0^\alpha(fg)(t,0)\geq D_0^\alpha f(t)D_0^\beta g(t)+D_0^\beta f(t)D_0^\alpha g(t).$$

Proof. Take $t>0$ and $\alpha,\beta\geq 1$ arbitrarily. As in the proof of Theorem 5.1.2, for any $\phi,\tau\geq 0$, we have

$$f(\tau)g(\tau)+f(\phi)g(\phi)\geq f(\tau)g(\phi)+f(\phi)g(\tau)$$

and

$$\begin{aligned}D_0^\alpha(fg)(t)&+f(\phi)g(\phi)h_\alpha(t,0)\\&\geq g(\phi)D_0^\alpha f(t)+f(\phi)D_0^\alpha g(t).\end{aligned}$$

Now we multiply both sides of the last inequality with $h_{\beta-1}(t,\sigma(\phi))$ and get

$$\begin{aligned}D_0^\alpha(fg)(t)h_{\beta-1}(t,\sigma(\phi))&+f(\phi)g(\phi)h_{\beta-1}(t,\sigma(\phi))h_\alpha(t,0)\\&\geq h_{\beta-1}(t,\sigma(\phi))g(\phi)D_0^\alpha f(t)+h_{\beta-1}(t,\sigma(\phi))f(\phi)D_0^\alpha g(t),\end{aligned}$$

which we integrate with respect to ϕ from 0 to t and find

$$\begin{aligned}&D_0^\alpha(fg)(t)\int_0^t h_{\beta-1}(t,\sigma(\phi))\Delta\phi+\left(\int_0^t f(\phi)g(\phi)h_{\beta-1}(t,\sigma(\phi))\Delta\phi\right)h_\alpha(t,0)\\&\geq\left(\int_0^t h_{\beta-1}(t,\sigma(\phi))g(\phi)\Delta\phi\right)D_0^\alpha f(t)+\left(\int_0^t h_{\beta-1}(t,\sigma(\phi))f(\phi)\Delta\phi\right)D_0^\alpha g(t)\end{aligned}$$

and

$$D_0^\alpha(fg)(t)h_\beta(t,0)+h_\alpha(t,0)D_0^\beta(fg)(t)\geq D_0^\alpha f(t)D_0^\beta g(t)+D_0^\beta f(t)D_0^\alpha g(t).$$

This completes the proof. □

Theorem 5.1.4. *Let $n\in\mathbb{N}$ be arbitrarily chosen and f_j, $j\in\{1,\ldots,n\}$, be positive increasing functions. Then for any $t>0$, $\alpha\geq 1$, we have*

$$D_0^\alpha\left(\prod_{j=1}^n f_j\right)(t)\geq(h_\alpha(t,0))^{1-n}\prod_{j=1}^n D_0^\alpha f_j(t). \tag{5.1}$$

Proof. We will use the principle of mathematical induction.

1. For $n=1$ the assertion is evident.
2. Assume that the assertion is true for some $n\in\mathbb{N}$.

3. We will prove

$$D_0^\alpha\left(\prod_{j=1}^{n+1} f_j\right)(t) \geq (h_\alpha(t,0))^{-n}\prod_{j=1}^{n+1} D_0^\alpha f_j(t).$$

We apply Theorem 5.1.2 for the functions

$$f(t) = f_{n+1}(t),$$
$$g(t) = \left(\prod_{j=1}^{n} f_j\right)(t)$$

and, using (5.1), get

$$\begin{aligned}
D_0^\alpha\left(\prod_{j=1}^{n+1} f_j\right)(t) &= D_0^\alpha(fg)(t)\\
&\geq (h_\alpha(t,0))^{-1} D_0^\alpha f_{n+1}(t) D_0^\alpha g(t)\\
&\geq (h_\alpha(t,0))^{-1} D_0^\alpha f_{n+1}(t)(h_\alpha(t,0))^{1-n}\prod_{j=1}^{n} D_0^\alpha f_j(t)\\
&= (h_\alpha(t,0))^{-n}\prod_{j=1}^{n+1} D_0^\alpha f_j(t).
\end{aligned}$$

Hence, by the principal of mathematical induction, we conclude that inequality (5.1) is true for any $n \in \mathbb{N}$. This completes the proof. □

5.2 A Poincaré-type inequality

Let $a \in \mathbb{T}$, $a > 0$. We will start with the following useful auxiliary results.

Lemma 5.2.1. *Let $\alpha, \beta > 1$, $f \in C_{rd}([0,a])$. Then*

$$\begin{aligned}
D_0^\alpha D_0^\beta f(t) &= D_0^{\alpha+\beta} f(t)\\
&\quad + \int_0^t f(u)\mu(u)h_{\alpha-1}(t,\sigma(u))h_{\beta-1}(u,\sigma(u))\Delta u, \quad t \in [0,a].
\end{aligned}$$

Proof. Let $t \in [0,a]$. Using Fubini's Theorem, we get

$$\begin{aligned}
D_0^\alpha D_0^\beta f(t) &= \int_0^t h_{\alpha-1}(t,\sigma(\tau))\int_0^\tau h_{\beta-1}(\tau,\sigma(u))f(u)\Delta u\Delta\tau\\
&= \int_0^t f(u)\left(\int_u^t h_{\alpha-1}(t,\sigma(\tau))h_{\beta-1}(\tau,\sigma(u))\Delta\tau\right)\Delta u
\end{aligned}$$

$$
\begin{aligned}
&= \int_0^t f(u)\left(\int_u^{\sigma(u)} h_{\alpha-1}(t,\sigma(\tau))h_{\beta-1}(\tau,\sigma(u))\Delta\tau \right.\\
&\quad \left. + \int_{\sigma(u)}^t h_{\alpha-1}(t,\sigma(\tau))h_{\beta-1}(\tau,\sigma(u))\Delta\tau \right)\Delta u\\
&= \int_0^t f(u)(h_{\alpha+\beta-1}(t,\sigma(u)) + \mu(u)h_{\alpha-1}(t,\sigma(u))h_{\beta-1}(u,\sigma(u)))\Delta u\\
&= \int_0^t h_{\alpha+\beta-1}(t,\sigma(u))f(u)\Delta u\\
&\quad + \int_0^t \mu(u)h_{\alpha-1}(t,\sigma(u))h_{\beta-1}(u,\sigma(u))\Delta u\\
&= D_0^{\alpha+\beta}f(t) + \int_0^t \mu(u)h_{\alpha-1}(t,\sigma(u))h_{\beta-1}(u,\sigma(u))\Delta u.
\end{aligned}
$$

This completes the proof. □

Definition 5.2.2. Let $\alpha,\beta > 1, f \in C_{rd}([0,a])$. The integral

$$
E(f,\alpha,\beta,t) = \int_0^t f(u)\mu(u)h_{\alpha-1}(t,\sigma(u))h_{\beta-1}(u,\sigma(u))\Delta u, \quad t \in [0,a],
$$

is called the forward graininess deviation functional of f.

By Lemma 5.2.1, we have

$$
D_0^\alpha D_0^\beta f(t) = D_0^{\alpha+\beta}f(t) + E(f,\alpha,\beta,t), \quad t \in [0,a].
$$

Definition 5.2.3. Let $\alpha > 2$ and $m-1 < \alpha < m$, $m \in \mathbb{N}$, $\nu = m-\alpha$. For a function $f \in C_{rd}^m([0,a])$, define

$$
\begin{aligned}
\Delta_0^{\alpha-1}f(t) &= D_0^{\nu+1}f^{\Delta^m}(t)\\
&= \int_0^t h_\nu(t,\sigma(u))f^{\Delta^m}(u)\Delta u, \quad t \in [0,a].
\end{aligned}
$$

Lemma 5.2.4. *Let $\alpha > 2$, $m-1 < \alpha < m$, $m \in \mathbb{N}$, $\nu = m-\alpha$, $f \in C_{rd}^m([0,a])$. Then*

$$
\begin{aligned}
\int_0^t h_{m-1}(t,\sigma(\tau))f^{\Delta^m}(\tau)\Delta\tau &= -\int_0^t f^{\Delta^m}(u)\mu(u)h_{\alpha-2}(t,\sigma(u))h_\nu(u,\sigma(u))\Delta u\\
&\quad + \int_0^t h_{\alpha-2}(t,\sigma(\tau))\Delta_0^{m-1}f(\tau)\Delta\tau, \quad t \in [0,a].
\end{aligned}
$$

Proof. We have

$$
\begin{aligned}
D_0^{\alpha-1}\Delta_0^{m-1}f(t) &= \int_0^t h_{\alpha-2}(t,\sigma(\tau))\Delta_0^{m-1}f(\tau)\Delta\tau \\
&= D_0^{\alpha-1}D_0^{\nu+1}f^{\Delta^m}(t) \\
&= D_0^{\alpha+\nu}f^{\Delta^m}(t) \\
&\quad + \int_0^t f^{\Delta^m}(u)\mu(u)h_{\alpha-1}(t,\sigma(u))h_{\nu-1}(u,\sigma(u))\Delta u \\
&= D_0^m f^{\Delta^m}(t) \\
&\quad + \int_0^t f^{\Delta^m}(u)\mu(u)h_{\alpha-1}(t,\sigma(u))h_{\nu-1}(u,\sigma(u))\Delta u \\
&= \int_0^t h_{m-1}(t,\sigma(u))f^{\Delta^m}(u)\Delta u \\
&\quad + \int_0^t f^{\Delta^m}(u)\mu(u)h_{\alpha-1}(t,\sigma(u))h_{\beta-1}(u,\sigma(u))\Delta u,
\end{aligned}
$$

$t \in [0,a]$. This completes the proof. □

Lemma 5.2.5 (Fractional Taylor Formula). *Let $\alpha > 2$, $m-1 < \alpha < m$, $m \in \mathbb{N}$, $\nu = m-\alpha$, $f \in C_{rd}^m([0,a])$. Then*

$$
\begin{aligned}
f(t) &= \sum_{k=0}^{m-1} h_k(t,0)f^{\Delta^k}(0) \\
&\quad - \int_0^t f^{\Delta^m}(u)\mu(u)h_{\alpha-2}(t,\sigma(u))h_\nu(u,\sigma(u))\Delta u \\
&\quad + \int_0^t h_{\alpha-2}(t,\sigma(\tau))\Delta_0^{\alpha-1}f(\tau)\Delta\tau, \quad t \in [0,a].
\end{aligned}
$$

Proof. By the Taylor formula on time scales and Lemma 5.2.4, we have

$$
\begin{aligned}
f(t) &= \sum_{k=0}^{m-1} h_k(t,0)f^{\Delta^k}(0) + \int_0^t h_{m-1}(t,\sigma(\tau))f^{\Delta^m}(\tau)\Delta\tau \\
&= \sum_{k=0}^{m-1} h_k(t,0)f^{\Delta^k}(0) \\
&\quad - \int_0^t f^{\Delta^m}(u)\mu(u)h_{\alpha-2}(t,\sigma(u))h_\nu(u,\sigma(u))\Delta u
\end{aligned}
$$

$$+\int_0^t h_{\alpha-2}(t,\sigma(\tau))\Delta_0^{\alpha-1}f(\tau)\Delta\tau, \quad t\in[0,a].$$

This completes the proof. □

Definition 5.2.6. For $\alpha>2$, $m-1<\alpha<m$, $m\in\mathbb{N}$, $\nu=m-\alpha$, $f\in C_{rd}^m([0,a])$, define

$$B(t)=f(t)+E(f^{\Delta^m},\alpha-1,\nu+1,t), \quad t\in[0,a].$$

Lemma 5.2.7. *Let $\alpha>2$, $m-1<\alpha<m$, $m\in\mathbb{N}$, $\nu=m-\alpha$, $f\in C_{rd}^m([0,a])$, $f^{\Delta^k}(0)=0$, $k\in\{0,\dots,m-1\}$. Then*

$$B(t)=\int_0^t h_{\alpha-2}(t,\sigma(\tau))\Delta_0^{\alpha-1}f(\tau)\Delta\tau, \quad t\in[0,a].$$

Proof. By the fractional Taylor formula, we get

$$\begin{aligned}f(t)&=\sum_{k=0}^{m-1}h_k(t,0)f^{\Delta^k}(0)\\&\quad-\int_0^t f^{\Delta^m}(u)\mu(u)h_{\alpha-2}(t,\sigma(u))h_\nu(u,\sigma(u))\Delta u\\&\quad+\int_0^t h_{\alpha-2}(t,\sigma(\tau))\Delta_0^{\alpha-1}f(\tau)\Delta\tau\\&=-\int_0^t f^{\Delta^m}(u)\mu(u)h_{\alpha-2}(t,\sigma(u))h_\nu(u,\sigma(u))\Delta u\\&\quad+\int_0^t h_{\alpha-2}(t,\sigma(\tau))\Delta_0^{\alpha-1}f(\tau)\Delta\tau, \quad t\in[0,a].\end{aligned}$$

Hence,

$$\begin{aligned}B(t)&=f(t)+E(f^{\Delta^m},\alpha-1,\nu+1,t)\\&=f(t)+\int_0^t \mu(u)h_{\alpha-2}(t,\sigma(u))h_\nu(u,\sigma(u))f^{\Delta^m}(u)\Delta u\\&=\int_0^t h_{\alpha-2}(t,\sigma(\tau))\Delta_0^{\alpha-1}f(\tau)\Delta\tau, \quad t\in[0,a].\end{aligned}$$

This completes the proof. □

Theorem 5.2.8 (Poincaré-type Inequality). *Let $\alpha > 2$, $m-1 < \alpha < m$, $m \in \mathbb{N}$, $h_{\alpha-2}(s,\sigma(\tau)), h_\nu(s,\sigma(\tau)) \in \mathcal{C}([0,a]\times[0,a])$, and $f^{\Delta^k}(0) = 0$, $k \in \{0,\dots,m-1\}$. Let also $p,q > 1$, $\frac{1}{p}+\frac{1}{q} = 1$. Then*

$$\int_0^a |B(t)|^q \Delta t \le \left(\int_0^a \left(\int_0^t |h_{\alpha-2}(t,\sigma(\tau))|^p \Delta\tau\right)^{\frac{q}{p}} \Delta t\right)\left(\int_0^a |\Delta_0^{\alpha-1} f(t)|^q \Delta t\right).$$

Proof. By Lemma 5.2.7, we have

$$B(t) = \int_0^t h_{\alpha-2}(t,\sigma(\tau))\Delta_0^{\alpha-1} f(\tau)\Delta\tau, \quad t\in[0,a].$$

Hence, by Hölder's inequality, we obtain

$$\begin{aligned}|B(t)| &= \left|\int_0^t h_{\alpha-2}(t,\sigma(\tau))\Delta_0^{\alpha-1} f(\tau)\Delta\tau\right| \\ &\le \int_0^t |h_{\alpha-2}(t,\sigma(\tau))||\Delta_0^{\alpha-1} f(\tau)|\Delta\tau \\ &\le \left(\int_0^t |h_{\alpha-2}(t,\sigma(\tau))|^p\Delta\tau\right)^{\frac{1}{p}}\left(\int_0^t |\Delta_0^{\alpha-1} f(\tau)|^q\Delta\tau\right)^{\frac{1}{q}}, \quad t\in[0,a].\end{aligned}$$

Hence,

$$\begin{aligned}|B(t)|^q &\le \left(\int_0^t |h_{\alpha-2}(t,\sigma(\tau))|^p\Delta\tau\right)^{\frac{q}{p}}\left(\int_0^t |\Delta_0^{\alpha-1} f(\tau)|^q\Delta\tau\right) \\ &\le \left(\int_0^t |h_{\alpha-2}(t,\sigma(\tau))|^p\Delta\tau\right)^{\frac{q}{p}}\left(\int_0^a |\Delta_0^{\alpha-1} f(\tau)|^q\Delta\tau\right), \quad t\in[0,a],\end{aligned}$$

and

$$\int_0^a |B(t)|^q \Delta t \le \left(\int_0^a \left(\int_0^t |h_{\alpha-2}(t,\sigma(\tau))|^p \Delta\tau\right)^{\frac{q}{p}} \Delta t\right)\left(\int_0^a |\Delta_0^{\alpha-1} f(t)|^q \Delta t\right).$$

This completes the proof. □

5.3 A Sobolev-type inequality

Theorem 5.3.1 (Sobolev-type Inequality). *Let $\alpha > 2$, $m-1 < \alpha < m$, $m \in \mathbb{N}$, $h_{\alpha-2}(s,\sigma(\tau))$, $h_\nu(s,\sigma(\tau)) \in \mathcal{C}([0,a]\times[0,a])$, and $f^{\Delta^k}(0) = 0$, $k \in \{0,\dots,m-1\}$. Let also $p,q > 1$, $r \ge 1$,*

$\frac{1}{p}+\frac{1}{q}=1$. *Then*

$$\left(\int_0^a |B(t)|^r \Delta t\right)^{\frac{1}{r}} \le \left(\int_0^a \left(\int_0^t |h_{\alpha-2}(t,\sigma(\tau))|^p \Delta\tau\right)^{\frac{r}{p}} \Delta t\right)^{\frac{1}{r}} \left(\int_0^a |\Delta_0^{\alpha-1} f(t)|^q \Delta t\right)^{\frac{1}{q}}.$$

Proof. As in the proof of Theorem 5.2.8, we have

$$|B(t)| \le \left(\int_0^t |h_{\alpha-2}(t,\sigma(\tau))|^p \Delta\tau\right)^{\frac{1}{p}} \left(\int_0^t |\Delta_0^{\alpha-1} f(\tau)|^q \Delta\tau\right)^{\frac{1}{q}}, \quad t \in [0,a].$$

Hence,

$$|B(t)|^r \le \left(\int_0^t |h_{\alpha-2}(t,\sigma(\tau))|^p \Delta\tau\right)^{\frac{r}{p}} \left(\int_0^a |\Delta_0^{\alpha-1} f(\tau)|^q \Delta\tau\right)^{\frac{r}{q}}, \quad t \in [0,a],$$

and

$$\left(\int_0^a |B(t)|^r \Delta t\right)^{\frac{1}{r}} \le \left(\int_0^a \left(\int_0^t |h_{\alpha-2}(t,\sigma(\tau))|^p \Delta\tau\right)^{\frac{r}{p}} \Delta t\right)^{\frac{1}{r}} \left(\int_0^a |\Delta_0^{\alpha-1} f(t)|^q \Delta t\right)^{\frac{1}{q}}.$$

This completes the proof. □

5.4 An Opial-type inequality

Theorem 5.4.1. *Let $\alpha > 2$, $m-1 < \alpha < m$, $m \in \mathbb{N}$, $h_{\alpha-2}(s,\sigma(\tau)), h_\nu(s,\sigma(\tau)) \in \mathcal{C}([0,a] \times [0,a])$, and $f^{\Delta^k}(0) = 0$, $k \in \{0,\ldots,m-1\}$. Let also $p,q > 1$, $\frac{1}{p}+\frac{1}{q}=1$, and suppose $|\Delta_0^\alpha f|$ is an increasing function on $[0,a]$. Then*

$$\int_0^a |B(t)||\Delta_0^{\alpha-1} f(t)|\Delta t \le a^{\frac{1}{q}} \left(\int_0^a \left(\int_0^t |h_{\alpha-2}(t,\sigma(\tau))|^p \Delta\tau\right) \Delta t\right)^{\frac{1}{p}} \times \left(\int_0^a (\Delta_0^{\alpha-1} f(t))^{2q} \Delta t\right)^{\frac{1}{q}}.$$

Proof. By the proof of Theorem 5.2.8, we get

$$\begin{aligned} |B(t)| &\le \left(\int_0^t |h_{\alpha-2}(t,\sigma(\tau))|^p \Delta\tau\right)^{\frac{1}{p}} \left(\int_0^t |\Delta_0^{\alpha-1} f(\tau)|^q \Delta\tau\right)^{\frac{1}{q}} \\ &\le \left(\int_0^t |h_{\alpha-2}(t,\sigma(\tau))|^p \Delta\tau\right)^{\frac{1}{p}} |\Delta_0^{\alpha-1} f(t)| t^{\frac{1}{q}}, \quad t \in [0,a]. \end{aligned}$$

Hence,

$$|B(t)||\Delta_0^{\alpha-1}f(t)| \le \left(\int_0^t |h_{\alpha-2}(t,\sigma(\tau))|^p \Delta\tau\right)^{\frac{1}{p}} (\Delta_0^{\alpha-1}f(t))^2 t^{\frac{1}{q}}, \quad t \in [0,a].$$

Then

$$\begin{aligned}\int_0^a |B(t)||\Delta_0^{\alpha-1}f(t)|\Delta t &\le \int_0^a \left(\left(\int_0^t |h_{\alpha-2}(t,\sigma(\tau))|^p \Delta\tau\right)^{\frac{1}{p}} \right.\\ &\quad \left. \times (\Delta_0^{\alpha-1}f(t))^2 t^{\frac{1}{q}}\right)\Delta t\\ &\le \left(\int_0^a \left(\int_0^t |h_{\alpha-2}(t,\sigma(\tau))|^p \Delta\tau\right)\Delta t\right)^{\frac{1}{p}}\\ &\quad \times \left(\int_0^a (\Delta_0^{\alpha-1}f(t))^{2q} t\Delta t\right)^{\frac{1}{q}}\\ &\le a^{\frac{1}{q}} \left(\int_0^a \left(\int_0^t |h_{\alpha-2}(t,\sigma(\tau))|^p \Delta\tau\right)\Delta t\right)^{\frac{1}{p}}\\ &\quad \times \left(\int_0^a (\Delta_0^{\alpha-1}f(t))^{2q} \Delta t\right)^{\frac{1}{q}}.\end{aligned}$$

This completes the proof. □

5.5 Ostrowski-type inequalities

Theorem 5.5.1. *Let $\alpha > 2$, $m-1 < \alpha < m$, $m \in \mathbb{N}$, $h_{\alpha-2}(s,\sigma(\tau)), h_\nu(s,\sigma(\tau)) \in \mathcal{C}([0,a] \times [0,a])$, and $f^{\Delta^k}(0) = 0$, $k \in \{1,\dots,m-1\}$. Then*

$$\left|\frac{1}{a}\int_0^a B(t)\Delta t - f(0)\right| \le \frac{1}{a}\left(\int_0^a \left(\int_0^t |h_{\alpha-2}(t,\sigma(\tau))|\Delta\tau\right)\Delta t\right) \sup_{t\in[0,a]} |\Delta_0^{\alpha-1}f(t)|.$$

Proof. By the fractional Taylor formula, we get

$$\begin{aligned}f(t) &= f(0) - \int_0^t f^{\Delta^m}(u)\mu(u)h_{\alpha-2}(t,\sigma(u))h_\nu(u,\sigma(u))\Delta u\\ &\quad + \int_0^t h_{\alpha-2}(t,\sigma(\tau))\Delta_0^{\alpha-1}f(\tau)\Delta\tau, \quad t \in [0,a].\end{aligned}$$

Hence, by Lemma 5.2.7, we get

$$B(t) - f(0) = \int_0^t h_{\alpha-2}(t, \sigma(\tau)) \Delta_0^{\alpha-1} f(\tau) \Delta\tau, \quad t \in [0, a],$$

and

$$\begin{aligned} |B(t) - f(0)| &= \left| \int_0^t h_{\alpha-2}(t, \sigma(\tau)) \Delta_0^{\alpha-1} f(\tau) \Delta\tau \right| \\ &\leq \int_0^t |h_{\alpha-2}(t, \sigma(\tau))| |\Delta_0^{\alpha-1} f(\tau)| \Delta\tau \\ &\leq \left(\int_0^t |h_{\alpha-2}(t, \sigma(\tau))| \Delta\tau \right) \sup_{t \in [0,a]} |\Delta_0^{\alpha-1} f(t)| \end{aligned}$$

Thus,

$$\begin{aligned} \left| \frac{1}{a} \int_0^a B(t) \Delta t - f(0) \right| &= \left| \frac{1}{a} \int_0^a (B(t) - f(0)) \Delta t \right| \\ &\leq \frac{1}{a} \int_0^a |B(t) - f(0)| \Delta t \\ &\leq \left(\frac{1}{a} \int_0^a \left(\int_0^t |h_{\alpha-2}(t, \sigma(\tau))| \Delta\tau \right) \Delta t \right) \sup_{t \in [0,a]} |\Delta_0^{\alpha-1} f(t)|. \end{aligned}$$ □

Theorem 5.5.2. *Suppose all conditions of Theorem 5.5.1 hold. Let also $p, q > 1$, $\frac{1}{p} + \frac{1}{q} = 1$. Then*

$$\begin{aligned} \left| \frac{1}{a} \int_0^a B(t) \Delta t - f(0) \right| &\leq \frac{1}{a} \left(\int_0^a \left(\int_0^t |h_{\alpha-2}(t, \sigma(\tau))|^p \Delta\tau \right)^{\frac{1}{p}} \Delta t \right) \\ &\quad \times \left(\int_0^a |\Delta_0^{\alpha-1} f(t)|^q \Delta t \right)^{\frac{1}{q}}. \end{aligned}$$

Proof. By the proof of Theorem 5.5.1, it follows that

$$\begin{aligned} |B(t) - f(0)| &\leq \int_0^t |h_{\alpha-2}(t, \sigma(\tau))| |\Delta_0^{\alpha-1} f(\tau)| \Delta\tau \\ &\leq \left(\int_0^t |h_{\alpha-2}(t, \sigma(\tau))|^p \Delta\tau \right)^{\frac{1}{p}} \end{aligned}$$

$$\times \left(\int_0^t |\Delta_0^{\alpha-1} f(t)|^q \Delta t \right)^{\frac{1}{q}}$$

$$\le \left(\int_0^t |h_{\alpha-2}(t,\sigma(\tau))|^p \Delta\tau \right)^{\frac{1}{p}}$$

$$\times \left(\int_0^a |\Delta_0^{\alpha-1} f(t)|^q \Delta t \right)^{\frac{1}{q}},$$

$t \in [0,a]$. Thus

$$\begin{aligned}
\left| \frac{1}{a} \int_0^a B(t)\Delta t - f(0) \right| &= \left| \frac{1}{a} \int_0^a (B(t) - f(0))\Delta t \right| \\
&\le \frac{1}{a} \int_0^a |B(t) - f(0)|\Delta t \\
&\le \frac{1}{a} \left(\int_0^a \left(\int_0^t |h_{\alpha-2}(t,\sigma(\tau))|^p \Delta\tau \right)^{\frac{1}{p}} \Delta t \right) \\
&\quad \times \left(\int_0^a |\Delta_0^{\alpha-1} f(t)|^q \Delta t \right)^{\frac{1}{q}}.
\end{aligned}$$

This completes the proof. □

6 Two-dimensional linear integral inequalities

This chapter is devoted on two dimensional linear integral inequalities. They are investigated Wendroff type and Pachpatte type two dimensional linear integral inequalities. Some of the results in this chapter can be found in [6, 24] and [25].

Let $\mathbb{T}_1$ and $\mathbb{T}_2$ be time scales with forward jump operators and delta differentiation operators σ_1, σ_2 and Δ_1, Δ_2, respectively. Suppose that $0 \in \mathbb{T}_1$ and $0 \in \mathbb{T}_2$.

6.1 Wendroff's inequality

Theorem 6.1.1 (Wendroff's Inequality). *Let $u, c \in \mathcal{C}((\mathbb{R}_+ \cap \mathbb{T}_1) \times (\mathbb{R}_+ \cap \mathbb{T}_2))$, $a \in \mathcal{C}^1(\mathbb{R}_+ \cap \mathbb{T}_1)$ and $b \in \mathcal{C}^1(\mathbb{R}_+ \cap \mathbb{T}_2)$ be nonnegative functions such that*

$$\begin{aligned} a(t_1) + b(0) &\neq 0, \quad t_1 \in \mathbb{R}_+ \cap \mathbb{T}_1, \\ a^{\Delta_1}(t_1) &\geq 0, \quad t_1 \in \mathbb{R}_+ \cap \mathbb{T}_1, \\ b^{\Delta_2}(t_2) &\geq 0, \quad t_2 \in \mathbb{R}_+ \cap \mathbb{T}_2. \end{aligned}$$

If

$$u(t_1, t_2) \leq a(t_1) + b(t_2) + \int_0^{t_1}\int_0^{t_2} c(s_1, s_2) u(s_1, s_2) \Delta_2 s_2 \Delta_1 s_1,$$

$(t_1, t_2) \in (\mathbb{R}_+ \cap \mathbb{T}_1) \times (\mathbb{R}_+ \cap \mathbb{T}_2)$, then

$$u(t_1, t_2) \leq (a(0) + b(t_2)) e_f(t_1, 0), \quad (t_1, t_2) \in (\mathbb{R}_+ \cap \mathbb{T}_1) \times (\mathbb{R}_+ \cap \mathbb{T}_2),$$

where

$$f(t_1, t_2) = \frac{a^{\Delta_1}(t_1)}{a(t_1) + b(0)} + \int_0^{t_2} c(t_1, s_2) \Delta_2 s_2, \quad (t_1, t_2) \in (\mathbb{R}_+ \cap \mathbb{T}_1) \times (\mathbb{R}_+ \cap \mathbb{T}_2).$$

Proof. Let

$$z(t_1, t_2) = a(t_1) + b(t_2) + \int_0^{t_1}\int_0^{t_2} c(s_1, s_2) u(s_1, s_2) \Delta_2 s_2 \Delta_1 s_1,$$

$(t_1, t_2) \in (\mathbb{R}_+ \cap \mathbb{T}_1) \times (\mathbb{R}_+ \cap \mathbb{T}_2)$. We have that

$$\begin{aligned} &u(t_1, t_2) \leq z(t_1, t_2), \\ &z(t_1, t_2) \geq 0, \\ &z(t_1, 0) = a(t_1) + b(0), \end{aligned}$$

https://doi.org/10.1515/9783110705553-006

$$\begin{aligned} z_{t_1}^{\Delta_1}(t_1,t_2) &= a^{\Delta_1}(t_1) + \int_0^{t_2} c(t_1,s_2)u(t_1,s_2)\Delta_2 s_2, \\ z_{t_1}^{\Delta_1}(t_1,0) &= a^{\Delta_1}(t_1), \\ z(0,t_2) &= a(0) + b(t_2), \\ z_{t_2}^{\Delta_2}(t_1,t_2) &= b^{\Delta_2}(t_2) + \int_0^{t_1} c(s_1,t_2)u(s_1,t_2)\Delta_1 s_1, \\ z_{t_2}^{\Delta_2}(0,t_2) &= b^{\Delta_2}(t_2), \quad (t_1,t_2) \in (\mathbb{R}_+ \cap \mathbb{T}_1) \times (\mathbb{R}_+ \cap \mathbb{T}_2). \end{aligned}$$

Next,

$$\begin{aligned} z_{t_1t_2}^{\Delta_1\Delta_2}(t_1,t_2) &= c(t_1,t_2)u(t_1,t_2) \\ &\le c(t_1,t_2)z(t_1,t_2), \quad (t_1,t_2) \in (\mathbb{R}_+ \cap \mathbb{T}_1) \times (\mathbb{R}_+ \cap \mathbb{T}_2). \end{aligned}$$

Hence,

$$\frac{z_{t_1t_2}^{\Delta_1\Delta_2}(t_1,t_2)}{z(t_1,t_2)} \le c(t_1,t_2), \quad (t_1,t_2) \in (\mathbb{R}_+ \cap \mathbb{T}_1) \times (\mathbb{R}_+ \cap \mathbb{T}_2),$$

or

$$\frac{z_{t_1t_2}^{\Delta_1\Delta_2}(t_1,t_2)z(t_1,t_2)}{(z(t_1,t_2))^2} \le c(t_1,t_2), \quad (t_1,t_2) \in (\mathbb{R}_+ \cap \mathbb{T}_1) \times (\mathbb{R}_+ \cap \mathbb{T}_2). \tag{6.1}$$

Since

$$a^{\Delta_1}(t_1) \ge 0 \quad \text{and} \quad b^{\Delta_2}(t_2) \ge 0 \quad \text{for} \quad (t_1,t_2) \in (\mathbb{R}_+ \cap \mathbb{T}_1) \times (\mathbb{R}_+ \cap \mathbb{T}_2),$$

we conclude that

$$z_{t_1}^{\Delta_1}(t_1,t_2) \ge 0 \quad \text{and} \quad z_{t_2}^{\Delta_2}(t_1,t_2) \ge 0 \quad \text{for} \quad (t_1,t_2) \in (\mathbb{R}_+ \cap \mathbb{T}_1) \times (\mathbb{R}_+ \cap \mathbb{T}_2).$$

Therefore

$$z(t_1,\sigma_2(t_2)) \ge z(t_1,t_2) \quad \text{for} \quad (t_1,t_2) \in (\mathbb{R}_+ \cap \mathbb{T}_1) \times (\mathbb{R}_+ \cap \mathbb{T}_2).$$

Also,

$$z_{t_1t_2}^{\Delta_1\Delta_2}(t_1,t_2) \ge 0, \quad (t_1,t_2) \in (\mathbb{R}_+ \cap \mathbb{T}_1) \times (\mathbb{R}_+ \cap \mathbb{T}_2).$$

From here and (6.1), we get

$$\frac{z_{t_1t_2}^{\Delta_1\Delta_2}(t_1,t_2)z(t_1,t_2)}{z(t_1,t_2)z(t_1,\sigma_2(t_2))} \le \frac{z_{t_1t_2}^{\Delta_1\Delta_2}(t_1,t_2)z(t_1,t_2)}{(z(t_1,t_2))^2}$$

$$\leq c(t_1, t_2)$$
$$\leq c(t_1, t_2) + \frac{z_{t_1}^{\Delta_1}(t_1, t_2) z_{t_2}^{\Delta_2}(t_1, t_2)}{z(t_1, t_2) z(t_1, \sigma_2(t_2))},$$

$(t_1, t_2) \in (\mathbb{R}_+ \cap \mathbb{T}_1) \times (\mathbb{R}_+ \cap \mathbb{T}_2)$. Hence,

$$\left(\frac{z_{t_1}^{\Delta_1}(t_1, t_2)}{z(t_1, t_2)}\right)_{t_2}^{\Delta_2} \leq c(t_1, t_2), \quad (t_1, t_2) \in (\mathbb{R}_+ \cap \mathbb{T}_1) \times (\mathbb{R}_+ \cap \mathbb{T}_2).$$

Therefore

$$\frac{z_{t_1}^{\Delta_1}(t_1, t_2)}{z(t_1, t_2)} - \frac{z_{t_1}^{\Delta_1}(t_1, 0)}{z(t_1, 0)} \leq \int_0^{t_2} c(t_1, s_2) \Delta_2 s_2, \quad (t_1, t_2) \in (\mathbb{R}_+ \cap \mathbb{T}_1) \times (\mathbb{R}_+ \cap \mathbb{T}_2),$$

or

$$\frac{z_{t_1}^{\Delta_1}(t_1, t_2)}{z(t_1, t_2)} - \frac{a^{\Delta_1}(t_1)}{a(t_1) + b(0)} \leq \int_0^{t_2} c(t_1, s_2) \Delta_2 s_2, \quad (t_1, t_2) \in (\mathbb{R}_+ \cap \mathbb{T}_1) \times (\mathbb{R}_+ \cap \mathbb{T}_2),$$

or

$$\frac{z_{t_1}^{\Delta_1}(t_1, t_2)}{z(t_1, t_2)} \leq \frac{a^{\Delta_1}(t_1)}{a(t_1) + b(0)} + \int_0^{t_2} c(t_1, s_2) \Delta_2 s_2$$
$$= f(t_1, t_2), \quad (t_1, t_2) \in (\mathbb{R}_+ \cap \mathbb{T}_1) \times (\mathbb{R}_+ \cap \mathbb{T}_2),$$

or

$$z_{t_1}^{\Delta_1}(t_1, t_2) \leq f(t_1, t_2) z(t_1, t_2), \quad (t_1, t_2) \in (\mathbb{R}_+ \cap \mathbb{T}_1) \times (\mathbb{R}_+ \cap \mathbb{T}_2).$$

Applying Lemma 2.1.1, we obtain

$$z(t_1, t_2) \leq z(0, t_2) e_f(t_1, 0)$$
$$= (a(0) + b(t_2)) e_f(t_1, 0), \quad (t_1, t_2) \in (\mathbb{R}_+ \cap \mathbb{T}_1) \times (\mathbb{R}_+ \cap \mathbb{T}_2).$$

Consequently,

$$u(t_1, t_2) \leq z(t_1, t_2)$$
$$\leq (a(0) + b(t_2)) e_f(t_1, 0), \quad (t_1, t_2) \in (\mathbb{R}_+ \cap \mathbb{T}_1) \times (\mathbb{R}_+ \cap \mathbb{T}_2).$$

This completes the proof. □

Theorem 6.1.2. *Let $u, a, b \in \mathcal{C}((\mathbb{R}_+ \cap \mathbb{T}_1) \times (\mathbb{R}_+ \cap \mathbb{T}_2))$ be nonnegative functions, $a(t_1, t_2)$ be nondecreasing in each variable t_1, t_2, $(t_1, t_2) \in (\mathbb{R}_+ \cap \mathbb{T}_1) \times (\mathbb{R}_+ \cap \mathbb{T}_2)$. If*

$$u(t_1, t_2) \leq a(t_1, t_2) + \int_0^{t_1} \int_0^{t_2} b(s_1, s_2) u(s_1, s_2) \Delta_2 s_2 \Delta_1 s_1,$$

$(t_1, t_2) \in (\mathbb{R}_+ \cap \mathbb{T}_1) \times (\mathbb{R}_+ \cap \mathbb{T}_2)$, *then*

$$u(t_1, t_2) \le a(t_1, t_2) e_c(t_1, 0), \quad (t_1, t_2) \in (\mathbb{R}_+ \cap \mathbb{T}_1) \times (\mathbb{R}_+ \cap \mathbb{T}_2),$$

where

$$c(t_1, t_2) = \int_0^{t_2} b(t_1, s_2) \Delta_2 s_2, \quad (t_1, t_2) \in (\mathbb{R}_+ \cap \mathbb{T}_1) \times (\mathbb{R}_+ \cap \mathbb{T}_2).$$

Proof. We have

$$\begin{aligned} \frac{u(t_1, t_2)}{a(t_1, t_2)} &\le 1 + \frac{1}{a(t_1, t_2)} \int_0^{t_1} \int_0^{t_2} b(s_1, s_2) u(s_1, s_2) \Delta_2 s_2 \Delta_1 s_1 \\ &= 1 + \int_0^{t_1} \int_0^{t_2} b(s_1, s_2) \frac{u(s_1, s_2)}{a(t_1, t_2)} \Delta_2 s_2 \Delta_1 s_1 \\ &\le 1 + \int_0^{t_1} \int_0^{t_2} b(s_1, s_2) \frac{u(s_1, s_2)}{a(s_1, s_2)} \Delta_2 s_2 \Delta_1 s_1, \end{aligned}$$

$(t_1, t_2) \in (\mathbb{R}_+ \cap \mathbb{T}_1) \times (\mathbb{R}_+ \cap \mathbb{T}_2)$. Hence, by Theorem 6.1.1, we get

$$\frac{u(t_1, t_2)}{a(t_1, t_2)} \le e_c(t_1, 0), \quad (t_1, t_2) \in (\mathbb{R}_+ \cap \mathbb{T}_1) \times (\mathbb{R}_+ \cap \mathbb{T}_2),$$

whereupon

$$u(t_1, t_2) \le a(t_1, t_2) e_c(t_1, 0), \quad (t_1, t_2) \in (\mathbb{R}_+ \cap \mathbb{T}_1) \times (\mathbb{R}_+ \cap \mathbb{T}_2).$$

This completes the proof. □

6.2 Pachpatte's inequalities

Theorem 6.2.1 (Pachpatte's Inequality). *Let $u, p, q \in \mathcal{C}((\mathbb{R}_+ \cap \mathbb{T}_1) \times (\mathbb{R}_+ \cap \mathbb{T}_2))$ be nonnegative functions, $k(t_1, t_2, s_1, s_2)$ and its partial derivatives*

$$\begin{aligned} &k_{t_1}^{\Delta_1}(t_1, t_2, s_1, s_2), && k_{t_1}^{\Delta_1}(\sigma_1(t_1), t_2, s_1, s_2), \\ &k_{t_2}^{\Delta_2}(t_1, t_2, s_1, s_2), && k_{t_2}^{\Delta_2}(t_1, \sigma_2(t_2), s_1, s_2), \\ &k_{t_1 t_2}^{\Delta_1 \Delta_2}(t_1, t_2, s_1, s_2), && (t_1, t_2), (s_1, s_2) \in (\mathbb{R}_+ \cap \mathbb{T}_1) \times (\mathbb{R}_+ \cap \mathbb{T}_2), \end{aligned}$$

be nonnegative continuous functions. If

$$u(t_1, t_2) \le p(t_1, t_2) + q(t_1, t_2) \int_0^{t_1} \int_0^{t_2} k(t_1, t_2, s_1, s_2) u(s_1, s_2) \Delta_2 s_2 \Delta_1 s_1,$$

$(t_1, t_2) \in (\mathbb{R}_+ \cap \mathbb{T}_1) \times (\mathbb{R}_+ \cap \mathbb{T}_2)$, *then*

$$u(t_1, t_2) \le p(t_1, t_2) + q(t_1, t_2)A(t_1, t_2)e_c(t_2, 0),$$

$(t_1, t_2) \in (\mathbb{R}_+ \cap \mathbb{T}_1) \times (\mathbb{R}_+ \cap \mathbb{T}_2)$, *where*

$$\begin{aligned}
a(t_1, t_2) &= k(\sigma_1(t_1), \sigma_2(t_2), t_1, t_2)p(t_1, t_2) \\
&\quad + \int_0^{t_2} k_{t_2}^{\Delta_2}(\sigma_1(t_1), t_2, t_1, s_2)p(t_1, s_2)\Delta_2 s_2 \\
&\quad + \int_0^{t_1} k_{t_1}^{\Delta_1}(t_1, \sigma_2(t_2), s_1, t_2)p(s_1, t_2)\Delta_1 s_1 \\
&\quad + \int_0^{t_1}\int_0^{t_2} k_{t_1 t_2}^{\Delta_1 \Delta_2}(t_1, t_2, s_1, s_2)p(s_1, s_2)\Delta_2 s_2 \Delta_1 s_1, \\
b(t_1, t_2) &= k(\sigma_1(t_1), \sigma_2(t_2), t_1, t_2)q(t_1, t_2) \\
&\quad + \int_0^{t_2} k_{t_2}^{\Delta_2}(\sigma_1(t_1), t_2, t_1, s_2)q(t_1, s_2)\Delta_2 s_2 \\
&\quad + \int_0^{t_1} k_{t_1}^{\Delta_1}(t_1, \sigma_2(t_2), s_1, t_2)q(s_1, t_2)\Delta_1 s_1 \\
&\quad + \int_0^{t_1}\int_0^{t_2} k_{t_1 t_2}^{\Delta_1 \Delta_2}(t_1, t_2, s_1, s_2)q(s_1, s_2)\Delta_2 s_2 \Delta_1 s_1, \\
A(t_1, t_2) &= \int_0^{t_1}\int_0^{t_2} a(s_1, s_2)\Delta_2 s_2 \Delta_1 s_1, \\
c(t_1, t_2) &= \int_0^{t_2} b(t_1, s_2)\Delta_2 s_2,
\end{aligned}$$

$(t_1, t_2) \in (\mathbb{R}_+ \cap \mathbb{T}_1) \times (\mathbb{R}_+ \cap \mathbb{T}_2)$.

Proof. Let

$$z(t_1, t_2) = \int_0^{t_1}\int_0^{t_2} k(t_1, t_2, s_1, s_2)u(s_1, s_2)\Delta_2 s_2 \Delta_1 s_1, \quad (t_1, t_2) \in (\mathbb{R}_+ \cap \mathbb{T}_1) \times (\mathbb{R}_+ \cap \mathbb{T}_2).$$

Then

$$u(t_1, t_2) \le p(t_1, t_2) + q(t_1, t_2)z(t_1, t_2), \quad (t_1, t_2) \in (\mathbb{R}_+ \cap \mathbb{T}_1) \times (\mathbb{R}_+ \cap \mathbb{T}_2),$$

and $z(t_1,t_2)$ is a nondecreasing function in each variable t_1, t_2, $(t_1,t_2) \in (\mathbb{R}_+ \cap \mathbb{T}_1) \times (\mathbb{R}_+ \cap \mathbb{T}_2)$. Then

$$
\begin{aligned}
z(0,0) &= 0,\\
z_{t_1}^{\Delta_1}(t_1,t_2) &= \int_0^{t_2} k(\sigma_1(t_1),t_2,t_1,s_2)u(t_1,s_2)\Delta_2 s_2\\
&\quad + \int_0^{t_1}\int_0^{t_2} k_{t_1}^{\Delta_1}(t_1,t_2,s_1,s_2)u(s_1,s_2)\Delta_2 s_2\Delta_1 s_1,\\
z_{t_1t_2}^{\Delta_1\Delta_2}(t_1,t_2) &= k(\sigma_1(t_1),\sigma_2(t_2),t_1,t_2)u(t_1,t_2)\\
&\quad + \int_0^{t_2} k_{t_2}^{\Delta_2}(\sigma_1(t_1),t_2,t_1,s_2)u(t_1,s_2)\Delta_2 s_2\\
&\quad + \int_0^{t_1} k_{t_1}^{\Delta_1}(t_1,\sigma_2(t_2),s_1,t_2)u(s_1,t_2)\Delta_1 s_1\\
&\quad + \int_0^{t_1}\int_0^{t_2} k_{t_1t_2}^{\Delta_1\Delta_2}(t_1,t_2,s_1,s_2)u(s_1,s_2)\Delta_2 s_2\Delta_1 s_1\\
&\le k(\sigma_1(t_1),\sigma_2(t_2),t_1,t_2)p(t_1,t_2)\\
&\quad + k(\sigma_1(t_1),\sigma_2(t_2),t_1,t_2)q(t_1,t_2)z(t_1,t_2)\\
&\quad + \int_0^{t_2} k_{t_2}^{\Delta_2}(\sigma_1(t_1),t_2,t_1,s_2)p(t_1,s_2)\Delta_2 s_2\\
&\quad + \int_0^{t_2} k_{t_2}^{\Delta_2}(\sigma_1(t_1),t_2,t_1,s_2)q(t_1,s_2)z(t_1,s_2)\Delta_2 s_2\\
&\quad + \int_0^{t_1} k_{t_1}^{\Delta_1}(t_1,\sigma_2(t_2),s_1,t_2)p(s_1,t_2)\Delta_1 s_1\\
&\quad + \int_0^{t_1} k_{t_1}^{\Delta_1}(t_1,\sigma_2(t_2),s_1,t_2)q(s_1,t_2)z(s_1,t_2)\Delta_1 s_1\\
&\quad + \int_0^{t_1}\int_0^{t_2} k_{t_1t_2}^{\Delta_1\Delta_2}(t_1,t_2,s_1,s_2)p(s_1,s_2)\Delta_2 s_2\Delta_1 s_1\\
&\quad + \int_0^{t_1}\int_0^{t_2} k_{t_1t_2}^{\Delta_1\Delta_2}(t_1,t_2,s_1,s_2)q(s_1,s_2)z(s_1,s_2)\Delta_2 s_2\Delta_1 s_1
\end{aligned}
$$

$$
\begin{aligned}
&\le a(t_1,t_2)\\
&\quad + k(\sigma_1(t_1),\sigma_2(t_2),t_1,t_2)q(t_1,t_2)z(t_1,t_2)\\
&\quad + \left(\int_0^{t_2} k_{t_2}^{\Delta_2}(\sigma_1(t_1),t_2,t_1,s_2)q(t_1,s_2)\Delta_2 s_2\right)z(t_1,t_2)\\
&\quad + \left(\int_0^{t_1} k_{t_1}^{\Delta_1}(t_1,\sigma_2(t_2),s_1,t_2)q(s_1,t_2)\Delta_1 s_1\right)z(t_1,t_2)\\
&\quad + \left(\int_0^{t_1}\int_0^{t_2} k_{t_1t_2}^{\Delta_1\Delta_2}(t_1,t_2,s_1,s_2)q(s_1,s_2)\Delta_2 s_2\Delta_1 s_1\right)z(t_1,t_2)\\
&= a(t_1,t_2) + b(t_1,t_2)z(t_1,t_2), \quad (t_1,t_2)\in(\mathbb{R}_+\cap\mathbb{T}_1)\times(\mathbb{R}_+\cap\mathbb{T}_2).
\end{aligned}
$$

Hence, using that

$$
\begin{aligned}
z_{t_1}^{\Delta_1}(t_1,0) &= 0,\\
z(0,t_2) &= 0, \quad (t_1,t_2)\in(\mathbb{R}_+\cap\mathbb{T}_1)\times(\mathbb{R}_+\cap\mathbb{T}_2),
\end{aligned}
$$

we obtain

$$
\begin{aligned}
z_{t_1}^{\Delta_1}(t_1,t_2) - z_{t_1}^{\Delta_1}(t_1,0) &\le \int_0^{t_2} a(t_1,s_2)\Delta_2 s_2 + \int_0^{t_2} b(t_1,s_2)z(t_1,s_2)\Delta_2 s_2,\\
z_{t_1}^{\Delta_1}(t_1,t_2) &\le \int_0^{t_2} a(t_1,s_2)\Delta_2 s_2 + \int_0^{t_2} b(t_1,s_2)z(t_1,s_2)\Delta_2 s_2,\\
z(t_1,t_2) - z(0,t_2) &\le \int_0^{t_1}\int_0^{t_2} a(s_1,s_2)\Delta_2 s_2\Delta_1 s_1\\
&\quad + \int_0^{t_1}\int_0^{t_2} b(s_1,s_2)z(s_1,s_2)\Delta_2 s_2\Delta_1 s_1,\\
z(t_1,t_2) &\le A(t_1,t_2) + \int_0^{t_1}\int_0^{t_2} b(s_1,s_2)z(s_1,s_2)\Delta_2 s_2\Delta_1 s_1,
\end{aligned}
$$

$(t_1,t_2)\in(\mathbb{R}_+\cap\mathbb{T}_1)\times(\mathbb{R}_+\cap\mathbb{T}_2)$. Now we apply Theorem 6.1.2 and get

$$
z(t_1,t_2) \le A(t_1,t_2)e_c(t_1,0), \quad (t_1,t_2)\in(\mathbb{R}_+\cap\mathbb{T}_1)\times(\mathbb{R}_+\cap\mathbb{T}_2),
$$

whereupon

$$
\begin{aligned}
u(t_1,t_2) &\le p(t_1,t_2) + q(t_1,t_2)z(t_1,t_2)\\
&\le p(t_1,t_2) + q(t_1,t_2)A(t_1,t_2)e_c(t_2,0),
\end{aligned}
$$

$(t_1,t_2)\in(\mathbb{R}_+\cap\mathbb{T}_1)\times(\mathbb{R}_+\cap\mathbb{T}_2)$. This completes the proof. □

Corollary 6.2.2. *Let $u, p, q, k \in \mathcal{C}((\mathbb{R}_+ \cap \mathbb{T}_1) \times (\mathbb{R}_+ \cap \mathbb{T}_2))$ be nonnegative functions. If*

$$u(t_1, t_2) \le p(t_1, t_2) + q(t_1, t_2) \int_0^{t_1}\int_0^{t_2} k(s_1, s_2)u(s_1, s_2)\Delta_2 s_2 \Delta_1 s_1,$$

$(t_1, t_2) \in (\mathbb{R}_+ \cap \mathbb{T}_1) \times (\mathbb{R}_+ \cap \mathbb{T}_2)$, then

$$u(t_1, t_2) \le p(t_1, t_2) + q(t_1, t_2)A(t_1, t_2)e_c(t_1, 0), \quad (t_1, t_2) \in (\mathbb{R}_+ \cap \mathbb{T}_1) \times (\mathbb{R}_+ \cap \mathbb{T}_2),$$

where

$$\begin{aligned} a(t_1, t_2) &= k(t_1, t_2)p(t_1, t_2), \\ b(t_1, t_2) &= k(t_1, t_2)q(t_1, t_2), \\ A(t_1, t_2) &= \int_0^{t_1}\int_0^{t_2} a(s_1, s_2)\Delta_2 s_2 \Delta_1 s_1, \\ c(t_1, t_2) &= \int_0^{t_2} b(t_1, s_2)\Delta_2 s_2, \quad (t_1, t_2) \in (\mathbb{R}_+ \cap \mathbb{T}_1) \times (\mathbb{R}_+ \cap \mathbb{T}_2). \end{aligned}$$

Theorem 6.2.3. *Let c_1 and c_2 be nonnegative constants, $u, v, h_i \in \mathcal{C}((\mathbb{R}_+\cap\mathbb{T}_1)\times(\mathbb{R}_+\cap\mathbb{T}_2))$, $i \in \{1, 2, 3, 4\}$, be nonnegative functions. If*

$$\begin{aligned} u(t_1, t_2) \le c_1 &+ \int_0^{t_1}\int_0^{t_2} h_1(s_1, s_2)u(s_1, s_2)\Delta_2 s_2 \Delta_1 s_1 \\ &+ \int_0^{t_1}\int_0^{t_2} h_2(s_1, s_2)v(s_1, s_2)\Delta_2 s_2 \Delta_1 s_1, \\ v(t_1, t_2) \le c_2 &+ \int_0^{t_1}\int_0^{t_2} h_3(s_1, s_2)u(s_1, s_2)\Delta_2 s_2 \Delta_1 s_1 \\ &+ \int_0^{t_1}\int_0^{t_2} h_4(s_1, s_2)v(s_1, s_2)\Delta_2 s_2 \Delta_1 s_1, \end{aligned}$$

$(t_1, t_2) \in (\mathbb{R}_+ \cap \mathbb{T}_1) \times (\mathbb{R}_+ \cap \mathbb{T}_2)$, then

$$u(t_1, t_2) + v(t_1, t_2) \le c_3 + A(t_1, t_2)e_c(t_1, 0), \quad (t_1, t_2) \in (\mathbb{R}_+ \cap \mathbb{T}_1) \times (\mathbb{R}_+ \cap \mathbb{T}_2),$$

where

$$\begin{aligned} c_3 &= c_1 + c_2, \\ H(t_1, t_2) &= \max\{h_1(t_1, t_2) + h_3(t_1, t_2), h_2(t_1, t_2) + h_4(t_1, t_2)\}, \end{aligned}$$

$$A(t_1,t_2) = c_3 \int_0^{t_1}\int_0^{t_2} H(s_1,s_2)\Delta_2 s_2 \Delta_1 s_1,$$

$$c(t_1,t_2) = \int_0^{t_2} H(t_1,s_2)\Delta_2 s_2, \quad (t_1,t_2) \in (\mathbb{R}_+ \cap \mathbb{T}_1) \times (\mathbb{R}_+ \cap \mathbb{T}_2).$$

Proof. Let

$$f(t_1,t_2) = u(t_1,t_2) + v(t_1,t_2), \quad (t_1,t_2) \in (\mathbb{R}_+ \cap \mathbb{T}_1) \times (\mathbb{R}_+ \cap \mathbb{T}_2).$$

Then

$$\begin{aligned}
f(t_1,t_2) &= u(t_1,t_2) + v(t_1,t_2) \\
&\le c_1 + \int_0^{t_1}\int_0^{t_2} h_1(s_1,s_2)u(s_1,s_2)\Delta_2 s_2 \Delta_1 s_1 \\
&\quad + \int_0^{t_1}\int_0^{t_2} h_2(s_1,s_2)v(s_1,s_2)\Delta_2 s_2 \Delta_1 s_1 \\
&\quad + c_2 + \int_0^{t_1}\int_0^{t_2} h_3(s_1,s_2)u(s_1,s_2)\Delta_2 s_2 \Delta_1 s_1 \\
&\quad + \int_0^{t_1}\int_0^{t_2} h_4(s_1,s_2)v(s_1,s_2)\Delta_2 s_2 \Delta_1 s_1 \\
&= c_3 + \int_0^{t_1}\int_0^{t_2} (h_1(s_1,s_2) + h_3(s_1,s_2))u(s_1,s_2)\Delta_2 s_2 \Delta_1 s_1 \\
&\quad + \int_0^{t_1}\int_0^{t_2} (h_2(s_1,s_2) + h_4(s_1,s_2))v(s_1,s_2)\Delta_2 s_2 \Delta_1 s_1 \\
&\le c_3 + \int_0^{t_1}\int_0^{t_2} H(s_1,s_2)u(s_1,s_2)\Delta_2 s_2 \Delta_1 s_1 \\
&\quad + \int_0^{t_1}\int_0^{t_2} H(s_1,s_2)v(s_1,s_2)\Delta_2 s_2 \Delta_1 s_1 \\
&= c_3 + \int_0^{t_1}\int_0^{t_2} H(s_1,s_2)(u(s_1,s_2) + v(s_1,s_2))\Delta_2 s_2 \Delta_1 s_1 \\
&= c_3 + \int_0^{t_1}\int_0^{t_2} H(s_1,s_2)f(s_1,s_2)\Delta_2 s_2 \Delta_1 s_1,
\end{aligned}$$

$(t_1,t_2) \in (\mathbb{R}_+ \cap \mathbb{T}_1) \times (\mathbb{R}_+ \cap \mathbb{T}_2)$. Hence, using Corollary 6.2.2, we obtain

$$\begin{aligned} u(t_1,t_2) + v(t_1,t_2) &= f(t_1,t_2) \\ &\le c_3 + A(t_1,t_2)e_c(t_1,0), \quad (t_1,t_2) \in (\mathbb{R}_+ \cap \mathbb{T}_1) \times (\mathbb{R}_+ \cap \mathbb{T}_2). \end{aligned}$$

This completes the proof. □

Theorem 6.2.4 (Pachpatte's Inequality). *Let $u \in \mathcal{C}^2((\mathbb{R}_+ \cap \mathbb{T}_1) \times (\mathbb{R}_+ \cap \mathbb{T}_2))$, $u^{\Delta_1\Delta_2}_{t_1,t_2}(t_1,t_2)$ and $c(t_1,t_2)$ be nonnegative continuous functions for $(t_1,t_2) \in (\mathbb{R}_+ \cap \mathbb{T}_1) \times (\mathbb{R}_+ \cap \mathbb{T}_2)$, and*

$$u(0,t_2) = u(t_1,0) = 0 \quad \textit{for } (t_1,t_2) \in (\mathbb{R}_+ \cap \mathbb{T}_1) \times (\mathbb{R}_+ \cap \mathbb{T}_2).$$

Let also $a \in \mathcal{C}^1(\mathbb{R}_+ \cap \mathbb{T}_1)$, $b \in \mathcal{C}^1(\mathbb{R}_+ \cap \mathbb{T}_2)$ be positive functions having derivatives such that

$$a^{\Delta_1}(t_1) \ge 0, \quad t_1 \in \mathbb{R}_+ \cap \mathbb{T}_1, \quad b^{\Delta_2}(t_2) \ge 0, \quad t_2 \in \mathbb{R}_+ \cap \mathbb{T}_2.$$

If

$$u^{\Delta_1\Delta_2}_{t_1t_2}(t_1,t_2) \le a(t_1) + b(t_2) + \int_0^{t_1}\int_0^{t_2} c(s_1,s_2)(u(s_1,s_2) + u^{\Delta_1\Delta_2}_{t_1t_2}(s_1,s_2))\Delta_2 s_2 \Delta_1 s_1,$$

$(t_1,t_2) \in (\mathbb{R}_+ \cap \mathbb{T}_1) \times (\mathbb{R}_+ \cap \mathbb{T}_2)$, then

$$u(t_1,t_2) \le \int_0^{t_1}\int_0^{t_2} h(s_1,s_2)\Delta_2 s_2 \Delta_1 s_1, \quad (t_1,t_2) \in (\mathbb{R}_+ \cap \mathbb{T}_1) \times (\mathbb{R}_+ \cap \mathbb{T}_2),$$

where

$$\begin{aligned} p(t_1,t_2) &= \frac{a^{\Delta_1}(t_1)}{a(t_1) + b(0)} + \int_0^{t_2} (1 + c(t_1,s_2))\Delta_2 s_2, \\ q(t_1,t_2) &= (a(0) + b(t_2))e_p(t_1,0)c(t_1,t_2), \\ h(t_1,t_2) &= a(t_1) + b(t_2) + \int_0^{t_1}\int_0^{t_2} q(s_1,s_2)\Delta_2 s_2 \Delta_1 s_1, \end{aligned}$$

$(t_1,t_2) \in (\mathbb{R}_+ \cap \mathbb{T}_1) \times (\mathbb{R}_+ \cap \mathbb{T}_2)$.

Proof. Let

$$z(t_1,t_2) = a(t_1) + b(t_2) + \int_0^{t_1}\int_0^{t_2} c(s_1,s_2)(u(s_1,s_2) + u^{\Delta_1\Delta_2}_{t_1t_2}(s_1,s_2))\Delta_2 s_2 \Delta_1 s_1,$$

$(t_1, t_2) \in (\mathbb{R}_+ \cap \mathbb{T}_1) \times (\mathbb{R}_+ \cap \mathbb{T}_2)$. Then

$$\begin{aligned} u^{\Delta_1\Delta_2}_{t_1t_2}(t_1, t_2) &\le z(t_1, t_2), \quad (t_1, t_2) \in (\mathbb{R}_+ \cap \mathbb{T}_1) \times (\mathbb{R}_+ \cap \mathbb{T}_2), \\ z(t_1, 0) &= a(t_1) + b(0), \quad t_1 \in \mathbb{R}_+ \cap \mathbb{T}_1, \\ z(0, t_2) &= a(0) + b(t_2), \quad t_2 \in \mathbb{R}_+ \cap \mathbb{T}_2, \\ z^{\Delta_1}_{t_1}(t_1, t_2) &= a^{\Delta_1}(t_1) + \int_0^{t_2} c(t_1, s_2)(u(t_1, s_2) + u^{\Delta_1\Delta_2}_{t_1t_2}(t_1, s_2))\Delta_2 s_2, \end{aligned} \tag{6.2}$$

and

$$z^{\Delta_1\Delta_2}_{t_1t_2}(t_1, t_2) = c(t_1, t_2)(u(t_1, t_2) + u^{\Delta_1\Delta_2}_{t_1t_2}(t_1, t_2)), \quad (t_1, t_2) \in (\mathbb{R}_+ \cap \mathbb{T}_1) \times (\mathbb{R}_+ \cap \mathbb{T}_2). \tag{6.3}$$

Since

$$u(0, t_2) = u(t_1, 0) = 0, \quad (t_1, t_2) \in (\mathbb{R}_+ \cap \mathbb{T}_1) \times (\mathbb{R}_+ \cap \mathbb{T}_2),$$

we have

$$u^{\Delta_2}_{t_2}(0, t_2) = u^{\Delta_1}_{t_1}(t_1, 0) = 0, \quad (t_1, t_2) \in (\mathbb{R}_+ \cap \mathbb{T}_1) \times (\mathbb{R}_+ \cap \mathbb{T}_2).$$

Hence, by (6.2), we obtain

$$u^{\Delta_1}_{t_1}(t_1, t_2) - u^{\Delta_1}_{t_1}(t_1, 0) \le \int_0^{t_2} z(t_1, s_2)\Delta_2 s_2, \quad (t_1, t_2) \in (\mathbb{R}_+ \cap \mathbb{T}_1) \times (\mathbb{R}_+ \cap \mathbb{T}_2),$$

or

$$u^{\Delta_1}_{t_1}(t_1, t_2) \le \int_0^{t_2} z(t_1, s_2)\Delta_2 s_2, \quad (t_1, t_2) \in (\mathbb{R}_+ \cap \mathbb{T}_1) \times (\mathbb{R}_+ \cap \mathbb{T}_2),$$

whereupon

$$u(t_1, t_2) - u(0, t_2) \le \int_0^{t_1}\int_0^{t_2} z(s_1, s_2)\Delta_2 s_2 \Delta_1 s_1, \quad (t_1, t_2) \in (\mathbb{R}_+ \cap \mathbb{T}_1) \times (\mathbb{R}_+ \cap \mathbb{T}_2),$$

or

$$u(t_1, t_2) \le \int_0^{t_1}\int_0^{t_2} z(s_1, s_2)\Delta_2 s_2 \Delta_1 s_1, \quad (t_1, t_2) \in (\mathbb{R}_+ \cap \mathbb{T}_1) \times (\mathbb{R}_+ \cap \mathbb{T}_2).$$

Using the last inequality and (6.3), we get

$$z^{\Delta_1\Delta_2}_{t_1t_2}(t_1, t_2) \le c(t_1, t_2)\left(z(t_1, t_2) + \int_0^{t_1}\int_0^{t_2} z(s_1, s_2)\Delta_2 s_2 \Delta_1 s_1 \right),$$

$(t_1, t_2) \in (\mathbb{R}_+ \cap \mathbb{T}_1) \times (\mathbb{R}_+ \cap \mathbb{T}_2)$. Let

$$v(t_1, t_2) = z(t_1, t_2) + \int_0^{t_1} \int_0^{t_2} z(s_1, s_2) \Delta_2 s_2 \Delta_1 s_1, \quad (t_1, t_2) \in (\mathbb{R}_+ \cap \mathbb{T}_1) \times (\mathbb{R}_+ \cap \mathbb{T}_2).$$

Then

$$\begin{aligned} v(t_1, 0) &= z(t_1, 0), \\ v(0, t_2) &= z(0, t_2), \quad (t_1, t_2) \in (\mathbb{R}_+ \cap \mathbb{T}_1) \times (\mathbb{R}_+ \cap \mathbb{T}_2), \end{aligned}$$

and

$$\begin{aligned} z(t_1, t_2) &\le v(t_1, t_2), \\ z^{\Delta_1 \Delta_2}_{t_1 t_2}(t_1, t_2) &\le c(t_1, t_2) v(t_1, t_2), \quad (t_1, t_2) \in (\mathbb{R}_+ \cap \mathbb{T}_1) \times (\mathbb{R}_+ \cap \mathbb{T}_2). \end{aligned}$$

Next,

$$\begin{aligned} v^{\Delta_1}_{t_1}(t_1, t_2) &= z^{\Delta_1}_{t_1}(t_1, t_2) + \int_0^{t_2} z(t_1, s_2) \Delta_2 s_2, \\ v^{\Delta_1 \Delta_2}_{t_1 t_2}(t_1, t_2) &= z^{\Delta_1 \Delta_2}_{t_1 t_2}(t_1, t_2) + z(t_1, t_2) \\ &\le c(t_1, t_2) v(t_1, t_2) + v(t_1, t_2) \\ &= (1 + c(t_1, t_2)) v(t_1, t_2), \quad (t_1, t_2) \in (\mathbb{R}_+ \cap \mathbb{T}_1) \times (\mathbb{R}_+ \cap \mathbb{T}_2). \end{aligned}$$

From here,

$$\frac{v^{\Delta_1 \Delta_2}_{t_1 t_2}(t_1, t_2)}{v(t_1, t_2)} \le 1 + c(t_1, t_2), \quad (t_1, t_2) \in (\mathbb{R}_+ \cap \mathbb{T}_1) \times (\mathbb{R}_+ \cap \mathbb{T}_2),$$

or

$$\frac{v^{\Delta_1 \Delta_2}_{t_1 t_2}(t_1, t_2) v(t_1, t_2)}{(v(t_1, t_2))^2} \le 1 + c(t_1, t_2), \tag{6.4}$$

$(t_1, t_2) \in (\mathbb{R}_+ \cap \mathbb{T}_1) \times (\mathbb{R}_+ \cap \mathbb{T}_2)$. Since

$$z^{\Delta_2}_{t_2}(t_1, t_2) = b^{\Delta_2}(t_2) + \int_0^{t_1} c(s_1, t_2)(u(s_1, t_2) + u^{\Delta_1 \Delta_2}_{t_1 t_2}(s_1, t_2)) \Delta_1 s_1,$$
$$(t_1, t_2) \in (\mathbb{R}_+ \cap \mathbb{T}_1) \times (\mathbb{R}_+ \cap \mathbb{T}_2),$$

and because

$$\begin{aligned} &b^{\Delta_2}(t_2) \ge 0, \quad t_2 \in \mathbb{R}_+ \cap \mathbb{T}_2, \\ &c(t_1, t_2) \ge 0, \quad u(t_1, t_2) \ge 0, \quad u^{\Delta_1 \Delta_2}_{t_1 t_2}(t_1, t_2) \ge 0, \end{aligned}$$

$(t_1, t_2) \in (\mathbb{R}_+ \cap \mathbb{T}_1) \times (\mathbb{R}_+ \cap \mathbb{T}_2)$, we conclude that $z(t_1, t_2)$ is a nondecreasing function with respect to t_2. Therefore $v(t_1, t_2)$ is a nondecreasing function with respect to t_2. From here,

$$v(t_1, t_2) \le v(t_1, \sigma_2(t_2)), \quad (t_1, t_2) \in (\mathbb{R}_+ \cap \mathbb{T}_1) \times (\mathbb{R}_+ \cap \mathbb{T}_2).$$

Hence, using (6.4), we obtain

$$\frac{v^{\Delta_1\Delta_2}_{t_1t_2}(t_1, t_2)v(t_1, t_2)}{v(t_1, t_2)v(t_1, \sigma_2(t_2))} \le 1 + c(t_1, t_2), \quad (t_1, t_2) \in (\mathbb{R}_+ \cap \mathbb{T}_1) \times (\mathbb{R}_+ \cap \mathbb{T}_2). \tag{6.5}$$

Observe that

$$\begin{aligned} v^{\Delta_1}_{t_1}(t_1, t_2) &\ge 0, \\ v^{\Delta_2}_{t_2}(t_1, t_2) &= z^{\Delta_2}_{t_2}(t_1, t_2) + \int_0^{t_1} z(s_1, t_2)\Delta_1 s_1 \\ &\ge 0, \quad (t_1, t_2) \in (\mathbb{R}_+ \cap \mathbb{T}_1) \times (\mathbb{R}_+ \cap \mathbb{T}_2). \end{aligned}$$

From here and (6.5), we obtain

$$\begin{aligned} \frac{v^{\Delta_1\Delta_2}_{t_1t_2}(t_1, t_2)v(t_1, t_2)}{v(t_1, t_2)v(t_1, \sigma_2(t_2))} &\le 1 + c(t_1, t_2) \\ &\le 1 + c(t_1, t_2) + \frac{v^{\Delta_1}_{t_1}(t_1, t_2)v^{\Delta_2}_{t_2}(t_1, t_2)}{v(t_1, t_2)v(t_1, \sigma_2(t_2))}, \end{aligned}$$

$(t_1, t_2) \in (\mathbb{R}_+ \cap \mathbb{T}_1) \times (\mathbb{R}_+ \cap \mathbb{T}_2)$, or

$$\frac{v^{\Delta_1\Delta_2}_{t_1t_2}(t_1, t_2)v(t_1, t_2)}{v(t_1, t_2)v(t_1, \sigma_2(t_2))} - \frac{v^{\Delta_1}_{t_1}(t_1, t_2)v^{\Delta_2}_{t_2}(t_1, t_2)}{v(t_1, t_2)v(t_1, \sigma_2(t_2))} \le 1 + c(t_1, t_2),$$

$(t_1, t_2) \in (\mathbb{R}_+ \cap \mathbb{T}_1) \times (\mathbb{R}_+ \cap \mathbb{T}_2)$, or

$$\left(\frac{v^{\Delta_1}_{t_1}(t_1, t_2)}{v(t_1, t_2)}\right)^{\Delta_2}_{t_2} \le 1 + c(t_1, t_2), \quad (t_1, t_2) \in (\mathbb{R}_+ \cap \mathbb{T}_1) \times (\mathbb{R}_+ \cap \mathbb{T}_2).$$

From here,

$$\frac{v^{\Delta_1}_{t_1}(t_1, t_2)}{v(t_1, t_2)} - \frac{v^{\Delta_1}_{t_1}(t_1, 0)}{v(t_1, 0)} \le \int_0^{t_2} (1 + c(t_1, s_2))\Delta_2 s_2, \quad (t_1, t_2) \in (\mathbb{R}_+ \cap \mathbb{T}_1) \times (\mathbb{R}_+ \cap \mathbb{T}_2),$$

or

$$\frac{v^{\Delta_1}_{t_1}(t_1, t_2)}{v(t_1, t_2)} - \frac{z^{\Delta_1}_{t_1}(t_1, 0)}{z(t_1, 0)} \le \int_0^{t_2} (1 + c(t_1, s_2))\Delta_2 s_2, \quad (t_1, t_2) \in (\mathbb{R}_+ \cap \mathbb{T}_1) \times (\mathbb{R}_+ \cap \mathbb{T}_2),$$

or

$$\frac{v_{t_1}^{\Delta_1}(t_1,t_2)}{v(t_1,t_2)} - \frac{a^{\Delta_1}(t_1)}{a(t_1)+b(0)} \le \int_0^{t_2} (1+c(t_1,s_2))\Delta_2 s_2, \quad (t_1,t_2) \in (\mathbb{R}_+ \cap \mathbb{T}_1) \times (\mathbb{R}_+ \cap \mathbb{T}_2),$$

or

$$\begin{aligned}\frac{v_{t_1}^{\Delta_1}(t_1,t_2)}{v(t_1,t_2)} &\le \frac{a^{\Delta_1}(t_1)}{a(t_1)+b(0)} + \int_0^{t_2} (1+c(t_1,s_2))\Delta_2 s_2 \\ &= p(t_1,t_2), \quad (t_1,t_2) \in (\mathbb{R}_+ \cap \mathbb{T}_1) \times (\mathbb{R}_+ \cap \mathbb{T}_2),\end{aligned}$$

or

$$v_{t_1}^{\Delta_1}(t_1,t_2) \le p(t_1,t_2)v(t_1,t_2), \quad (t_1,t_2) \in (\mathbb{R}_+ \cap \mathbb{T}_1) \times (\mathbb{R}_+ \cap \mathbb{T}_2).$$

From the last inequality and Lemma 2.1.1, we obtain

$$\begin{aligned}v(t_1,t_2) &\le v(0,t_2)e_p(t_1,0) \\ &= z(0,t_2)e_p(t_1,0) \\ &= (a(0)+b(t_2))e_p(t_1,0), \quad (t_1,t_2) \in (\mathbb{R}_+ \cap \mathbb{T}_1) \times (\mathbb{R}_+ \cap \mathbb{T}_2).\end{aligned}$$

Therefore

$$\begin{aligned}z_{t_1t_2}^{\Delta_1\Delta_2}(t_1,t_2) &\le c(t_1,t_2)v(t_1,t_2) \\ &\le (a(0)+b(t_2))e_p(t_1,0)c(t_1,t_2) \\ &= q(t_1,t_2), \quad (t_1,t_2) \in (\mathbb{R}_+ \cap \mathbb{T}_1) \times (\mathbb{R}_+ \cap \mathbb{T}_2),\end{aligned}$$

$$z_{t_1}^{\Delta_1}(t_1,t_2) - z_{t_1}^{\Delta_1}(t_1,0) \le \int_0^{t_2} q(t_1,s_2)\Delta_2 s_2,$$

$$z_{t_1}^{\Delta_1}(t_1,t_2) \le a^{\Delta_1}(t_1) + \int_0^{t_2} q(t_1,s_2)\Delta_2 s_2,$$

$$z(t_1,t_2) - z(0,t_2) \le a(t_1) - a(0) + \int_0^{t_1}\int_0^{t_2} q(s_1,s_2)\Delta_2 s_2\Delta_1 s_1,$$

$$\begin{aligned}z(t_1,t_2) &\le a(0) + b(t_2) + a(t_1) - a(0) + \int_0^{t_1}\int_0^{t_2} q(s_1,s_2)\Delta_2 s_2\Delta_1 s_1 \\ &= a(t_1) + b(t_2) + \int_0^{t_1}\int_0^{t_2} q(s_1,s_2)\Delta_2 s_2\Delta_1 s_1 \\ &= h(t_1,t_2),\end{aligned}$$

$$
\begin{aligned}
u^{\Delta_1\Delta_2}_{t_1t_2}(t_1,t_2) &\le h(t_1,t_2),\\
u^{\Delta_1}_{t_1}(t_1,t_2) - u^{\Delta_1}_{t_1}(t_1,0) &\le \int_0^{t_2} h(t_1,s_2)\Delta_2 s_2,\\
u^{\Delta_1}_{t_1}(t_1,t_2) &\le \int_0^{t_2} h(t_1,s_2)\Delta_2 s_2,\\
u(t_1,t_2) - u(0,t_2) &\le \int_0^{t_1}\int_0^{t_2} h(s_1,s_2)\Delta_2 s_2\Delta_1 s_1,\\
u(t_1,t_2) &\le \int_0^{t_1}\int_0^{t_2} h(s_1,s_2)\Delta_2 s_2\Delta_1 s_1, \quad (t_1,t_2)\in(\mathbb{R}_+\cap\mathbb{T}_1)\times(\mathbb{R}_+\cap\mathbb{T}_2).
\end{aligned}
$$

This completes the proof. □

Theorem 6.2.5 (Pachpatte's Inequality). *Let* $u \in \mathcal{C}^2((\mathbb{R}_+\cap\mathbb{T}_1)\times(\mathbb{R}_+\cap\mathbb{T}_2))$, $u^{\Delta_1\Delta_2}_{t_1,t_2}(t_1,t_2)$ *and* $c(t_1,t_2)$ *be nonnegative continuous functions for* $(t_1,t_2)\in(\mathbb{R}_+\cap\mathbb{T}_1)\times(\mathbb{R}_+\cap\mathbb{T}_2)$, *and*

$$u(0,t_2) = u(t_1,0) = 0 \quad \textit{for} \quad (t_1,t_2)\in(\mathbb{R}_+\cap\mathbb{T}_1)\times(\mathbb{R}_+\cap\mathbb{T}_2).$$

Let also $a \in \mathcal{C}^1(\mathbb{R}_+\cap\mathbb{T}_1)$, $b\in\mathcal{C}^1(\mathbb{R}_+\cap\mathbb{T}_2)$ *be positive functions having derivatives such that*

$$a^{\Delta_1}(t_1)\ge 0,\quad t_1\in\mathbb{R}_+\cap\mathbb{T}_1,\quad b^{\Delta_2}(t_2)\ge 0,\quad t_2\in\mathbb{R}_+\cap\mathbb{T}_2,$$

and assume that M *is a nonnegative constant. If*

$$
\begin{aligned}
u^{\Delta_1\Delta_2}_{t_1t_2}(t_1,t_2) &\le a(t_1)+b(t_2)\\
&\quad + M\left(u(t_1,t_2) + \int_0^{t_1}\int_0^{t_2} c(s_1,s_2)\big(u(s_1,s_2)+u^{\Delta_1\Delta_2}_{t_1t_2}(s_1,s_2)\big)\Delta_2 s_2\Delta_1 s_1\right),
\end{aligned}
$$

$(t_1,t_2)\in(\mathbb{R}_+\cap\mathbb{T}_1)\times(\mathbb{R}_+\cap\mathbb{T}_2)$, *then*

$$u^{\Delta_1\Delta_2}_{t_1t_2}(t_1,t_2) \le q(t_1,t_2)e_{h_1}(t_1,0),\quad (t_1,t_2)\in(\mathbb{R}_+\cap\mathbb{T}_1)\times(\mathbb{R}_+\cap\mathbb{T}_2),$$

where

$$
\begin{aligned}
q(t_1,t_2) &= a(t_1)+b(t_2),\\
p(t_1,t_2) &= c(t_1,t_2)(1+M)+M,\\
h_1(t_1,t_2) &= \int_0^{t_2} p(t_1,s_2)\Delta_2 s_2,\quad (t_1,t_2)\in(\mathbb{R}_+\cap\mathbb{T}_1)\times(\mathbb{R}_+\cap\mathbb{T}_2).
\end{aligned}
$$

Proof. Let

$$z(t_1,t_2) = a(t_1) + b(t_2) + M\left(u(t_1,t_2) + \int_0^{t_1}\int_0^{t_2} c(s_1,s_2)(u(s_1,s_2) + u_{t_1t_2}^{\Delta_1\Delta_2}(s_1,s_2))\Delta_2 s_2\Delta_1 s_1\right),$$

$(t_1,t_2) \in (\mathbb{R}_+ \cap \mathbb{T}_1) \times (\mathbb{R}_+ \cap \mathbb{T}_2)$. Then

$$\begin{aligned}
Mu(t_1,t_2) &\le z(t_1,t_2),\\
z(0,t_2) &= a(0) + b(t_2) + Mu(0,t_2)\\
&= a(0) + b(t_2),\\
z(t_1,0) &= a(t_1) + b(0) + Mu(t_1,0)\\
&= a(t_1) + b(0),\\
u_{t_1t_2}^{\Delta_1\Delta_2}(t_1,t_2) &\le z(t_1,t_2), \quad (t_1,t_2) \in (\mathbb{R}_+ \cap \mathbb{T}_1) \times (\mathbb{R}_+ \cap \mathbb{T}_2).
\end{aligned}$$

Next,

$$\begin{aligned}
z_{t_1}^{\Delta_1}(t_1,t_2) &= a^{\Delta_1}(t_1) + M\left(u_{t_1}^{\Delta_1}(t_1,t_2) + \int_0^{t_2} c(t_1,s_2)(u(t_1,s_2) + u_{t_1t_2}^{\Delta_1\Delta_2}(t_1,s_2))\Delta_2 s_2\right),\\
z_{t_1t_2}^{\Delta_1\Delta_2}(t_1,t_2) &= M(u_{t_1t_2}^{\Delta_1\Delta_2}(t_1,t_2) + c(t_1,t_2)(u(t_1,t_2) + u_{t_1t_2}^{\Delta_1\Delta_2}(t_1,t_2)))\\
&= M(1 + c(t_1,t_2))u_{t_1t_2}^{\Delta_1\Delta_2}(t_1,t_2)\\
&\quad + Mc(t_1,t_2)u(t_1,t_2)\\
&\le M(1 + c(t_1,t_2))z(t_1,t_2) + c(t_1,t_2)z(t_1,t_2)\\
&= (c(t_1,t_2)(1 + M) + M)z(t_1,t_2)\\
&= p(t_1,t_2)z(t_1,t_2),\\
z_{t_1}^{\Delta_1}(t_1,t_2) - z_{t_1}^{\Delta_1}(t_1,0) &= \int_0^{t_2} p(t_1,s_2)z(t_1,s_2)\Delta_2 s_2,\\
z_{t_1}^{\Delta_1}(t_1,t_2) &= a^{\Delta_1}(t_1) + Mu_{t_1}^{\Delta_1}(t_1,0)\\
&\quad + \int_0^{t_2} p(t_1,s_2)z(t_1,s_2)\Delta_2 s_2,\\
&= a^{\Delta_1}(t_1) + \int_0^{t_2} p(t_1,s_2)z(t_1,s_2)\Delta_2 s_2,\\
z(t_1,t_2) - z(0,t_2) &= a(t_1) - a(0) + \int_0^{t_1}\int_0^{t_2} p(s_1,s_2)z(s_1,s_2)\Delta_2 s_2\Delta_1 s_1,
\end{aligned}$$

$$z(t_1,t_2) = a(0) + b(t_2) + a(t_1) - a(0) + \int_0^{t_1}\int_0^{t_2} p(s_1,s_2)z(s_1,s_2)\Delta_2 s_2 \Delta_1 s_1$$
$$= a(t_1) + b(t_2) + \int_0^{t_1}\int_0^{t_2} p(s_1,s_2)z(s_1,s_2)\Delta_2 s_2 \Delta_1 s_1$$
$$= q(t_1,t_2) + \int_0^{t_1}\int_0^{t_2} p(s_1,s_2)z(s_1,s_2)\Delta_2 s_2 \Delta_1 s_1,$$

$(t_1,t_2) \in (\mathbb{R}_+ \cap \mathbb{T}_1) \times (\mathbb{R}_+ \cap \mathbb{T}_2)$. Hence, by Theorem 6.1.2, we get

$$z(t_1,t_2) \le q(t_1,t_2)e_{h_1}(t_1,0), \quad (t_1,t_2) \in (\mathbb{R}_+ \cap \mathbb{T}_1) \times (\mathbb{R}_+ \cap \mathbb{T}_2).$$

Then

$$u^{\Delta_1\Delta_2}_{t_1 t_2}(t_1,t_2) \le q(t_1,t_2)e_{h_1}(t_1,0),$$

$(t_1,t_2) \in (\mathbb{R}_+ \cap \mathbb{T}_1) \times (\mathbb{R}_+ \cap \mathbb{T}_2)$. This completes the proof. □

Theorem 6.2.6 (Pachpatte's Inequality). *Let $u(t_1,t_2)$, $f(t_1,t_2)$, and $g(t_1,t_2)$ be nonnegative continuous functions for $(t_1,t_2) \in (\mathbb{R}_+ \cap \mathbb{T}_1) \times (\mathbb{R}_+ \cap \mathbb{T}_2)$, $a(t_1)$ and $b(t_2)$ be positive continuous functions for $(t_1,t_2) \in (\mathbb{R}_+ \cap \mathbb{T}_1) \times (\mathbb{R}_+ \cap \mathbb{T}_2)$ such that*

$$a^{\Delta_1}(t_1) \ge 0, \quad b^{\Delta_2}(t_2) \ge 0, \quad (t_1,t_2) \in (\mathbb{R}_+ \cap \mathbb{T}_1) \times (\mathbb{R}_+ \cap \mathbb{T}_2).$$

If

$$u(t_1,t_2) \le a(t_1) + b(t_2) + \int_0^{t_1}\int_0^{t_2} f(s_1,s_2)\left(u(s_1,s_2) + \int_0^{s_1}\int_0^{s_2} g(\tau_1,\tau_2)u(\tau_1,\tau_2)\Delta_2\tau_2\Delta_1\tau_1 \right)\Delta_2 s_2 \Delta_1 s_1,$$

$(t_1,t_2) \in (\mathbb{R}_+ \cap \mathbb{T}_1) \times (\mathbb{R}_+ \cap \mathbb{T}_2)$, then

$$u(t_1,t_2) \le a(t_1) + b(t_2) + \int_0^{t_1}\int_0^{t_2} h_3(s_1,s_2)\Delta_2 s_2 \Delta_1 s_1,$$

$(t_1,t_2) \in (\mathbb{R}_+ \cap \mathbb{T}_1) \times (\mathbb{R}_+ \cap \mathbb{T}_2)$, where

$$h_1(t_1,t_2) = f(t_1,t_2) + g(t_1,t_2),$$
$$h_2(t_1,t_2) = \frac{a^{\Delta_1}(t_1)}{a(t_1)+b(0)} + \int_0^{t_2} h_1(t_1,s_2)\Delta_2 s_2,$$
$$h_3(t_1,t_2) = f(t_1,t_2)(a(0)+b(t_2))e_{h_2}(t_1,0),$$

$(t_1,t_2) \in (\mathbb{R}_+ \cap \mathbb{T}_1) \times (\mathbb{R}_+ \cap \mathbb{T}_2)$.

Proof. Let

$$\begin{aligned} z(t_1,t_2) &= a(t_1)+b(t_2) \\ &\quad + \int_0^{t_1}\int_0^{t_2} f(s_1,s_2)\left(u(s_1,s_2)+\int_0^{s_1}\int_0^{s_2} g(\tau_1,\tau_2)u(\tau_1,\tau_2)\Delta_2\tau_2\Delta_1\tau_1\right)\Delta_2 s_2\Delta_1 s_1, \\ &\quad (t_1,t_2)\in(\mathbb{R}_+\cap\mathbb{T}_1)\times(\mathbb{R}_+\cap\mathbb{T}_2). \end{aligned}$$

Then

$$\begin{aligned} u(t_1,t_2) &\le z(t_1,t_2), \\ z(0,t_2) &= a(0)+b(t_2), \\ z(t_1,0) &= a(t_1)+b(0), \\ z_{t_1}^{\Delta_1}(t_1,t_2) &= a^{\Delta_1}(t_1) \\ &\quad + \int_0^{t_2} f(t_1,s_2)\left(u(t_1,s_2)+\int_0^{t_1}\int_0^{s_2} g(\tau_1,\tau_2)u(\tau_1,\tau_2)\Delta_2\tau_2\Delta_1\tau_1\right)\Delta_2 s_2\Delta_1 s_1, \\ z_{t_1t_2}^{\Delta_1\Delta_2}(t_1,t_2) &= f(t_1,t_2)\left(u(t_1,t_2)+\int_0^{t_1}\int_0^{t_2} g(s_1,s_2)u(s_1,s_2)\Delta_2 s_2\Delta_1 s_1\right) \\ &\le f(t_1,t_2)\left(z(t_1,t_2)+\int_0^{t_1}\int_0^{t_2} g(s_1,s_2)z(s_1,s_2)\Delta_2 s_2\Delta_1 s_1\right), \end{aligned}$$

$(t_1,t_2)\in(\mathbb{R}_+\cap\mathbb{T}_1)\times(\mathbb{R}_+\cap\mathbb{T}_2)$. Let

$$v(t_1,t_2) = z(t_1,t_2)+\int_0^{t_1}\int_0^{t_2} g(s_1,s_2)z(s_1,s_2)\Delta_2 s_2\Delta_1 s_1,$$

$(t_1,t_2)\in(\mathbb{R}_+\cap\mathbb{T}_1)\times(\mathbb{R}_+\cap\mathbb{T}_2)$. Then

$$\begin{aligned} z_{t_1t_2}^{\Delta_1\Delta_2}(t_1,t_2) &\le f(t_1,t_2)v(t_1,t_2), \\ z(t_1,t_2) &\le v(t_1,t_2), \\ v(t_1,0) &= z(t_1,0) \\ &= a(t_1)+b(0), \\ v(0,t_2) &= z(0,t_2) \\ &= a(0)+b(t_2), \quad (t_1,t_2)\in(\mathbb{R}_+\cap\mathbb{T}_1)\times(\mathbb{R}_+\cap\mathbb{T}_2). \end{aligned}$$

Next,

$$v_{t_1}^{\Delta_1}(t_1,t_2) = z_{t_1}^{\Delta_1}(t_1,t_2)+\int_0^{t_2} g(t_1,s_2)z(t_1,s_2)\Delta_2 s_2, \tag{6.6}$$

$(t_1, t_2) \in (\mathbb{R}_+ \cap \mathbb{T}_1) \times (\mathbb{R}_+ \cap \mathbb{T}_2)$, and

$$\begin{aligned} v_{t_1 t_2}^{\Delta_1 \Delta_2}(t_1, t_2) &= z_{t_1 t_2}^{\Delta_1 \Delta_2}(t_1, t_2) + g(t_1, t_2) z(t_1, t_2) \\ &\leq f(t_1, t_2) v(t_1, t_2) + g(t_1, t_2) v(t_1, t_2) \\ &= \big(f(t_1, t_2) + g(t_1, t_2)\big) v(t_1, t_2) \\ &= h_1(t_1, t_2) v(t_1, t_2), \end{aligned}$$

$(t_1, t_2) \in (\mathbb{R}_+ \cap \mathbb{T}_1) \times (\mathbb{R}_+ \cap \mathbb{T}_2)$. Since $a(t_1)$ and $b(t_2)$ are positive continuous functions for $(t_1, t_2) \in (\mathbb{R}_+ \cap \mathbb{T}_1) \times (\mathbb{R}_+ \cap \mathbb{T}_2)$, we get that $z(t_1, t_2)$ is a positive function for $(t_1, t_2) \in (\mathbb{R}_+ \cap \mathbb{T}_1) \times (\mathbb{R}_+ \cap \mathbb{T}_2)$, and then $v(t_1, t_2)$ is a positive function for $(t_1, t_2) \in (\mathbb{R}_+ \cap \mathbb{T}_1) \times (\mathbb{R}_+ \cap \mathbb{T}_2)$. Hence,

$$\begin{aligned} \frac{v_{t_1 t_2}^{\Delta_1 \Delta_2}(t_1, t_2) v(t_1, t_2)}{(v(t_1, t_2))^2} &= \frac{v_{t_1 t_2}^{\Delta_1 \Delta_2}(t_1, t_2)}{v(t_1, t_2)} \\ &\leq h_1(t_1, t_2), \quad (t_1, t_2) \in (\mathbb{R}_+ \cap \mathbb{T}_1) \times (\mathbb{R}_+ \cap \mathbb{T}_2). \end{aligned} \tag{6.7}$$

Also since

$$v_{t_2}^{\Delta_2}(t_1, t_2) = z_{t_2}^{\Delta_2}(t_1, t_2) + \int_0^{t_1} g(s_1, t_2) z(s_1, t_2) \Delta_1 s_1,$$

$(t_1, t_2) \in (\mathbb{R}_+ \cap \mathbb{T}_1) \times (\mathbb{R}_+ \cap \mathbb{T}_2)$, and

$$\begin{aligned} z_{t_2}^{\Delta_2}(t_1, t_2) &= b^{\Delta_2}(t_2) \\ &\quad + \int_0^{t_1} f(s_1, t_2) \left(u(s_1, t_2) + \int_0^{s_1} \int_0^{t_2} g(\tau_1, \tau_2) u(\tau_1, \tau_2) \Delta_2 \tau_2 \Delta_1 \tau_1 \right) \Delta_2 s_2 \Delta_1 s_1 \\ &\geq 0, \quad (t_1, t_2) \in (\mathbb{R}_+ \cap \mathbb{T}_1) \times (\mathbb{R}_+ \cap \mathbb{T}_2), \end{aligned}$$

we get

$$v_{t_2}^{\Delta_2}(t_1, t_2) \geq 0, \quad (t_1, t_2) \in (\mathbb{R}_+ \cap \mathbb{T}_1) \times (\mathbb{R}_+ \cap \mathbb{T}_2).$$

By (6.6), we obtain that

$$v_{t_1}^{\Delta_1}(t_1, t_2) \geq 0, \quad (t_1, t_2) \in (\mathbb{R}_+ \cap \mathbb{T}_1) \times (\mathbb{R}_+ \cap \mathbb{T}_2).$$

Consequently, $v(t_1, t_2)$ is a nondecreasing function with respect to t_1 and t_2, $(t_1, t_2) \in (\mathbb{R}_+ \cap \mathbb{T}_1) \times (\mathbb{R}_+ \cap \mathbb{T}_2)$. Then

$$v(t_1, t_2) \leq v(t_1, \sigma_2(t_2)), \quad (t_1, t_2) \in (\mathbb{R}_+ \cap \mathbb{T}_1) \times (\mathbb{R}_+ \cap \mathbb{T}_2).$$

Hence, by (6.7), we obtain

$$\frac{v_{t_1 t_2}^{\Delta_1 \Delta_2}(t_1, t_2) v(t_1, t_2)}{v(t_1, t_2) v(t_1, \sigma_2(t_2))} \leq \frac{v_{t_1 t_2}^{\Delta_1 \Delta_2}(t_1, t_2) v(t_1, t_2)}{(v(t_1, t_2))^2}$$

$$\begin{aligned}&\leq h_1(t_1,t_2)\\ &\leq h_1(t_1,t_2)\\ &\quad+\frac{v_{t_1}^{\Delta_1}(t_1,t_2)v_{t_2}^{\Delta_2}(t_1,t_2)}{v(t_1,t_2)v(t_1,\sigma_2(t_2))},\end{aligned}$$

$(t_1,t_2)\in(\mathbb{R}_+\cap\mathbb{T}_1)\times(\mathbb{R}_+\cap\mathbb{T}_2)$, or

$$\frac{v_{t_1t_2}^{\Delta_1\Delta_2}(t_1,t_2)v(t_1,t_2)-v_{t_1}^{\Delta_1}(t_1,t_2)v_{t_2}^{\Delta_2}(t_1,t_2)}{v(t_1,t_2)v(t_1,\sigma_2(t_2))}\leq h_1(t_1,t_2),$$

$(t_1,t_2)\in(\mathbb{R}_+\cap\mathbb{T}_1)\times(\mathbb{R}_+\cap\mathbb{T}_2)$, or

$$\left(\frac{v_{t_1}^{\Delta_1}(t_1,t_2)}{v(t_1,t_2)}\right)_{t_2}^{\Delta_2}\leq h_1(t_1,t_2),$$

$(t_1,t_2)\in(\mathbb{R}_+\cap\mathbb{T}_1)\times(\mathbb{R}_+\cap\mathbb{T}_2)$. From here,

$$\frac{v_{t_1}^{\Delta_1}(t_1,t_2)}{v(t_1,t_2)}-\frac{v_{t_1}^{\Delta_1}(t_1,0)}{v(t_1,0)}\leq\int_0^{t_2}h_1(t_1,s_2)\Delta_2s_2,$$

$(t_1,t_2)\in(\mathbb{R}_+\cap\mathbb{T}_1)\times(\mathbb{R}_+\cap\mathbb{T}_2)$, or

$$\frac{v_{t_1}^{\Delta_1}(t_1,t_2)}{v(t_1,t_2)}-\frac{z_{t_1}^{\Delta_1}(t_1,0)}{z(t_1,0)}\leq\int_0^{t_2}h_1(t_1,s_2)\Delta_2s_2,$$

$(t_1,t_2)\in(\mathbb{R}_+\cap\mathbb{T}_1)\times(\mathbb{R}_+\cap\mathbb{T}_2)$, or

$$\frac{v_{t_1}^{\Delta_1}(t_1,t_2)}{v(t_1,t_2)}-\frac{a^{\Delta_1}(t_1)}{a(t_1)+b(0)}\leq\int_0^{t_2}h_1(t_1,s_2)\Delta_2s_2,$$

$(t_1,t_2)\in(\mathbb{R}_+\cap\mathbb{T}_1)\times(\mathbb{R}_+\cap\mathbb{T}_2)$, or

$$\begin{aligned}\frac{v_{t_1}^{\Delta_1}(t_1,t_2)}{v(t_1,t_2)}&\leq\frac{a^{\Delta_1}(t_1)}{a(t_1)+b(0)}+\int_0^{t_2}h_1(t_1,s_2)\Delta_2s_2\\ &=h_2(t_1,t_2),\end{aligned}$$

$(t_1,t_2)\in(\mathbb{R}_+\cap\mathbb{T}_1)\times(\mathbb{R}_+\cap\mathbb{T}_2)$, or

$$v_{t_1}^{\Delta_1}(t_1,t_2)\leq h_2(t_1,t_2)v(t_1,t_2),$$

$(t_1,t_2)\in(\mathbb{R}_+\cap\mathbb{T}_1)\times(\mathbb{R}_+\cap\mathbb{T}_2)$. Hence, by Lemma 2.1.1, we obtain

$$v(t_1,t_2)\leq v(0,t_2)e_{h_2}(t_1,0)$$

$$
\begin{aligned}
&= z(0,t_2)e_{h_2}(t_1,0)\\
&= \big(a(0)+b(t_2)\big)e_{h_2}(t_1,0),\\
z^{\Delta_1\Delta_2}_{t_1t_2}(t_1,t_2) &\le f(t_1,t_2)\big(a(0)+b(t_2)\big)e_{h_2}(t_1,0)\\
&= h_3(t_1,t_2),\\
z^{\Delta_1}_{t_1}(t_1,t_2)-z^{\Delta_1}_{t_1}(t_1,0) &\le \int_0^{t_2} h_3(t_1,s_2)\Delta_2 s_2,\\
z^{\Delta_1}_{t_1}(t_1,t_2) &\le a^{\Delta_1}(t_1)+\int_0^{t_2} h_3(t_1,s_2)\Delta_2 s_2,\\
z(t_1,t_2)-z(0,t_2) &\le a(t_1)-a(0)+\int_0^{t_1}\int_0^{t_2} h_3(s_1,s_2)\Delta_2 s_2\Delta_1 s_1,\\
z(t_1,t_2) &\le a(0)+b(t_2)+a(t_1)-a(0)\\
&\quad+\int_0^{t_1}\int_0^{t_2} h_3(s_1,s_2)\Delta_2 s_2\Delta_1 s_1\\
&= a(t_1)+b(t_2)+\int_0^{t_1}\int_0^{t_2} h_3(s_1,s_2)\Delta_2 s_2\Delta_1 s_1,\\
u(t_1,t_2) &\le z(t_1,t_2)\\
&\le a(t_1)+b(t_2)+\int_0^{t_1}\int_0^{t_2} h_3(s_1,s_2)\Delta_2 s_2\Delta_1 s_1,
\end{aligned}
$$

$(t_1,t_2)\in(\mathbb{R}_+\cap\mathbb{T}_1)\times(\mathbb{R}_+\cap\mathbb{T}_2)$. This completes the proof. □

Theorem 6.2.7. *Let $u(t_1,t_2)$, $f(t_1,t_2)$, and $g(t_1,t_2)$ be nonnegative continuous functions for $(t_1,t_2)\in(\mathbb{R}_+\cap\mathbb{T}_1)\times(\mathbb{R}_+\cap\mathbb{T}_2)$ and let $c(t_1,t_2)$ be a positive continuous function for $(t_1,t_2)\in(\mathbb{R}_+\cap\mathbb{T}_1)\times(\mathbb{R}_+\cap\mathbb{T}_2)$ and nondecreasing in each variable t_1, t_2, $(t_1,t_2)\in(\mathbb{R}_+\cap\mathbb{T}_1)\times(\mathbb{R}_+\cap\mathbb{T}_2)$. If*

$$
\begin{aligned}
u(t_1,t_2) \le c(t_1,t_2)+\int_0^{t_1}\int_0^{t_2} f(s_1,s_2)\Bigg(&u(s_1,s_2)\\
&+\int_0^{s_1}\int_0^{s_2} g(\tau_1,\tau_2)u(\tau_1,\tau_2)\Delta_2\tau_2\Delta_1\tau_1\Bigg)\Delta_2 s_2\Delta_1 s_1,
\end{aligned}
$$

$(t_1,t_2)\in(\mathbb{R}_+\cap\mathbb{T}_1)\times(\mathbb{R}_+\cap\mathbb{T}_2)$, then

$$
u(t_1,t_2)\le c(t_1,t_2)\left(1+\int_0^{t_1}\int_0^{t_2} h_3(s_1,s_2)\Delta_2 s_2\Delta_1 s_1\right),
$$

$(t_1, t_2) \in (\mathbb{R}_+ \cap \mathbb{T}_1) \times (\mathbb{R}_+ \cap \mathbb{T}_2)$, *where*

$$h_1(t_1, t_2) = f(t_1, t_2) + g(t_1, t_2),$$
$$h_2(t_1, t_2) = \int_0^{t_2} h_1(t_1, s_2)\Delta_2 s_2,$$
$$h_3(t_1, t_2) = f(t_1, t_2)e_{h_2}(t_1, 0), \quad (t_1, t_2) \in (\mathbb{R}_+ \cap \mathbb{T}_1) \times (\mathbb{R}_+ \cap \mathbb{T}_2).$$

Proof. Since $c(t_1, t_2)$ is positive for $(t_1, t_2) \in (\mathbb{R}_+ \cap \mathbb{T}_1) \times (\mathbb{R}_+ \cap \mathbb{T}_2)$ and nondecreasing in each variable t_1 and t_2, $(t_1, t_2) \in (\mathbb{R}_+ \cap \mathbb{T}_1) \times (\mathbb{R}_+ \cap \mathbb{T}_2)$, we get

$$\begin{aligned}\frac{u(t_1, t_2)}{c(t_1, t_2)} &\le 1 + \int_0^{t_1}\int_0^{t_2} f(s_1, s_2)\Bigg(\frac{u(s_1, s_2)}{c(t_1, t_2)} \\ &\quad + \int_0^{s_1}\int_0^{s_2} g(\tau_1, \tau_2)\frac{u(\tau_1, \tau_2)}{c(t_1, t_2)}\Delta_2\tau_2\Delta_1\tau_1\Bigg)\Delta_2 s_2\Delta_1 s_1 \\ &\le 1 + \int_0^{t_1}\int_0^{t_2} f(s_1, s_2)\Bigg(\frac{u(s_1, s_2)}{c(s_1, s_2)} \\ &\quad + \int_0^{s_1}\int_0^{s_2} g(\tau_1, \tau_2)\frac{u(\tau_1, \tau_2)}{c(\tau_1, \tau_2)}\Delta_2\tau_2\Delta_1\tau_1\Bigg)\Delta_2 s_2\Delta_1 s_1,\end{aligned}$$

$(t_1, t_2) \in (\mathbb{R}_+ \cap \mathbb{T}_1) \times (\mathbb{R}_+ \cap \mathbb{T}_2)$. Let

$$z(t_1, t_2) = \frac{u(t_1, t_2)}{c(t_1, t_2)}, \quad (t_1, t_2) \in (\mathbb{R}_+ \cap \mathbb{T}_1) \times (\mathbb{R}_+ \cap \mathbb{T}_2).$$

Then

$$\begin{aligned}z(t_1, t_2) &\le 1 + \int_0^{t_1}\int_0^{t_2} f(s_1, s_2)\Bigg(z(s_1, s_2) \\ &\quad + \int_0^{s_1}\int_0^{s_2} g(\tau_1, \tau_2)z(\tau_1, \tau_2)\Delta_2\tau_2\Delta_1\tau_1\Bigg)\Delta_2 s_2\Delta_1 s_1,\end{aligned}$$

$(t_1, t_2) \in (\mathbb{R}_+ \cap \mathbb{T}_1) \times (\mathbb{R}_+ \cap \mathbb{T}_2)$. Hence, by Theorem 6.2.6, we obtain

$$z(t_1, t_2) \le 1 + \int_0^{t_1}\int_0^{t_2} h_3(s_1, s_2)\Delta_2 s_2\Delta_1 s_1,$$

$(t_1, t_2) \in (\mathbb{R}_+ \cap \mathbb{T}_1) \times (\mathbb{R}_+ \cap \mathbb{T}_2)$, whereupon

$$\frac{u(t_1, t_2)}{c(t_1, t_2)} \le 1 + \int_0^{t_1}\int_0^{t_2} h_3(s_1, s_2)\Delta_2 s_2\Delta_1 s_1, \quad (t_1, t_2) \in (\mathbb{R}_+ \cap \mathbb{T}_1) \times (\mathbb{R}_+ \cap \mathbb{T}_2),$$

or

$$u(t_1,t_2) \le c(t_1,t_2)\left(1+\int_0^{t_1}\int_0^{t_2} h_3(s_1,s_2)\Delta_2 s_2\Delta_1 s_1\right),$$

$(t_1,t_2) \in (\mathbb{R}_+ \cap \mathbb{T}_1) \times (\mathbb{R}_+ \cap \mathbb{T}_2)$. This completes the proof. □

Theorem 6.2.8. *Let $u(t_1,t_2)$, $f(t_1,t_2)$, $g(t_1,t_2)$, and $p(t_1,t_2)$ be nonnegative continuous functions for $(t_1,t_2) \in (\mathbb{R}_+ \cap \mathbb{T}_1) \times (\mathbb{R}_+ \cap \mathbb{T}_2)$, and u_0 be a positive constant. If*

$$\begin{aligned} u(t_1,t_2) &\le u_0 + \int_0^{t_1}\int_0^{t_2} (f(s_1,s_2)u(s_1,s_2) + p(s_1,s_2))\Delta_2 s_2\Delta_1 s_1 \\ &\quad + \int_0^{t_1}\int_0^{t_2} f(s_1,s_2)\left(\int_0^{s_1}\int_0^{s_2} g(\tau_1,\tau_2)u(\tau_1,\tau_2)\Delta_2\tau_2\Delta_1\tau_1\right)\Delta_2 s_2\Delta_1 s_1, \end{aligned}$$

$(t_1,t_2) \in (\mathbb{R}_+ \cap \mathbb{T}_1) \times (\mathbb{R}_+ \cap \mathbb{T}_2)$, then

$$u(t_1,t_2) \le \left(u_0 + \int_0^{t_1}\int_0^{t_2} p(s_1,s_2)\Delta_2 s_2\Delta_1 s_1\right)\left(1+\int_0^{t_1}\int_0^{t_2} h_3(s_1,s_2)\Delta_2 s_2\Delta_1 s_1\right),$$

$(t_1,t_2) \in (\mathbb{R}_+ \cap \mathbb{T}_1) \times (\mathbb{R}_+ \cap \mathbb{T}_2)$, where $h_1(t_1,t_2)$, $h_2(t_1,t_2)$ and $h_3(t_1,t_2)$ are defined as in Theorem 6.2.7.

Proof. Let

$$c(t_1,t_2) = u_0 + \int_0^{t_1}\int_0^{t_2} p(s_1,s_2)\Delta_2 s_2\Delta_1 s_1, \quad (t_1,t_2) \in (\mathbb{R}_+ \cap \mathbb{T}_1) \times (\mathbb{R}_+ \cap \mathbb{T}_2).$$

Then $c(t_1,t_2)$ is a positive continuous function for $(t_1,t_2) \in (\mathbb{R}_+ \cap \mathbb{T}_1) \times (\mathbb{R}_+ \cap \mathbb{T}_2)$, and $c(t_1,t_2)$ is a nondecreasing function in each variable t_1 and t_2, $(t_1,t_2) \in (\mathbb{R}_+ \cap \mathbb{T}_1) \times (\mathbb{R}_+ \cap \mathbb{T}_2)$. Then

$$\begin{aligned} u(t_1,t_2) &\le c(t_1,t_2) + \int_0^{t_1}\int_0^{t_2} f(s_1,s_2)\Bigg(u(s_1,s_2) \\ &\quad + \int_0^{s_1}\int_0^{s_2} g(\tau_1,\tau_2)u(\tau_1,\tau_2)\Delta_2\tau_2\Delta_1\tau_1\Bigg)\Delta_2 s_2\Delta_1 s_1, \end{aligned}$$

$(t_1,t_2) \in (\mathbb{R}_+ \cap \mathbb{T}_1) \times (\mathbb{R}_+ \cap \mathbb{T}_2)$. Hence, by Theorem 6.2.7, we get

$$u(t_1,t_2) \le c(t_1,t_2)\left(1+\int_0^{t_1}\int_0^{t_2} h_3(s_1,s_2)\Delta_2 s_2\Delta_1 s_1\right),$$

$(t_1,t_2) \in (\mathbb{R}_+ \cap \mathbb{T}_1) \times (\mathbb{R}_+ \cap \mathbb{T}_2)$. This completes the proof. □

Theorem 6.2.9. *Let $u(t_1,t_2)$, $f(t_1,t_2)$, $g(t_1,t_2)$, and $p(t_1,t_2)$ be nonnegative continuous functions for $(t_1,t_2) \in (\mathbb{R}_+ \cap \mathbb{T}_1) \times (\mathbb{R}_+ \cap \mathbb{T}_2)$, and u_0 be a positive constant. If*

$$u(t_1,t_2) \leq u_0 + \int_0^{t_1}\int_0^{t_2} f(s_1,s_2)u(s_1,s_2)\Delta_2 s_2 \Delta_1 s_1 + \int_0^{t_1}\int_0^{t_2} f(s_1,s_2)\left(\int_0^{s_1}\int_0^{s_2}(g(\tau_1,\tau_2)u(\tau_1,\tau_2) + p(\tau_1,\tau_2))\Delta_2\tau_2\Delta_1\tau_1\right)\Delta_2 s_2\Delta_1 s_1,$$

$(t_1,t_2) \in (\mathbb{R}_+ \cap \mathbb{T}_1) \times (\mathbb{R}_+ \cap \mathbb{T}_2)$, then

$$u(t_1,t_2) \leq \left(u_0 + \int_0^{t_1}\int_0^{t_2} f(s_1,s_2)\int_0^{s_1}\int_0^{s_2} p(\sigma_1,\sigma_2)\Delta_2\sigma_2\Delta_1\sigma_1\Delta_2 s_2\Delta_1 s_1\right) \times \left(1 + \int_0^{t_1}\int_0^{t_2} h_3(s_1,s_2)\Delta_2 s_2\Delta_1 s_1\right),$$

$(t_1,t_2) \in (\mathbb{R}_+ \cap \mathbb{T}_1) \times (\mathbb{R}_+ \cap \mathbb{T}_2)$, where $h_1(t_1,t_2)$, $h_2(t_1,t_2)$, and $h_3(t_1,t_2)$ are defined as in Theorem 6.2.7.

Proof. Let

$$c(t_1,t_2) = u_0 + \int_0^{t_1}\int_0^{t_2} f(s_1,s_2)\int_0^{s_1}\int_0^{s_2} p(\tau_1,\tau_2)\Delta_2\tau_2\Delta_1\tau_1\Delta_2 s_2\Delta_1 s_1,$$

$(t_1,t_2) \in (\mathbb{R}_+ \cap \mathbb{T}_1) \times (\mathbb{R}_+ \cap \mathbb{T}_2)$. Then $c(t_1,t_2)$ is a positive continuous function for $(t_1,t_2) \in (\mathbb{R}_+ \cap \mathbb{T}_1) \times (\mathbb{R}_+ \cap \mathbb{T}_2)$, and $c(t_1,t_2)$ is a nondecreasing function in each variable t_1 and t_2, $(t_1,t_2) \in (\mathbb{R}_+ \cap \mathbb{T}_1) \times (\mathbb{R}_+ \cap \mathbb{T}_2)$. Then

$$u(t_1,t_2) \leq c(t_1,t_2) + \int_0^{t_1}\int_0^{t_2} f(s_1,s_2)\left(u(s_1,s_2) + \int_0^{s_1}\int_0^{s_2} g(\tau_1,\tau_2)u(\tau_1,\tau_2)\Delta_2\tau_2\Delta_1\tau_1\right)\Delta_2 s_2\Delta_1 s_1,$$

$(t_1,t_2) \in (\mathbb{R}_+ \cap \mathbb{T}_1) \times (\mathbb{R}_+ \cap \mathbb{T}_2)$. Hence, by Theorem 6.2.7, we get

$$u(t_1,t_2) \leq c(t_1,t_2)\left(1 + \int_0^{t_1}\int_0^{t_2} h_3(s_1,s_2)\Delta_2 s_2\Delta_1 s_1\right),$$

$(t_1,t_2) \in (\mathbb{R}_+ \cap \mathbb{T}_1) \times (\mathbb{R}_+ \cap \mathbb{T}_2)$. This completes the proof. □

Theorem 6.2.10. *Let $u(t_1,t_2)$, $f(t_1,t_2)$, $g(t_1,t_2)$, and $h(t_1,t_2)$ be nonnegative continuous functions for $(t_1,t_2) \in (\mathbb{R}_+ \cap \mathbb{T}_1) \times (\mathbb{R}_+ \cap \mathbb{T}_2)$, u_0 be a positive constant. If*

$$\begin{aligned} u(t_1,t_2) &\le u_0 + \int_0^{t_1}\int_0^{t_2} h(s_1,s_2)u(s_1,s_2)\Delta_2 s_2\Delta_1 s_1 \\ &\quad + \int_0^{t_1}\int_0^{t_2} f(s_1,s_2)\left(u(s_1,s_2) \right. \\ &\quad \left. + \int_0^{s_1}\int_0^{s_2} g(\tau_1,\tau_2)u(\tau_1,\tau_2)\Delta_2\tau_2\Delta_1\tau_1 \right)\Delta_2 s_2\Delta_1 s_1, \end{aligned}$$

$(t_1,t_2) \in (\mathbb{R}_+ \cap \mathbb{T}_1) \times (\mathbb{R}_+ \cap \mathbb{T}_2)$. Then

$$u(t_1,t_2) \le u_0\left(1 + \int_0^{t_1}\int_0^{t_2} h_3(s_1,s_2)\Delta_2 s_2\Delta_1 s_1 \right),$$

$(t_1,t_2) \in (\mathbb{R}_+ \cap \mathbb{T}_1) \times (\mathbb{R}_+ \cap \mathbb{T}_2)$, where

$$\begin{aligned} f_1(t_1,t_2) &= 2(f(t_1,t_2) + g(t_1,t_2)), \\ h_1(t_1,t_2) &= f_1(t_1,t_2) + g(t_1,t_2), \\ h_2(t_1,t_2) &= \int_0^{t_2} h_1(t_1,s_2)\Delta_2 s_2, \\ h_3(t_1,t_2) &= f_1(t_1,t_2)e_{h_2}(t_1,0), \quad (t_1,t_2) \in (\mathbb{R}_+ \cap \mathbb{T}_1) \times (\mathbb{R}_+ \cap \mathbb{T}_2). \end{aligned}$$

Proof. We have

$$\begin{aligned} u(t_1,t_2) &\le u_0 + \int_0^{t_1}\int_0^{t_2} (h(s_1,s_2) + f(s_1,s_2))u(s_1,s_2)\Delta_2 s_2\Delta_1 s_1 \\ &\quad + \int_0^{t_1}\int_0^{t_2} (f(s_1,s_2) + h(s_1,s_2))\left(u(s_1,s_2) \right. \\ &\quad \left. + \int_0^{s_1}\int_0^{s_2} g(\tau_1,\tau_2)u(\tau_1,\tau_2)\Delta_2\tau_2\Delta_1\tau_1 \right)\Delta_2 s_2\Delta_1 s_1 \\ &\le u_0 + \int_0^{t_1}\int_0^{t_2} 2(h(s_1,s_2) + f(s_1,s_2))\left(u(s_1,s_2) \right. \\ &\quad \left. + \int_0^{s_1}\int_0^{s_2} g(\tau_1,\tau_2)u(\tau_1,\tau_2)\Delta_2\tau_2\Delta_1\tau_1 \right)\Delta_2 s_2\Delta_1 s_1 \end{aligned}$$

$$= u_0 + \int_0^{t_1}\int_0^{t_2} f_1(s_1,s_2)\Bigg(u(s_1,s_2) + \int_0^{s_1}\int_0^{s_2} g(\tau_1,\tau_2)u(\tau_1,\tau_2)\Delta_2\tau_2\Delta_1\tau_1 \Bigg)\Delta_2 s_2\Delta_1 s_1,$$

$(t_1,t_2) \in (\mathbb{R}_+ \cap \mathbb{T}_1) \times (\mathbb{R}_+ \cap \mathbb{T}_2)$. Hence, by Theorem 6.2.7, we get

$$u(t_1,t_2) \le u_0\Bigg(1 + \int_0^{t_1}\int_0^{t_2} h_3(s_1,s_2)\Delta_2 s_2\Delta_1 s_1 \Bigg),$$

$(t_1,t_2) \in (\mathbb{R}_+ \cap \mathbb{T}_1) \times (\mathbb{R}_+ \cap \mathbb{T}_2)$. This completes the proof. □

Theorem 6.2.11. *Let $u(t_1,t_2)$, $h(t_1,t_2)$, $p(t_1,t_2)$, $f(t_1,t_2)$, and $g(t_1,t_2)$ be nonnegative continuous functions for $(t_1,t_2) \in (\mathbb{R}_+ \cap \mathbb{T}_1) \times (\mathbb{R}_+ \cap \mathbb{T}_2)$, and*

$$u(t_1,t_2) \le h(t_1,t_2) + p(t_1,t_2)\int_0^{t_1}\int_0^{t_2} f(s_1,s_2)\Bigg(u(s_1,s_2) + p(s_1,s_2) \times \int_0^{s_1}\int_0^{s_2} g(\tau_1,\tau_2)u(\tau_1,\tau_2)\Delta_2\tau_2\Delta_1\tau_1 \Bigg)\Delta_2 s_2\Delta_1 s_1,$$

$(t_1,t_2) \in (\mathbb{R}_+ \cap \mathbb{T}_1) \times (\mathbb{R}_+ \cap \mathbb{T}_2)$. Then

$$u(t_1,t_2) \le h(t_1,t_2) + p(t_1,t_2)(1 + h(t_1,t_2)) \times \Bigg(1 + \int_0^{t_1}\int_0^{t_2} h_4(s_1,s_2)\Delta_2 s_2\Delta_1 s_1 \Bigg),$$

$(t_1,t_2) \in (\mathbb{R}_+ \cap \mathbb{T}_1) \times (\mathbb{R}_+ \cap \mathbb{T}_2)$, where

$$h_1(t_1,t_2) = \int_0^{t_1}\int_0^{t_2}\Bigg(f(s_1,s_2)h(s_1,s_2) + p(s_1,s_2)\int_0^{s_1}\int_0^{s_2} g(\tau_1,\tau_2)h(\tau_1,\tau_2)\Delta_2\tau_2\Delta_1\tau_1 \Bigg)\Delta_2 s_2\Delta_1 s_1,$$

$$p_1(t_1,t_2) = g(t_1,t_2)p(t_1,t_2),$$

$$f_1(t_1,t_2) = (1 + f(t_1,t_2))p(t_1,t_2),$$

$$h_2(t_1,t_2) = f_1(t_1,t_2) + p_1(t_1,t_2),$$

$$h_3(t_1,t_2) = \int_0^{t_2} h_2(t_1,s_2)\Delta_2 s_2,$$

$$h_4(t_1,t_2) = f_1(t_1,t_2)e_{h_3}(t_1,0),$$

$(t_1,t_2) \in (\mathbb{R}_+ \cap \mathbb{T}_1) \times (\mathbb{R}_+ \cap \mathbb{T}_2)$.

Proof. Let

$$z(t_1,t_2) = \int_0^{t_1}\int_0^{t_2} f(s_1,s_2)\Bigg(u(s_1,s_2) + p(s_1,s_2) \times \int_0^{s_1}\int_0^{s_2} g(\tau_1,\tau_2)u(\tau_1,\tau_2)\Delta_2\tau_2\Delta_1\tau_1 \Bigg)\Delta_2 s_2\Delta_1 s_1,$$

$(t_1,t_2) \in (\mathbb{R}_+ \cap \mathbb{T}_1) \times (\mathbb{R}_+ \cap \mathbb{T}_2)$. Then

$$u(t_1,t_2) \le h(t_1,t_2) + p(t_1,t_2)z(t_1,t_2), \tag{6.8}$$

$(t_1,t_2) \in (\mathbb{R}_+ \cap \mathbb{T}_1) \times (\mathbb{R}_+ \cap \mathbb{T}_2)$. Hence,

$$\begin{aligned} z(t_1,t_2) &\le \int_0^{t_1}\int_0^{t_2} f(s_1,s_2)\Bigg(h(s_1,s_2) + p(s_1,s_2)z(s_1,s_2) \\ &\quad + p(s_1,s_2)\int_0^{s_1}\int_0^{s_2} g(\tau_1,\tau_2)(h(\tau_1,\tau_2) \\ &\quad + p(\tau_1,\tau_2)z(\tau_1,\tau_2))\Delta_2\tau_2\Delta_1\tau_1 \Bigg)\Delta_2 s_2\Delta_1 s_1 \\ &= \int_0^{t_1}\int_0^{t_2}\Bigg(f(s_1,s_2)h(s_1,s_2) + p(s_1,s_2) \\ &\quad \times \int_0^{s_1}\int_0^{s_2} g(\tau_1,\tau_2)h(\tau_1,\tau_2)\Delta_2\tau_2\Delta_1\tau_1 \Bigg) \\ &\quad + \int_0^{t_1}\int_0^{t_2}\Bigg(f(s_1,s_2)p(s_1,s_2)z(s_1,s_2) + p(s_1,s_2) \\ &\quad \times \int_0^{s_1}\int_0^{s_2} g(\tau_1,\tau_2)p(\tau_1,\tau_2)z(\tau_1,\tau_2)\Delta_2\tau_2\Delta_1\tau_1 \Bigg)\Delta_2 s_2\Delta_1 s_1 \\ &= h_1(t_1,t_2) + \int_0^{t_1}\int_0^{t_2} (f(s_1,s_2)+1)p(s_1,s_2)\Bigg(z(s_1,s_2) \\ &\quad + \int_0^{s_1}\int_0^{s_2} p_1(\tau_1,\tau_2)z(\tau_1,\tau_2)\Delta_2\tau_2\Delta_1\tau_1 \Bigg)\Delta_2 s_2\Delta_1 s_1 \end{aligned}$$

$$= 1 + h_1(t_1,t_2) + \int_0^{t_1}\int_0^{t_2} f_1(s_1,s_2)\Bigg(z(s_1,s_2) + \int_0^{s_1}\int_0^{s_2} p_1(\tau_1,\tau_2)z(\tau_1,\tau_2)\Delta_2\tau_2\Delta_1\tau_1 \Bigg)\Delta_2 s_2\Delta_1 s_1,$$

$(t_1,t_2) \in (\mathbb{R}_+\cap\mathbb{T}_1)\times(\mathbb{R}_+\cap\mathbb{T}_2)$. Since the function $1+h_1(t_1,t_2)$ is a nondecreasing function for $(t_1,t_2) \in (\mathbb{R}_+\cap\mathbb{T}_1)\times(\mathbb{R}_+\cap\mathbb{T}_2)$, we can apply Theorem 6.2.7 and obtain

$$z(t_1,t_2) \le (1 + h_1(t_1,t_2))\Bigg(1 + \int_0^{t_1}\int_0^{t_2} h_4(s_1,s_2)\Delta_2 s_2\Delta_1 s_1 \Bigg),$$

$(t_1,t_2) \in (\mathbb{R}_+\cap\mathbb{T}_1)\times(\mathbb{R}_+\cap\mathbb{T}_2)$. From here and (6.8), we get

$$u(t_1,t_2) \le h(t_1,t_2) + p(t_1,t_2)(1 + h(t_1,t_2)) \times \Bigg(1 + \int_0^{t_1}\int_0^{t_2} h_4(s_1,s_2)\Delta_2 s_2\Delta_1 s_1 \Bigg),$$

$(t_1,t_2) \in (\mathbb{R}_+\cap\mathbb{T}_1)\times(\mathbb{R}_+\cap\mathbb{T}_2)$. This completes the proof. □

Theorem 6.2.12 (Pachpatte's Inequality). *Let $u(t_1,t_2)$, $h(t_1,t_2)$, and $p(t_1,t_2)$ be nonnegative continuous functions for $(t_1,t_2) \in (\mathbb{R}_+\cap\mathbb{T}_1)\times(\mathbb{R}_+\cap\mathbb{T}_2)$, $f(t_1,t_2)$ be a positive continuous function for $(t_1,t_2) \in (\mathbb{R}_+\cap\mathbb{T}_1)\times(\mathbb{R}_+\cap\mathbb{T}_2)$, and suppose u_0 is a nonnegative constant. If*

$$u(t_1,t_2) \le u_0 + \int_0^{t_1}\int_0^{t_2} f(s_1,s_2)\Bigg(h(s_1,s_2) + \int_0^{s_1}\int_0^{s_2} p(\tau_1,\tau_2)u(\tau_1,\tau_2)\Delta_2\tau_2\Delta_1\tau_1 \Bigg)\Delta_2 s_2\Delta_1 s_1,$$

$(t_1,t_2) \in (\mathbb{R}_+\cap\mathbb{T}_1)\times(\mathbb{R}_+\cap\mathbb{T}_2)$, then

$$u(t_1,t_2) \le q(t_1,t_2)e_{q_2}(t_1,0), \quad (t_1,t_2) \in (\mathbb{R}_+\cap\mathbb{T}_1)\times(\mathbb{R}_+\cap\mathbb{T}_2),$$

where

$$q(t_1,t_2) = u_0 + \int_0^{t_1}\int_0^{t_2} f(s_1,s_2)h(s_1,s_2)\Delta_2 s_2\Delta_1 s_1,$$

$$q_1(t_1,t_2) = f(t_1,t_2)\int_0^{t_1}\int_0^{t_2} p(\tau_1,\tau_2)\Delta_2\tau_2\Delta_1\tau_1,$$

$$q_2(t_1,t_2) = \int_0^{t_2} q_1(t_1,s_2)\Delta_2 s_2, \quad (t_1,t_2) \in (\mathbb{R}_+\cap\mathbb{T}_1)\times(\mathbb{R}_+\cap\mathbb{T}_2).$$

Proof. We can rewrite the given inequality in the form

$$\begin{aligned} u(t_1,t_2) &\le u_0 + \int_0^{t_1}\int_0^{t_2} f(s_1,s_2)h(s_1,s_2)\Delta_2 s_2\Delta_1 s_1 \\ &\quad + \int_0^{t_1}\int_0^{t_2} f(s_1,s_2)\int_0^{s_1}\int_0^{s_2} p(\tau_1,\tau_2)u(\tau_1,\tau_2)\Delta_2\tau_2\Delta_1\tau_1\Delta_2 s_2\Delta_1 s_1 \\ &= q(t_1,t_2) \\ &\quad + \int_0^{t_1}\int_0^{t_2} f(s_1,s_2)\int_0^{s_1}\int_0^{s_2} p(\tau_1,\tau_2)u(\tau_1,\tau_2)\Delta_2\tau_2\Delta_1\tau_1\Delta_2 s_2\Delta_1 s_1, \end{aligned}$$

$(t_1,t_2) \in (\mathbb{R}_+ \cap \mathbb{T}_1) \times (\mathbb{R}_+ \cap \mathbb{T}_2)$. Note that $q(t_1,t_2)$ is a positive nondecreasing function in each variable t_1, t_2, $(t_1,t_2) \in (\mathbb{R}_+ \cap \mathbb{T}_1) \times (\mathbb{R}_+ \cap \mathbb{T}_2)$. Then

$$\begin{aligned} \frac{u(t_1,t_2)}{q(t_1,t_2)} &\le 1 + \int_0^{t_1}\int_0^{t_2} f(s_1,s_2)\int_0^{s_1}\int_0^{s_2} p(\tau_1,\tau_2)\frac{u(\tau_1,\tau_2)}{q(t_1,t_2)}\Delta_2\tau_2\Delta_1\tau_1\Delta_2 s_2\Delta_1 s_1 \\ &\le 1 + \int_0^{t_1}\int_0^{t_2} f(s_1,s_2)\int_0^{s_1}\int_0^{s_2} p(\tau_1,\tau_2)\frac{u(\tau_1,\tau_2)}{q(\tau_1,\tau_2)}\Delta_2\tau_2\Delta_1\tau_1\Delta_2 s_2\Delta_1 s_1, \end{aligned}$$

$(t_1,t_2) \in (\mathbb{R}_+ \cap \mathbb{T}_1) \times (\mathbb{R}_+ \cap \mathbb{T}_2)$. Let

$$z(t_1,t_2) = \frac{u(t_1,t_2)}{q(t_1,t_2)}, \quad (t_1,t_2) \in (\mathbb{R}_+ \cap \mathbb{T}_1) \times (\mathbb{R}_+ \cap \mathbb{T}_2).$$

Then

$$z(t_1,t_2) \le 1 + \int_0^{t_1}\int_0^{t_2} f(s_1,s_2)\int_0^{s_1}\int_0^{s_2} p(\tau_1,\tau_2)z(\tau_1,\tau_2)\Delta_2\tau_2\Delta_1\tau_1\Delta_2 s_2\Delta_1 s_1,$$

$(t_1,t_2) \in (\mathbb{R}_+ \cap \mathbb{T}_1) \times (\mathbb{R}_+ \cap \mathbb{T}_2)$. We set

$$v(t_1,t_2) = 1 + \int_0^{t_1}\int_0^{t_2} f(s_1,s_2)\int_0^{s_1}\int_0^{s_2} p(\tau_1,\tau_2)z(\tau_1,\tau_2)\Delta_2\tau_2\Delta_1\tau_1\Delta_2 s_2\Delta_1 s_1,$$

$(t_1,t_2) \in (\mathbb{R}_+ \cap \mathbb{T}_1) \times (\mathbb{R}_+ \cap \mathbb{T}_2)$. Hence,

$$\begin{aligned} z(t_1,t_2) &\le v(t_1,t_2), \\ v(t_1,0) &= v(0,t_2) \\ &= v(0,0) \\ &= 1, \end{aligned}$$

$$
\begin{aligned}
v_{t_1}^{\Delta_1}(t_1,t_2) &= \int_0^{t_2} f(t_1,s_2)\int_0^{t_1}\int_0^{s_2} p(\tau_1,\tau_2)z(\tau_1,\tau_2)\Delta_2\tau_2\Delta_1\tau_1\Delta_2 s_2,\\
v_{t_1}^{\Delta_1}(t_1,0) &= 0,\\
v_{t_2}^{\Delta_2}(t_1,t_2) &= \int_0^{t_1} f(s_1,t_2)\int_0^{s_1}\int_0^{t_2} p(\tau_1,\tau_2)z(\tau_1,\tau_2)\Delta_2\tau_2\Delta_1\tau_1\Delta_1 s_1,\\
v_{t_2}^{\Delta_2}(0,t_2) &= 0,\\
v_{t_1t_2}^{\Delta_1\Delta_2}(t_1,t_2) &= f(t_1,t_2)\int_0^{t_1}\int_0^{t_2} p(s_1,s_2)z(s_1,s_2)\Delta_2 s_2\Delta_1 s_1,\\
\frac{v_{t_1t_2}^{\Delta_1\Delta_2}(t_1,t_2)}{f(t_1,t_2)} &= \int_0^{t_1}\int_0^{t_2} p(s_1,s_2)z(s_1,s_2)\Delta_2 s_2\Delta_1 s_1,\\
\left(\frac{v_{t_1t_2}^{\Delta_1\Delta_2}(t_1,t_2)}{f(t_1,t_2)}\right)_{t_1}^{\Delta_1} &= \int_0^{t_2} p(t_1,s_2)z(t_1,s_2)\Delta_2 s_2,\\
\left(\frac{v_{t_1t_2}^{\Delta_1\Delta_2}(t_1,t_2)}{f(t_1,t_2)}\right)_{t_1t_2}^{\Delta_1\Delta_2} &= p(t_1,t_2)z(t_1,t_2)\\
&\le p(t_1,t_2)v(t_1,t_2),\\
\frac{\left(\frac{v_{t_1t_2}^{\Delta_1\Delta_2}(t_1,t_2)}{f(t_1,t_2)}\right)_{t_1t_2}^{\Delta_1\Delta_2}}{v(t_1,t_2)} &\le p(t_1,t_2),\\
\frac{\left(\frac{v_{t_1t_2}^{\Delta_1\Delta_2}(t_1,t_2)}{f(t_1,t_2)}\right)_{t_1t_2}^{\Delta_1\Delta_2}v(t_1,t_2)}{v(t_1,t_2)v(t_1,\sigma_2(t_2))} &\le \frac{\left(\frac{v_{t_1t_2}^{\Delta_1\Delta_2}(t_1,t_2)}{f(t_1,t_2)}\right)_{t_1t_2}^{\Delta_1\Delta_2}v(t_1,t_2)}{(v(t_1,t_2))^2}\\
&\le p(t_1,t_2)
\end{aligned}
$$

$$
\begin{aligned}
&\le p(t_1,t_2)\\
&\qquad + \frac{\left(\frac{v_{t_1t_2}^{\Delta_1\Delta_2}(t_1,t_2)}{f(t_1,t_2)}\right)_{t_1}^{\Delta_1}v_{t_2}^{\Delta_2}(t_1,t_2)}{v(t_1,t_2)v(t_1,\sigma_2(t_2))},\\
\left(\frac{\left(\frac{v_{t_1t_2}^{\Delta_1\Delta_2}(t_1,t_2)}{f(t_1,t_2)}\right)_{t_1}^{\Delta_1}}{v(t_1,t_2)}\right)_{t_2}^{\Delta_2} &\le p(t_1,t_2),\\
\frac{\left(\frac{v_{t_1t_2}^{\Delta_1\Delta_2}(t_1,t_2)}{f(t_1,t_2)}\right)_{t_1}^{\Delta_1}}{v(t_1,t_2)} &\le \int_0^{t_2} p(t_1,s_2)\Delta_2 s_2,\\
\frac{\left(\frac{v_{t_1t_2}^{\Delta_1\Delta_2}(t_1,t_2)}{f(t_1,t_2)}\right)_{t_1}^{\Delta_1}v(t_1,t_2)}{v(t_1,t_2)v(\sigma_1(t_1),t_2)} &\le \frac{\left(\frac{v_{t_1t_2}^{\Delta_1\Delta_2}(t_1,t_2)}{f(t_1,t_2)}\right)_{t_1}^{\Delta_1}v(t_1,t_2)}{(v(t_1,t_2))^2}
\end{aligned}
$$

$$
\begin{aligned}
&\le \int_0^{t_2} p(t_1,s_2)\Delta_2 s_2 \\
&\le \int_0^{t_2} p(t_1,s_2)\Delta_2 s_2 \\
&\quad + \frac{\frac{v_{t_1t_2}^{\Delta_1\Delta_2}(t_1,t_2)}{f(t_1,t_2)} v_{t_1}^{\Delta_1}(t_1,t_2)}{v(t_1,t_2)v(\sigma_1(t_1),t_2)}, \\
\left(\frac{v_{t_1t_2}^{\Delta_1\Delta_2}(t_1,t_2)}{f(t_1,t_2)v(t_1,t_2)}\right)_{t_1}^{\Delta_1} &\le \int_0^{t_2} p(t_1,s_2)\Delta_2 s_2, \\
\frac{v_{t_1t_2}^{\Delta_1\Delta_2}(t_1,t_2)}{f(t_1,t_2)v(t_1,t_2)} &\le \int_0^{t_1}\int_0^{t_2} p(s_1,s_2)\Delta_2 s_2\Delta_1 s_1, \\
\frac{v_{t_1t_2}^{\Delta_1\Delta_2}(t_1,t_2)}{v(t_1,t_2)} &\le f(t_1,t_2)\int_0^{t_1}\int_0^{t_2} p(s_1,s_2)\Delta_2 s_2\Delta_1 s_1, \\
\frac{v_{t_1t_2}^{\Delta_1\Delta_2}(t_1,t_2)v(t_1,t_2)}{v(t_1,t_2)v(t_1,\sigma_2(t_2))} &\le \frac{v_{t_1t_2}^{\Delta_1\Delta_2}(t_1,t_2)v(t_1,t_2)}{(v(t_1,t_2))^2} \\
&\le f(t_1,t_2)\int_0^{t_1}\int_0^{t_2} p(s_1,s_2)\Delta_2 s_2\Delta_1 s_1 \\
&\le f(t_1,t_2)\int_0^{t_1}\int_0^{t_2} p(s_1,s_2)\Delta_2 s_2\Delta_1 s_1 \\
&\quad + \frac{v_{t_1}^{\Delta_1}(t_1,t_2)v_{t_2}^{\Delta_2}(t_1,t_2)}{v(t_1,t_2)v(t_1,\sigma_2(t_2))}, \\
\left(\frac{v_{t_1}^{\Delta_1}(t_1,t_2)}{v(t_1,t_2)}\right)_{t_2}^{\Delta_2} &\le q_1(t_1,t_2), \\
\frac{v_{t_1}^{\Delta_1}(t_1,t_2)}{v(t_1,t_2)} &\le \int_0^{t_2} q_1(t_1,s_2)\Delta_2 s_2 \\
&= q_2(t_1,t_2), \\
v_{t_1}^{\Delta_1}(t_1,t_2) &\le q_2(t_1,t_2)v(t_1,t_2),
\end{aligned}
$$

$(t_1,t_2) \in (\mathbb{R}_+ \cap \mathbb{T}_1) \times (\mathbb{R}_+ \cap \mathbb{T}_2)$. From the last inequality and Lemma 2.1.1, we conclude that

$$
v(t_1,t_2) \le e_{q_2}(t_1,0), \quad (t_1,t_2) \in (\mathbb{R}_+ \cap \mathbb{T}_1) \times (\mathbb{R}_+ \cap \mathbb{T}_2).
$$

Therefore

$$z(t_1,t_2) \le e_{q_2}(t_1,0),$$
$$u(t_1,t_2) \le q(t_1,t_2)e_{q_2}(t_1,0), \quad (t_1,t_2) \in (\mathbb{R}_+ \cap \mathbb{T}_1) \times (\mathbb{R}_+ \cap \mathbb{T}_2).$$

This completes the proof. □

Theorem 6.2.13. *Let $u(t_1,t_2)$, $h(t_1,t_2)$, and $p(t_1,t_2)$ be nonnegative continuous functions for $(t_1,t_2) \in (\mathbb{R}_+ \cap \mathbb{T}_1) \times (\mathbb{R}_+ \cap \mathbb{T}_2)$, $f(t_1,t_2)$ and $g(t_1,t_2)$ be positive continuous functions for $(t_1,t_2) \in (\mathbb{R}_+ \cap \mathbb{T}_1) \times (\mathbb{R}_+ \cap \mathbb{T}_2)$, and suppose u_0 is a positive constant. If*

$$u(t_1,t_2) \le u_0 + \int_0^{t_1}\int_0^{t_2} f(s_1,s_2)\Bigg(h(s_1,s_2) + \int_0^{s_1}\int_0^{s_2} g(r_1,r_2) \times \left(\int_0^{r_1}\int_0^{r_2} p(\tau_1,\tau_2)u(\tau_1,\tau_2)\Delta_2\tau_2\Delta_1\tau_1\right) \Delta_2 r_2\Delta_1 r_1 \Bigg)\Delta_2 s_2\Delta_1 s_1,$$

$(t_1,t_2) \in (\mathbb{R}_+ \cap \mathbb{T}_1) \times (\mathbb{R}_+ \cap \mathbb{T}_2)$, then

$$u(t_1,t_2) \le q(t_1,t_2)e_{h_3}(t_1,0), \quad (t_1,t_2) \in (\mathbb{R}_+ \cap \mathbb{T}_1) \times (\mathbb{R}_+ \cap \mathbb{T}_2),$$

where

$$q(t_1,t_2) = u_0 + \int_0^{t_1}\int_0^{t_2} f(s_1,s_2)h(s_1,s_2)\Delta_2 s_2\Delta_1 s_1,$$
$$h_1(t_1,t_2) = g(t_1,t_2)\int_0^{t_1}\int_0^{t_2} p(s_1,s_2)\Delta_2 s_2\Delta_1 s_1,$$
$$h_2(t_1,t_2) = f(t_1,t_2)\int_0^{t_1}\int_0^{t_2} h_1(s_1,s_2)\Delta_2 s_2\Delta_1 s_1,$$
$$h_3(t_1,t_2) = \int_0^{t_2} h_2(t_1,s_2)\Delta_2 s_2,$$

$(t_1,t_2) \in (\mathbb{R}_+ \cap \mathbb{T}_1) \times (\mathbb{R}_+ \cap \mathbb{T}_2)$.

Proof. We can rewrite the given inequality in the following form:

$$u(t_1,t_2) \le u_0 + \int_0^{t_1}\int_0^{t_2} f(s_1,s_2)h(s_1,s_2)\Delta_2 s_2\Delta_1 s_1$$

$$
\begin{aligned}
&+\int_0^{t_1}\int_0^{t_2} f(s_1,s_2)\int_0^{s_1}\int_0^{s_2} g(r_1,r_2)\left(\int_0^{r_1}\int_0^{r_2} p(\tau_1,\tau_2)u(\tau_1,\tau_2)\right.\\
&\left.\Delta_2\tau_2\Delta_1\tau_1\right)\Delta_2 r_2\Delta_1 r_1\Delta_2 s_2\Delta_1 s_1\\
&= q(t_1,t_2)\\
&+\int_0^{t_1}\int_0^{t_2} f(s_1,s_2)\int_0^{s_1}\int_0^{s_2} g(r_1,r_2)\left(\int_0^{r_1}\int_0^{r_2} p(\tau_1,\tau_2)u(\tau_1,\tau_2)\right.\\
&\left.\Delta_2\tau_2\Delta_1\tau_1\right)\Delta_2 r_2\Delta_1 r_1\Delta_2 s_2\Delta_1 s_1,
\end{aligned}
$$

$(t_1,t_2)\in(\mathbb{R}_+\cap\mathbb{T}_1)\times(\mathbb{R}_+\cap\mathbb{T}_2)$. Note that $q(t_1,t_2)$ is a positive nondecreasing function in each variable t_1, t_2, $(t_1,t_2)\in(\mathbb{R}_+\cap\mathbb{T}_1)\times(\mathbb{R}_+\cap\mathbb{T}_2)$. Then

$$
\begin{aligned}
\frac{u(t_1,t_2)}{q(t_1,t_2)} &\le 1\\
&+\int_0^{t_1}\int_0^{t_2} f(s_1,s_2)\int_0^{s_1}\int_0^{s_2} g(r_1,r_2)\left(\int_0^{r_1}\int_0^{r_2} p(\tau_1,\tau_2)\frac{u(\tau_1,\tau_2)}{q(t_1,t_2)}\right.\\
&\left.\Delta_2\tau_2\Delta_1\tau_1\right)\Delta_2 r_2\Delta_1 r_1\Delta_2 s_2\Delta_1 s_1\\
&\le 1\\
&+\int_0^{t_1}\int_0^{t_2} f(s_1,s_2)\int_0^{s_1}\int_0^{s_2} g(r_1,r_2)\left(\int_0^{r_1}\int_0^{r_2} p(\tau_1,\tau_2)\frac{u(\tau_1,\tau_2)}{q(\tau_1,\tau_2)}\right.\\
&\left.\Delta_2\tau_2\Delta_1\tau_1\right)\Delta_2 r_2\Delta_1 r_1\Delta_2 s_2\Delta_1 s_1,
\end{aligned}
$$

$(t_1,t_2)\in(\mathbb{R}_+\cap\mathbb{T}_1)\times(\mathbb{R}_+\cap\mathbb{T}_2)$. Let

$$
z(t_1,t_2)=\frac{u(t_1,t_2)}{q(t_1,t_2)},\quad (t_1,t_2)\in(\mathbb{R}_+\cap\mathbb{T}_1)\times(\mathbb{R}_+\cap\mathbb{T}_2).
$$

Hence,

$$
\begin{aligned}
z(t_1,t_2) &\le 1\\
&+\int_0^{t_1}\int_0^{t_2} f(s_1,s_2)\int_0^{s_1}\int_0^{s_2} g(r_1,r_2)\left(\int_0^{r_1}\int_0^{r_2} p(\tau_1,\tau_2)z(\tau_1,\tau_2)\right.\\
&\left.\Delta_2\tau_2\Delta_1\tau_1\right)\Delta_2 r_2\Delta_1 r_1\Delta_2 s_2\Delta_1 s_1,
\end{aligned}
$$

$(t_1, t_2) \in (\mathbb{R}_+ \cap \mathbb{T}_1) \times (\mathbb{R}_+ \cap \mathbb{T}_2)$. Now we set

$$\begin{aligned} v(t_1, t_2) &= 1 \\ &\quad + \int_0^{t_1}\int_0^{t_2} f(s_1, s_2) \int_0^{s_1}\int_0^{s_2} g(r_1, r_2)\left(\int_0^{r_1}\int_0^{r_2} p(\tau_1, \tau_2) z(\tau_1, \tau_2)\right. \\ &\quad \left. \Delta_2\tau_2\Delta_1\tau_1\right)\Delta_2\sigma_2\Delta_1 r_1\Delta_2 s_2\Delta_1 s_1, \end{aligned}$$

$(t_1, t_2) \in (\mathbb{R}_+ \cap \mathbb{T}_1) \times (\mathbb{R}_+ \cap \mathbb{T}_2)$. Then

$$\begin{aligned} z(t_1, t_2) &\le v(t_1, t_2), \\ v(t_1, 0) &= v(0, t_2) \\ &= v(0, 0) \\ &= 1, \\ v_{t_1}^{\Delta_1}(t_1, t_2) &= \int_0^{t_2} f(t_1, s_2) \int_0^{t_1}\int_0^{s_2} g(r_1, r_2)\left(\int_0^{r_1}\int_0^{r_2} p(\tau_1, \tau_2) z(\tau_1, \tau_2)\right. \\ &\quad \left. \Delta_2\tau_2\Delta_1\tau_1\right)\Delta_2 r_2\Delta_1 r_1\Delta_2 s_2, \\ v_{t_1}^{\Delta_1}(t_1, 0) &= 0, \\ v_{t_2}^{\Delta_2}(t_1, t_2) &= \int_0^{t_1} f(s_1, t_2) \int_0^{s_1}\int_0^{t_2} g(r_1, r_2)\left(\int_0^{r_1}\int_0^{r_2} p(\tau_1, \tau_2) z(\tau_1, \tau_2)\right. \\ &\quad \left. \Delta_2\tau_2\Delta_1\tau_1\right)\Delta_2 r_2\Delta_1 r_1\Delta_1 s_1, \\ v_{t_2}^{\Delta_2}(0, t_2) &= 0, \quad (t_1, t_2) \in (\mathbb{R}_+ \cap \mathbb{T}_1) \times (\mathbb{R}_+ \cap \mathbb{T}_2). \end{aligned}$$

Note that

$$v_{t_1}^{\Delta_1}(t_1, t_2) \ge 0, \quad v_{t_2}^{\Delta_2}(t_1, t_2) \ge 0, \quad (t_1, t_2) \in (\mathbb{R}_+ \cap \mathbb{T}_1) \times (\mathbb{R}_+ \cap \mathbb{T}_2).$$

Therefore $v(t_1, t_2)$ is a positive nondecreasing function in each variable t_1, t_2, $(t_1, t_2) \in (\mathbb{R}_+ \cap \mathbb{T}_1) \times (\mathbb{R}_+ \cap \mathbb{T}_2)$. Consequently,

$$\begin{aligned} v(t_1, t_2) &\le v(t_1, \sigma_2(t_2)), \\ v(t_1, t_2) &\le v(\sigma_1(t_1), t_2), \quad (t_1, t_2) \in (\mathbb{R}_+ \cap \mathbb{T}_1) \times (\mathbb{R}_+ \cap \mathbb{T}_2). \end{aligned} \tag{6.9}$$

Next we get

$$\begin{aligned} v_{t_1 t_2}^{\Delta_1\Delta_2}(t_1, t_2) &= f(t_1, t_2) \int_0^{t_1}\int_0^{t_2} g(s_1, s_2) \int_0^{s_1}\int_0^{s_2} p(r_1, r_2) z(r_1, r_2) \\ &\quad \Delta_2 r_2\Delta_1 r_1\Delta_2 s_2\Delta_1 s_1, \end{aligned}$$

$$
\begin{aligned}
\frac{v^{\Delta_1\Delta_2}_{t_1t_2}(t_1,t_2)}{f(t_1,t_2)} &= \int_0^{t_1}\int_0^{t_2} g(s_1,s_2)\int_0^{s_1}\int_0^{s_2} p(r_1,r_2)z(r_1,r_2)\\
&\quad \Delta_2r_2\Delta_1r_1\Delta_2s_2\Delta_1s_1,\\
\frac{v^{\Delta_1\Delta_2}_{t_1t_2}(t_1,0)}{f(t_1,0)} &= \frac{v^{\Delta_1\Delta_2}_{t_1t_2}(0,t_2)}{f(0,t_2)}\\
&= \frac{v^{\Delta_1\Delta_2}_{t_1t_2}(0,0)}{f(0,0)}\\
&= 0,\\
\left(\frac{v^{\Delta_1\Delta_2}_{t_1t_2}(t_1,t_2)}{f(t_1,t_2)}\right)^{\Delta_1}_{t_1} &= \int_0^{t_2} g(t_1,s_2)\int_0^{t_1}\int_0^{s_2} p(r_1,r_2)z(r_1,r_2)\\
&\quad \Delta_2r_2\Delta_1r_1\Delta_2s_2,\\
\left(\frac{v^{\Delta_1\Delta_2}_{t_1t_2}(t_1,t_2)}{f(t_1,t_2)}\right)^{\Delta_1\Delta_2}_{t_1t_2} &= g(t_1,t_2)\int_0^{t_1}\int_0^{t_2} p(r_1,r_2)z(r_1,r_2)\Delta_2r_2\Delta_1r_1,\\
\frac{\left(\frac{v^{\Delta_1\Delta_2}_{t_1t_2}(t_1,t_2)}{f(t_1,t_2)}\right)^{\Delta_1\Delta_2}_{t_1t_2}}{g(t_1,t_2)} &= \int_0^{t_1}\int_0^{t_2} p(r_1,r_2)z(r_1,r_2)\Delta_2r_2\Delta_1r_1,\\
\left(\frac{\left(\frac{v^{\Delta_1\Delta_2}_{t_1t_2}(t_1,t_2)}{f(t_1,t_2)}\right)^{\Delta_1\Delta_2}_{t_1t_2}}{g(t_1,t_2)}\right)^{\Delta_1}_{t_1} &= \int_0^{t_2} p(t_1,r_2)z(t_1,r_2)\Delta_2r_2,\\
\left(\frac{\left(\frac{v^{\Delta_1\Delta_2}_{t_1t_2}(t_1,t_2)}{f(t_1,t_2)}\right)^{\Delta_1\Delta_2}_{t_1t_2}}{g(t_1,t_2)}\right)^{\Delta_1\Delta_2}_{t_1t_2} &= p(t_1,t_2)z(t_1,t_2)\\
&\le p(t_1,t_2)v(t_1,t_2),\\
\frac{\left(\frac{\left(\frac{v^{\Delta_1\Delta_2}_{t_1t_2}(t_1,t_2)}{f(t_1,t_2)}\right)^{\Delta_1\Delta_2}_{t_1t_2}}{g(t_1,t_2)}\right)^{\Delta_1\Delta_2}_{t_1t_2}}{v(t_1,t_2)} &\le p(t_1,t_2),
\end{aligned}
$$

$(t_1,t_2) \in (\mathbb{R}_+ \cap \mathbb{T}_1) \times (\mathbb{R}_+ \cap \mathbb{T}_2)$. Hence, using (6.9), we get

$$
\begin{aligned}
\frac{\left(\frac{\left(\frac{v^{\Delta_1\Delta_2}_{t_1t_2}(t_1,t_2)}{f(t_1,t_2)}\right)^{\Delta_1\Delta_2}_{t_1t_2}}{g(t_1,t_2)}\right)^{\Delta_1\Delta_2}_{t_1t_2} v(t_1,t_2)}{v(t_1,t_2)v(t_1,\sigma_2(t_2))} &\le \frac{\left(\frac{\left(\frac{v^{\Delta_1\Delta_2}_{t_1t_2}(t_1,t_2)}{f(t_1,t_2)}\right)^{\Delta_1\Delta_2}_{t_1t_2}}{g(t_1,t_2)}\right)^{\Delta_1\Delta_2}_{t_1t_2}}{v(t_1,t_2)}\\
&\le p(t_1,t_2)\\
&\le p(t_1,t_2)\\
&\quad + \frac{\left(\frac{\left(\frac{v^{\Delta_1\Delta_2}_{t_1t_2}(t_1,t_2)}{f(t_1,t_2)}\right)^{\Delta_1\Delta_2}_{t_1t_2}}{g(t_1,t_2)}\right)^{\Delta_1}_{t_1} v^{\Delta_2}_{t_2}(t_1,t_2)}{v(t_1,t_2)v(t_1,\sigma_2(t_2))},
\end{aligned}
$$

$(t_1, t_2) \in (\mathbb{R}_+ \cap \mathbb{T}_1) \times (\mathbb{R}_+ \cap \mathbb{T}_2)$, whereupon

$$\left(\frac{\left(\frac{\left(\frac{v_{t_1 t_2}^{\Delta_1 \Delta_2}(t_1,t_2)}{f(t_1,t_2)} \right)_{t_1 t_2}^{\Delta_1 \Delta_2}}{g(t_1,t_2)} \right)_{t_1}^{\Delta_1}}{v(t_1,t_2)} \right)_{t_2}^{\Delta_2} \le p(t_1,t_2),$$

$$\frac{\left(\frac{\left(\frac{v_{t_1 t_2}^{\Delta_1 \Delta_2}(t_1,t_2)}{f(t_1,t_2)} \right)_{t_1 t_2}^{\Delta_1 \Delta_2}}{g(t_1,t_2)} \right)_{t_1}^{\Delta_1}}{v(t_1,t_2)} \le \int_0^{t_2} p(t_1,s_2)\Delta_2 s_2,$$

$(t_1, t_2) \in (\mathbb{R}_+ \cap \mathbb{T}_1) \times (\mathbb{R}_+ \cap \mathbb{T}_2)$. From here, using (6.9), we obtain

$$\begin{aligned}
\frac{\left(\frac{\left(\frac{v_{t_1 t_2}^{\Delta_1 \Delta_2}(t_1,t_2)}{f(t_1,t_2)} \right)_{t_1 t_2}^{\Delta_1 \Delta_2}}{g(t_1,t_2)} \right)_{t_1}^{\Delta_1} v(t_1,t_2)}{v(t_1,t_2)v(r_1(t_1),t_2)} &\le \frac{\left(\frac{\left(\frac{v_{t_1 t_2}^{\Delta_1 \Delta_2}(t_1,t_2)}{f(t_1,t_2)} \right)_{t_1 t_2}^{\Delta_1 \Delta_2}}{g(t_1,t_2)} \right)_{t_1}^{\Delta_1}}{v(t_1,t_2)} \\
&\le \int_0^{t_2} p(t_1,s_2)\Delta_2 s_2 \\
&\le \int_0^{t_2} p(t_1,s_2)\Delta_2 s_2 \\
&\quad + \frac{\left(\frac{\left(\frac{v_{t_1 t_2}^{\Delta_1 \Delta_2}(t_1,t_2)}{f(t_1,t_2)} \right)_{t_1 t_2}^{\Delta_1 \Delta_2}}{g(t_1,t_2)} \right) v_{t_1}^{\Delta_1}(t_1,t_2)}{v(t_1,t_2)v(\sigma_1(t_1),t_2)},
\end{aligned}$$

$(t_1, t_2) \in (\mathbb{R}_+ \cap \mathbb{T}_1) \times (\mathbb{R}_+ \cap \mathbb{T}_2)$. Therefore

$$\left(\frac{\left(\frac{v_{t_1 t_2}^{\Delta_1 \Delta_2}(t_1,t_2)}{f(t_1,t_2)} \right)_{t_1 t_2}^{\Delta_1 \Delta_2}}{g(t_1,t_2)v(t_1,t_2)} \right)_{t_1}^{\Delta_1} \le \int_0^{t_2} p(t_1,s_2)\Delta_2 s_2,$$

$$\frac{\left(\frac{v_{t_1 t_2}^{\Delta_1 \Delta_2}(t_1,t_2)}{f(t_1,t_2)} \right)_{t_1 t_2}^{\Delta_1 \Delta_2}}{g(t_1,t_2)v(t_1,t_2)} \le \int_0^{t_1}\int_0^{t_2} p(s_1,s_2)\Delta_2 s_2 \Delta_1 s_1,$$

$$\begin{aligned}
\frac{\left(\frac{v_{t_1 t_2}^{\Delta_1 \Delta_2}(t_1,t_2)}{f(t_1,t_2)} \right)_{t_1 t_2}^{\Delta_1 \Delta_2}}{v(t_1,t_2)} &\le g(t_1,t_2) \int_0^{t_1}\int_0^{t_2} p(s_1,s_2)\Delta_2 s_2 \Delta_1 s_1 \\
&= h_1(t_1,t_2),
\end{aligned}$$

$(t_1, t_2) \in (\mathbb{R}_+ \cap \mathbb{T}_1) \times (\mathbb{R}_+ \cap \mathbb{T}_2)$. Using (6.9), we obtain

$$\frac{\left(\frac{v_{t_1 t_2}^{\Delta_1 \Delta_2}(t_1,t_2)}{f(t_1,t_2)} \right)_{t_1 t_2}^{\Delta_1 \Delta_2} v(t_1,t_2)}{v(t_1,t_2)v(t_1,\sigma_2(t_2))} \le \frac{\left(\frac{v_{t_1 t_2}^{\Delta_1 \Delta_2}(t_1,t_2)}{f(t_1,t_2)} \right)_{t_1 t_2}^{\Delta_1 \Delta_2}}{v(t_1,t_2)}$$

$$
\begin{aligned}
&\le h_1(t_1,t_2)\\
&\le h_1(t_1,t_2)\\
&\quad+\frac{\left(\frac{v_{t_1t_2}^{\Delta_1\Delta_2}(t_1,t_2)}{f(t_1,t_2)}\right)_{t_1}^{\Delta_1} v_{t_2}^{\Delta_2}(t_1,t_2)}{v(t_1,t_2)v(t_1,\sigma_2(t_2))},
\end{aligned}
$$

$(t_1,t_2)\in(\mathbb{R}_+\cap\mathbb{T}_1)\times(\mathbb{R}_+\cap\mathbb{T}_2)$. Hence,

$$
\left(\frac{\left(\frac{v_{t_1t_2}^{\Delta_1\Delta_2}(t_1,t_2)}{f(t_1,t_2)}\right)_{t_1}^{\Delta_1}}{v(t_1,t_2)}\right)_{t_2}^{\Delta_2}\le h_1(t_1,t_2),
$$

$$
\frac{\left(\frac{v_{t_1t_2}^{\Delta_1\Delta_2}(t_1,t_2)}{f(t_1,t_2)}\right)_{t_1}^{\Delta_1}}{v(t_1,t_2)}\le\int_0^{t_2}h_1(t_1,s_2)\Delta_2s_2,
$$

$(t_1,t_2)\in(\mathbb{R}_+\cap\mathbb{T}_1)\times(\mathbb{R}_+\cap\mathbb{T}_2)$. Now we use (6.9) and find

$$
\begin{aligned}
\frac{\left(\frac{v_{t_1t_2}^{\Delta_1\Delta_2}(t_1,t_2)}{f(t_1,t_2)}\right)_{t_1}^{\Delta_1}v(t_1,t_2)}{v(t_1,t_2)v(\sigma_1(t_1),t_2)}&\le\frac{\left(\frac{v_{t_1t_2}^{\Delta_1\Delta_2}(t_1,t_2)}{f(t_1,t_2)}\right)_{t_1}^{\Delta_1}}{v(t_1,t_2)}\\
&\le\int_0^{t_2}h_1(t_1,s_2)\Delta_2s_2\\
&\le\int_0^{t_2}h_1(t_1,s_2)\Delta_2s_2\\
&\quad+\frac{\left(\frac{v_{t_1t_2}^{\Delta_1\Delta_2}(t_1,t_2)}{f(t_1,t_2)}\right)v_{t_1}^{\Delta_1}(t_1,t_2)}{v(t_1,t_2)v(\sigma_1(t_1),t_2)},
\end{aligned}
$$

$(t_1,t_2)\in(\mathbb{R}_+\cap\mathbb{T}_1)\times(\mathbb{R}_+\cap\mathbb{T}_2)$. Then

$$
\left(\frac{v_{t_1t_2}^{\Delta_1\Delta_2}(t_1,t_2)}{v(t_1,t_2)f(t_1,t_2)}\right)_{t_1}^{\Delta_1}\le\int_0^{t_2}h_1(t_1,s_2)\Delta_2s_2,
$$

$$
\frac{v_{t_1t_2}^{\Delta_1\Delta_2}(t_1,t_2)}{f(t_1,t_2)v(t_1,t_2)}\le\int_0^{t_1}\int_0^{t_2}h_1(s_1,s_2)\Delta_2s_2\Delta_1s_1,
$$

$$
\begin{aligned}
\frac{v_{t_1t_2}^{\Delta_1\Delta_2}(t_1,t_2)}{v(t_1,t_2)}&\le f(t_1,t_2)\int_0^{t_1}\int_0^{t_2}h_1(s_1,s_2)\Delta_2s_2\Delta_1s_1\\
&=h_2(t_1,t_2),
\end{aligned}
$$

$(t_1,t_2)\in(\mathbb{R}_+\cap\mathbb{T}_1)\times(\mathbb{R}_+\cap\mathbb{T}_2)$. Using (6.9), we get

$$
\frac{v_{t_1t_2}^{\Delta_1\Delta_2}(t_1,t_2)v(t_1,t_2)}{v(t_1,t_2)v(t_1,\sigma_2(t_2))}\le\frac{v_{t_1t_2}^{\Delta_1\Delta_2}(t_1,t_2)}{v(t_1,t_2)}
$$

$$\leq h_2(t_1,t_2)$$
$$\leq h_2(t_1,t_2) + \frac{v_{t_1}^{\Delta_1}(t_1,t_2)v_{t_2}^{\Delta_2}(t_1,t_2)}{v(t_1,t_2)v(t_1,\sigma_2(t_2))},$$

$(t_1,t_2) \in (\mathbb{R}_+ \cap \mathbb{T}_1) \times (\mathbb{R}_+ \cap \mathbb{T}_2)$. Therefore

$$\left(\frac{v_{t_1}^{\Delta_1}(t_1,t_2)}{v(t_1,t_2)}\right)_{t_2}^{\Delta_2} \leq h_2(t_1,t_2),$$
$$\frac{v_{t_1}^{\Delta_1}(t_1,t_2)}{v(t_1,t_2)} \leq \int_0^{t_2} h_2(t_1,s_2)\Delta_2 s_2$$
$$= h_3(t_1,t_2), \quad (t_1,t_2) \in (\mathbb{R}_+ \cap \mathbb{T}_1) \times (\mathbb{R}_+ \cap \mathbb{T}_2),$$

or

$$v_{t_1}^{\Delta_1}(t_1,t_2) \leq h_3(t_1,t_2)v(t_1,t_2), \quad (t_1,t_2) \in (\mathbb{R}_+ \cap \mathbb{T}_1) \times (\mathbb{R}_+ \cap \mathbb{T}_2).$$

Hence, by Lemma 2.1.1, we obtain

$$v(t_1,t_2) \leq e_{h_3}(t_1,0), \quad (t_1,t_2) \in (\mathbb{R}_+ \cap \mathbb{T}_1) \times (\mathbb{R}_+ \cap \mathbb{T}_2),$$

and

$$\frac{u(t_1,t_2)}{q(t_1,t_2)} = z(t_1,t_2)$$
$$\leq v(t_1,t_2)$$
$$\leq e_{h_3}(t_1,0), \quad (t_1,t_2) \in (\mathbb{R}_+ \cap \mathbb{T}_1) \times (\mathbb{R}_+ \cap \mathbb{T}_2),$$

or

$$u(t_1,t_2) \leq q(t_1,t_2)e_{h_3}(t_1,0), \quad (t_1,t_2) \in (\mathbb{R}_+ \cap \mathbb{T}_1) \times (\mathbb{R}_+ \cap \mathbb{T}_2).$$

This completes the proof. □

Theorem 6.2.14. *Let $u(t_1,t_2)$ and $h(t_1,t_2)$ be positive continuous functions for $(t_1,t_2) \in (\mathbb{R}_+ \cap \mathbb{T}_1) \times (\mathbb{R}_+ \cap \mathbb{T}_2)$, $b(t_2)$, $p(t_1)$ and $q(t_2)$ be positive twice continuously-differentiable functions for $(t_1,t_2) \in (\mathbb{R}_+ \cap \mathbb{T}_1) \times (\mathbb{R}_+ \cap \mathbb{T}_2)$, $a^{\Delta_1}(t_1)$, $p^{\Delta_1}(t_1)$ and $b^{\Delta_2}(t_2)$, $q^{\Delta_2}(t_2)$ be nonnegative functions for $(t_1,t_2) \in (\mathbb{R}_+ \cap \mathbb{T}_1) \times (\mathbb{R}_+ \cap \mathbb{T}_2)$, and $a^{\Delta_1\Delta_1}(t_1)$ and $p^{\Delta_1\Delta_1}(t_1)$ be positive functions for $t_1 \in \mathbb{R}_+ \cap \mathbb{T}_1$. If*

$$u(t_1,t_2) \leq a(t_1) + b(t_2) + t_2 p(t_1) + t_1 q(t_2)$$
$$+ \int_0^{t_1}\int_0^{s_1}\int_0^{t_2}\int_0^{s_2} h(r_1,r_2)u(r_1,r_2)\Delta_2 r_2 \Delta_2 s_2 \Delta_1 r_1 \Delta_1 s_1,$$

$(t_1, t_2) \in (\mathbb{R}_+ \cap \mathbb{T}_1) \times (\mathbb{R}_+ \cap \mathbb{T}_2)$, *then*

$$u(t_1, t_2) \le (a(0) + b(t_2) + t_2 p(0)) e_{h_3}(t_1, 0), \quad (t_1, t_2) \in (\mathbb{R}_+ \cap \mathbb{T}_1) \times (\mathbb{R}_+ \cap \mathbb{T}_2),$$

where

$$\begin{aligned} h_1(t_1, t_2) &= \frac{p^{\Delta_1 \Delta_1}(t_1)}{a(t_1) + b(0) + t_1 q(0)} + \int_0^{t_2} h(t_1, s_2) \Delta_2 s_2, \\ h_2(t_1, t_2) &= \int_0^{t_2} h_1(t_1, s_2) \Delta_2 s_2 + \frac{a^{\Delta_1 \Delta_1}(t_1)}{a(t_1) + b(0) + t_1 q(0)}, \\ h_3(t_1, t_2) &= \frac{a^{\Delta_1}(0) + t_2 p^{\Delta_1}(0) + q(t_2)}{a(0) + b(t_2) + t_1 q(0)} \\ &\quad + \int_0^{t_1} h_2(s_1, t_2) \Delta_1 s_1, \end{aligned}$$

$(t_1, t_2) \in (\mathbb{R}_+ \cap \mathbb{T}_1) \times (\mathbb{R}_+ \cap \mathbb{T}_2)$.

Proof. Let

$$\begin{aligned} z(t_1, t_2) &= a(t_1) + b(t_2) + t_2 p(t_1) + t_1 q(t_2) \\ &\quad + \int_0^{t_1} \int_0^{s_1} \int_0^{t_2} \int_0^{s_2} h(r_1, r_2) u(r_1, r_2) \Delta_2 r_2 \Delta_2 s_2 \Delta_1 r_1 \Delta_1 s_1, \end{aligned}$$

$(t_1, t_2) \in (\mathbb{R}_+ \cap \mathbb{T}_1) \times (\mathbb{R}_+ \cap \mathbb{T}_2)$. Then

$$\begin{aligned} u(t_1, t_2) &\le z(t_1, t_2), \\ u(t_1, 0) &= a(t_1) + b(0) + t_1 q(0), \\ u(0, t_2) &= a(0) + b(t_2) + t_2 p(0), \\ z_{t_1}^{\Delta_1}(t_1, t_2) &= a^{\Delta_1}(t_1) + t_2 p^{\Delta_1}(t_1) + q(t_2) \\ &\quad + \int_0^{t_1} \int_0^{t_2} \int_0^{s_2} h(r_1, r_2) u(r_1, r_2) \Delta_2 r_2 \Delta_2 s_2 \Delta_1 r_1, \\ z_{t_1}^{\Delta_1}(t_1, 0) &= a^{\Delta_1}(t_1) + q(0), \\ z_{t_1}^{\Delta_1}(0, t_2) &= a^{\Delta_1}(0) + t_2 p^{\Delta_1}(0) + q(t_2), \\ z_{t_1 t_1}^{\Delta_1 \Delta_1}(t_1, t_2) &= a^{\Delta_1 \Delta_1}(t_1) + t_2 p^{\Delta_1 \Delta_1}(t_1) \\ &\quad + \int_0^{t_2} \int_0^{s_2} h(t_1, r_2) u(t_1, r_2) \Delta_2 r_2 \Delta_2 s_2, \end{aligned}$$

$$\begin{aligned}
z^{\Delta_1\Delta_1}_{t_1t_1}(t_1,0) &= a^{\Delta_1\Delta_1}(t_1),\\
z^{\Delta_1\Delta_1}_{t_1t_1}(0,t_2) &= a^{\Delta_1\Delta_1}(0) + t_2 p^{\Delta_1\Delta_1}(0)\\
&\quad + \int_0^{t_2}\int_0^{s_2} h(0,r_2)u(0,r_2)\Delta_2 r_2\Delta_2 s_2,\\
z^{\Delta_1\Delta_1\Delta_2}_{t_1t_1t_2}(t_1,t_2) &= p^{\Delta_1\Delta_1}(t_1) + \int_0^{t_2} h(t_1,r_2)u(t_1,r_2)\Delta_2 r_2,\\
z^{\Delta_1\Delta_1\Delta_2}_{t_1t_1t_2}(t_1,0) &= p^{\Delta_1\Delta_1}(t_1),\\
z^{\Delta_1\Delta_1\Delta_2\Delta_2}_{t_1t_1t_2t_2}(t_1,t_2) &= h(t_1,t_2)u(t_1,t_2)\\
&\le h(t_1,t_2)z(t_1,t_2),\\
\frac{z^{\Delta_1\Delta_1\Delta_2\Delta_2}_{t_1t_1t_2t_2}(t_1,t_2)}{z(t_1,t_2)} &\le h(t_1,t_2),
\end{aligned}$$

$(t_1,t_2) \in (\mathbb{R}_+ \cap \mathbb{T}_1) \times (\mathbb{R}_+ \cap \mathbb{T}_2)$. Since $z(t_1,t_2)$ is a positive nondecreasing function in each variable t_1, t_2, $(t_1,t_2) \in (\mathbb{R}_+ \cap \mathbb{T}_1) \times (\mathbb{R}_+ \cap \mathbb{T}_2)$, we get

$$\begin{aligned}
\frac{z^{\Delta_1\Delta_1\Delta_2\Delta_2}_{t_1t_1t_2t_2}(t_1,t_2)z(t_1,t_2)}{z(t_1,t_2)z(t_1,\sigma_2(t_2))} &\le \frac{z^{\Delta_1\Delta_1\Delta_2\Delta_2}_{t_1t_1t_2t_2}(t_1,t_2)}{z(t_1,t_2)}\\
&\le h(t_1,t_2)\\
&\le h(t_1,t_2)\\
&\quad + \frac{z^{\Delta_1\Delta_1\Delta_2}_{t_1t_1t_2}(t_1,t_2)z^{\Delta_2}_{t_2}(t_1,t_2)}{z(t_1,t_2)z(t_1,\sigma_2(t_2))},
\end{aligned}$$

$(t_1,t_2) \in (\mathbb{R}_+ \cap \mathbb{T}_1) \times (\mathbb{R}_+ \cap \mathbb{T}_2)$. Hence,

$$\left(\frac{z^{\Delta_1\Delta_1\Delta_2}_{t_1t_1t_2}(t_1,t_2)}{z(t_1,t_2)}\right)^{\Delta_2}_{t_2} \le h(t_1,t_2),$$

$(t_1,t_2) \in (\mathbb{R}_+ \cap \mathbb{T}_1) \times (\mathbb{R}_+ \cap \mathbb{T}_2)$. Then

$$\frac{z^{\Delta_1\Delta_1\Delta_2}_{t_1t_1t_2}(t_1,t_2)}{z(t_1,t_2)} - \frac{z^{\Delta_1\Delta_1\Delta_2}_{t_1t_1t_2}(t_1,0)}{z(t_1,0)} \le \int_0^{t_2} h(t_1,s_2)\Delta_2 s_2,$$

$(t_1,t_2) \in (\mathbb{R}_+ \cap \mathbb{T}_1) \times (\mathbb{R}_+ \cap \mathbb{T}_2)$, or

$$\frac{z^{\Delta_1\Delta_1\Delta_2}_{t_1t_1t_2}(t_1,t_2)}{z(t_1,t_2)} - \frac{p^{\Delta_1\Delta_1}(t_1)}{a(t_1)+b(0)+t_1 q(0)} \le \int_0^{t_2} h(t_1,s_2)\Delta_2 s_2,$$

$(t_1, t_2) \in (\mathbb{R}_+ \cap \mathbb{T}_1) \times (\mathbb{R}_+ \cap \mathbb{T}_2)$, or

$$\begin{aligned}\frac{z_{t_1 t_1 t_2}^{\Delta_1 \Delta_1 \Delta_2}(t_1, t_2)}{z(t_1, t_2)} &\leq \frac{p^{\Delta_1 \Delta_1}(t_1)}{a(t_1) + b(0) + t_1 q(0)} + \int_0^{t_2} h(t_1, s_2)\Delta_2 s_2 \\ &= h_1(t_1, t_2),\end{aligned}$$

$(t_1, t_2) \in (\mathbb{R}_+ \cap \mathbb{T}_1) \times (\mathbb{R}_+ \cap \mathbb{T}_2)$. Now, since $z(t_1, t_2)$ is a positive nondecreasing function in t_2, $t_2 \in \mathbb{R}_+ \cap \mathbb{T}_2$, we get

$$\begin{aligned}\frac{z_{t_1 t_1 t_2}^{\Delta_1 \Delta_1 \Delta_2}(t_1, t_2) z(t_1, t_2)}{z(t_1, t_2) z(t_1, \sigma_2(t_2))} &\leq \frac{z_{t_1 t_1 t_2}^{\Delta_1 \Delta_1 \Delta_2}(t_1, t_2)}{z(t_1, t_2)} \\ &\leq h_1(t_1, t_2) \\ &\leq h_1(t_1, t_2) \\ &\quad + \frac{z_{t_1 t_1}^{\Delta_1 \Delta_1}(t_1, t_2) z_{t_2}^{\Delta_2}(t_1, t_2)}{z(t_1, t_2) z(t_1, \sigma_2(t_2))},\end{aligned}$$

$(t_1, t_2) \in (\mathbb{R}_+ \cap \mathbb{T}_1) \times (\mathbb{R}_+ \cap \mathbb{T}_2)$, whereupon

$$\left(\frac{z_{t_1 t_1}^{\Delta_1 \Delta_1}(t_1, t_2)}{z(t_1, t_2)} \right)_{t_2}^{\Delta_2} \leq h_1(t_1, t_2),$$

$(t_1, t_2) \in (\mathbb{R}_+ \cap \mathbb{T}_1) \times (\mathbb{R}_+ \cap \mathbb{T}_2)$, or

$$\frac{z_{t_1 t_1}^{\Delta_1 \Delta_1}(t_1, t_2)}{z(t_1, t_2)} - \frac{z_{t_1 t_1}^{\Delta_1 \Delta_1}(t_1, 0)}{z(t_1, 0)} \leq \int_0^{t_2} h_1(t_1, s_2)\Delta_2 s_2,$$

$(t_1, t_2) \in (\mathbb{R}_+ \cap \mathbb{T}_1) \times (\mathbb{R}_+ \cap \mathbb{T}_2)$, or

$$\frac{z_{t_1 t_1}^{\Delta_1 \Delta_1}(t_1, t_2)}{z(t_1, t_2)} - \frac{a^{\Delta_1 \Delta_1}(t_1)}{a(t_1) + b(0) + t_1 q(0)} \leq \int_0^{t_2} h_1(t_1, s_2)\Delta_2 s_2,$$

$(t_1, t_2) \in (\mathbb{R}_+ \cap \mathbb{T}_1) \times (\mathbb{R}_+ \cap \mathbb{T}_2)$, or

$$\begin{aligned}\frac{z_{t_1 t_1}^{\Delta_1 \Delta_1}(t_1, t_2)}{z(t_1, t_2)} &\leq \int_0^{t_2} h_1(t_1, s_2)\Delta_2 s_2 + \frac{a^{\Delta_1 \Delta_1}(t_1)}{a(t_1) + b(0) + t_1 q(0)} \\ &= h_2(t_1, t_2),\end{aligned}$$

$(t_1, t_2) \in (\mathbb{R}_+ \cap \mathbb{T}_1) \times (\mathbb{R}_+ \cap \mathbb{T}_2)$. Now we use that $z(t_1, t_2)$ is a positive nondecreasing function in t_1, $t_1 \in \mathbb{R}_+ \cap \mathbb{T}_1$, and find

$$\frac{z_{t_1 t_1}^{\Delta_1 \Delta_1}(t_1, t_2) z(t_1, t_2)}{z(t_1, t_2) z(\sigma_1(t_1), t_2)} \leq \frac{z_{t_1 t_1}^{\Delta_1 \Delta_1}(t_1, t_2)}{z(t_1, t_2)}$$

$$\begin{aligned} &\leq h_2(t_1,t_2) \\ &\leq h_2(t_1,t_2) \\ &\quad + \frac{(z_{t_1}^{\Delta_1}(t_1,t_2))^2}{z(t_1,t_2)z(\sigma_1(t_1),t_2)}, \end{aligned}$$

$(t_1,t_2) \in (\mathbb{R}_+ \cap \mathbb{T}_1) \times (\mathbb{R}_+ \cap \mathbb{T}_2)$, whereupon

$$\left(\frac{z_{t_1}^{\Delta_1}(t_1,t_2)}{z(t_1,t_2)}\right)_{t_1}^{\Delta_1} \leq h_2(t_1,t_2),$$

$(t_1,t_2) \in (\mathbb{R}_+ \cap \mathbb{T}_1) \times (\mathbb{R}_+ \cap \mathbb{T}_2)$, or

$$\frac{z_{t_1}^{\Delta_1}(t_1,t_2)}{z(t_1,t_2)} - \frac{z_{t_1}^{\Delta_1}(0,t_2)}{z(0,t_2)} \leq \int_0^{t_1} h_2(s_1,t_2)\Delta_1 s_1,$$

$(t_1,t_2) \in (\mathbb{R}_+ \cap \mathbb{T}_1) \times (\mathbb{R}_+ \cap \mathbb{T}_2)$, or

$$\frac{z_{t_1}^{\Delta_1}(t_1,t_2)}{z(t_1,t_2)} - \frac{a^{\Delta_1}(0) + t_2 p^{\Delta_1}(0) + q(t_2)}{a(0) + b(t_2) + t_1 q(0)} \leq \int_0^{t_1} h_2(s_1,t_2)\Delta_1 s_1,$$

$(t_1,t_2) \in (\mathbb{R}_+ \cap \mathbb{T}_1) \times (\mathbb{R}_+ \cap \mathbb{T}_2)$

$$\begin{aligned} \frac{z_{t_1}^{\Delta_1}(t_1,t_2)}{z(t_1,t_2)} &\leq \frac{a^{\Delta_1}(0) + t_2 p^{\Delta_1}(0) + q(t_2)}{a(0) + b(t_2) + t_1 q(0)} + \int_0^{t_1} h_2(s_1,t_2)\Delta_1 s_1 \\ &= h_3(t_1,t_2), \end{aligned}$$

$(t_1,t_2) \in (\mathbb{R}_+ \cap \mathbb{T}_1) \times (\mathbb{R}_+ \cap \mathbb{T}_2)$, or

$$z_{t_1}^{\Delta_1}(t_1,t_2) \leq h_3(t_1,t_2)z(t_1,t_2),$$

$(t_1,t_2) \in (\mathbb{R}_+ \cap \mathbb{T}_1) \times (\mathbb{R}_+ \cap \mathbb{T}_2)$. By the last inequality and Lemma 2.1.1, we obtain

$$\begin{aligned} z(t_1,t_2) &\leq z(0,t_2)e_{h_3}(t_1,0) \\ &= (a(0) + b(t_2) + t_2 p(0))e_{h_3}(t_1,0), \end{aligned}$$

$(t_1,t_2) \in (\mathbb{R}_+ \cap \mathbb{T}_1) \times (\mathbb{R}_+ \cap \mathbb{T}_2)$. Consequently,

$$\begin{aligned} u(t_1,t_2) &\leq z(t_1,t_2) \\ &= (a(0) + b(t_2) + t_2 p(0))e_{h_3}(t_1,0), \end{aligned}$$

$(t_1,t_2) \in (\mathbb{R}_+ \cap \mathbb{T}_1) \times (\mathbb{R}_+ \cap \mathbb{T}_2)$. This completes the proof. □

7 Snow's inequalities

This chapter deals with Snow type two dimensional linear integral inequalities. The material in this chapter is based on some results in [9, 24] and [25].

Let $\mathbb{T}_1$ and $\mathbb{T}_2$ be time scales with forward jump operators and delta differentiation operators σ_1, σ_2 and Δ_1, Δ_2, respectively.

7.1 Existence of solutions of some partial dynamic equations

Let D and $\tilde{D}$ be open bounded domains in $\mathbb{T}_1 \times \mathbb{T}_2$ such that $D \subset \tilde{D}$ and

$$\{(\sigma_1(t_1), \sigma_2(t_2)) : (t_1, t_2) \in D\} \subset \tilde{D}.$$

By ∂D we will denote the boundary of D and let $\overline{D} = D \cup \partial D$. Let also $(t_1^0, t_2^0) \in D$, $(t_1^0, t_2^0) \notin \partial D$. By $\mathcal{C}^2_{1,2}(D)$ we will denote the space of all functions $u \in \mathcal{C}(D)$ such that $u^{\Delta_1}_{t_1}(t_1, t_2)$, $u^{\Delta_2}_{t_2}(t_1, t_2)$, and $u^{\Delta_1\Delta_2}_{t_1t_2}(t_1, t_2)$ exist and are continuous for $(t_1, t_2) \in D$. Consider the problem

$$v^{\Delta_1\Delta_2}_{t_1t_2} - b(t_1, t_2)v = 0, \quad (t_1, t_2) \in D,\ t_1 \geq t_1^0,\ t_2 \geq t_2^0, \tag{7.1}$$

$$v(t_1, t_2^0) = 1, \quad v^{\Delta_2}_{t_2}(t_1^0, t_2) = 0, \quad (t_1, t_2) \in D,\ t_1 \geq t_1^0,\ t_2 \geq t_2^0. \tag{7.2}$$

Theorem 7.1.1. *Let b be a continuous function on D. Then $v \in \mathcal{C}^2_{1,2}(D)$ is a solution of the problem* (7.1), (7.2) *if and only if it satisfies the integral equation*

$$v(t_1, t_2) = 1 + \int_{t_1^0}^{t_1} \int_{t_2^0}^{t_2} b(s_1, s_2)v(s_1, s_2)\Delta_2 s_2 \Delta_1 s_1, \tag{7.3}$$

$(t_1, t_2) \in D$, $t_1 \geq t_1^0$, $t_2 \geq t_2^0$.

Proof.

1. Let $v \in \mathcal{C}^2_{1,2}(D)$ be a solution of the problem (7.1), (7.2). We integrate (7.1) from t_1^0 to t_1 and using that $v^{\Delta_2}_{t_2}(t_1^0, t_2) = 0$, $t_2 \geq t_2^0$, $(t_1^0, t_2) \in D$, get

$$v^{\Delta_2}_{t_2}(t_1, t_2) - v^{\Delta_2}_{t_2}(t_1^0, t_2) = \int_{t_1^0}^{t_1} b(s_1, t_2)v(s_1, t_2)\Delta_1 s_1,$$

$(t_1, t_2) \in D$, $t_1 \geq t_1^0$, $t_2 \geq t_2^0$, or

$$v^{\Delta_2}_{t_2}(t_1, t_2) = \int_{t_1^0}^{t_1} b(s_1, t_2)v(s_1, t_2)\Delta_1 s_1, \quad (t_1, t_2) \in D,\ t_1 \geq t_1^0,\ t_2 \geq t_2^0.$$

https://doi.org/10.1515/9783110705553-007

Now we integrate the last equation from t_2^0 to t_2 and using that $v(t_1, t_2^0) = 1$, $t_1 \geq t_1^0$, $(t_1, t_2^0) \in D$, get

$$v(t_1, t_2) - v(t_1, t_2^0) = \int_{t_1^0}^{t_1} \int_{t_2^0}^{t_2} b(s_1, s_2) v(s_1, s_2) \Delta_2 s_2 \Delta_1 s_1,$$

$(t_1, t_2) \in D$, $t_1 \geq t_1^0$, $t_2 \geq t_2^0$, or

$$v(t_1, t_2) = 1 + \int_{t_1^0}^{t_1} \int_{t_2^0}^{t_2} b(s_1, s_2) v(s_1, s_2) \Delta_2 s_2 \Delta_1 s_1,$$

$(t_1, t_2) \in D$, $t_1 \geq t_1^0$, $t_2 \geq t_2^0$, i. e., v satisfies equation (7.3).

2. Let $v \in \mathcal{C}^2_{1,2}(D)$ satisfy equation (7.3). Then

$$v(t_1, t_2^0) = 1, \quad (t_1, t_2^0) \in D,\ t_1 \geq t_1^0.$$

We differentiate (7.3) with respect to t_2 and find

$$v_{t_2}^{\Delta_2}(t_1, t_2) = \int_{t_1^0}^{t_1} b(s_1, t_2) v(s_1, t_2) \Delta_1 s_1, \quad (t_1, t_2) \in D,\ t_1 \geq t_1^0,\ t_2 \geq t_2^0, \tag{7.4}$$

whereupon

$$v_{t_2}^{\Delta_2}(t_1^0, t_2) = 0, \quad (t_1^0, t_2) \in D,\ t_2 \geq t_2^0.$$

Now we differentiate (7.4) with respect to t_1 and get

$$v_{t_1 t_2}^{\Delta_1 \Delta_2}(t_1, t_2) = b(t_1, t_2) v(t_1, t_2), \quad (t_1, t_2) \in D,\ t_1 \geq t_1^0,\ t_2 \geq t_2^0,$$

i. e., v satisfies equation (7.1). Therefore v is a solution of the problem (7.1), (7.2). This completes the proof. □

Lemma 7.1.2. *Let $b \in \mathcal{C}(D)$ and u be a nonnegative continuous function on $\overline{D}$. If*

$$u(t_1, t_2) \leq \int_{t_1^0}^{t_1} \int_{t_2^0}^{t_2} b(s_1, s_2) u(s_1, s_2) \Delta_2 s_2 \Delta_1 s_1, \quad (t_1, t_2) \in D,\ t_1 \geq t_1^0,\ t_2 \geq t_2^0,$$

then

$$u(t_1, t_2) = 0, \quad (t_1, t_2) \in D,\ t_1 \geq t_1^0,\ t_2 \geq t_2^0.$$

Proof. Let

$$z(t_1,t_2)=\int_{t_1^0}^{t_1}\int_{t_2^0}^{t_2} b(s_1,s_2)u(s_1,s_2)\Delta_2 s_2\Delta_1 s_1,\quad (t_1,t_2)\in D,\ t_1\geq t_1^0,\ t_2\geq t_2^0.$$

Then

$$\begin{aligned}
u(t_1,t_2)&\leq z(t_1,t_2),\\
z(t_1^0,t_2)&=0,\\
z_{t_1}^{\Delta_1}(t_1,t_2)&=\int_{t_2^0}^{t_2} b(t_1,s_2)u(t_1,s_2)\Delta_2 s_2\\
&\leq \int_{t_2^0}^{t_2} b(t_1,s_2)z(t_1,s_2)\Delta_2 s_2,\quad (t_1,t_2)\in D,\ t_1\geq t_1^0,\ t_2\geq t_2^0.
\end{aligned}$$

From here, using Gronwall's inequality, we get

$$z(t_1,t_2)=0,\quad (t_1,t_2)\in D,\ t_1\geq t_1^0,\ t_2\geq t_2^0.$$

Hence,

$$u(t_1,t_2)=0,\quad (t_1,t_2)\in D,\ t_1\geq t_1^0,\ t_2\geq t_2^0.$$

This completes the proof. □

Theorem 7.1.3. *Let $b\in \mathcal{C}(\overline{D})$. Then the problem* (7.1), (7.2) *has a unique solution $v\in \mathcal{C}^2_{1,2}(D)$.*

Proof. Define the sequence $\{v^l\}_{l\in\mathbb{N}_0}$ as follows:

$$\begin{aligned}
v^0(t_1,t_2)&=1,\\
v^l(t_1,t_2)&=1+\int_{t_1^0}^{t_1}\int_{t_2^0}^{t_2} b(s_1,s_2)v^{l-1}(s_1,s_2)\Delta_2 s_2\Delta_1 s_1,
\end{aligned}$$

$(t_1,t_2)\in D$, $t_1\geq t_1^0$, $t_2\geq t_2^0$, $l\in\mathbb{N}$. Let

$$M=\sup_{(t_1,t_2)\in\overline{D}}|b(t_1,t_2)|.$$

Then

$$|v^1(t_1,t_2)-v^0(t_1,t_2)|=\left|\int_{t_1^0}^{t_1}\int_{t_2^0}^{t_2} b(s_1,s_2)\Delta_2 s_2\Delta_1 s_1\right|$$

$$\leq \int_{t_1^0}^{t_1}\int_{t_2^0}^{t_2} |b(s_1,s_2)|\Delta_2 s_2 \Delta_1 s_1$$
$$\leq M h_1(t_1,t_1^0)h_1(t_2,t_2^0),$$

$(t_1,t_2) \in D$, $t_1 \geq t_1^0$, $t_2 \geq t_2^0$. Assume that

$$|v^l(t_1,t_2) - v^{l-1}(t_1,t_2)| \leq M^l h_l(t_1,t_1^0)h_l(t_2,t_2^0), \tag{7.5}$$

$(t_1,t_2) \in D$, $t_1 \geq t_1^0$, $t_2 \geq t_2^0$, for some $l \in \mathbb{N}$. We will prove that

$$|v^{l+1}(t_1,t_2) - v^l(t_1,t_2)| \leq M^{l+1} h_{l+1}(t_1,t_1^0)h_{l+1}(t_2,t_2^0),$$

$(t_1,t_2) \in D$, $t_1 \geq t_1^0$, $t_2 \geq t_2^0$. In fact,

$$|v^{l+1}(t_1,t_2) - v^l(t_1,t_2)| = \Bigg|1 + \int_{t_1^0}^{t_1}\int_{t_2^0}^{t_2} b(s_1,s_2)v^l(s_1,s_2)\Delta_2 s_2\Delta_1 s_1$$
$$- 1 - \int_{t_1^0}^{t_1}\int_{t_2^0}^{t_2} b(s_1,s_2)v^{l-1}(s_1,s_2)\Delta_2 s_2\Delta_1 s_1\Bigg|$$
$$= \left|\int_{t_1^0}^{t_1}\int_{t_2^0}^{t_2} b(s_1,s_2)(v^l(s_1,s_2) - v^{l-1}(s_1,s_2))\Delta_2 s_2\Delta_1 s_1\right|$$
$$\leq \int_{t_1^0}^{t_1}\int_{t_2^0} |b(s_1,s_2)||v^l(s_1,s_2) - v^{l-1}(s_1,s_2)|\Delta_2 s_2\Delta_1 s_1$$
$$\leq M^{l+1}\int_{t_1^0}^{t_1}\int_{t_2^0}^{t_2} h_l(s_1,t_1^0)h_l(s_2,t_2^0)\Delta_2 s_2\Delta_1 s_1$$
$$= M^{l+1} h_{l+1}(t_1,t_1^0)h_{l+1}(t_2,t_2^0),$$

$(t_1,t_2) \in D$, $t_1 \geq t_1^0$, $t_2 \geq t_2^0$. Therefore (7.5) holds for any $l \in \mathbb{N}$ and $(t_1,t_2) \in D$, $t_1 \geq t_1^0$, $t_2 \geq t_2^0$. Since D is a bounded domain, there is a constant $P > 0$ such that

$$(\sigma_1(t_1) - t_1^0)(\sigma_2(t_2) - t_2^0) \leq P, \quad (t_1,t_2) \in D,\ t_1 \geq t_1^0,\ t_2 \geq t_2^0.$$

We have

$$\lim_{l\to\infty}|v^l(t_1,t_2) - 1| = \left|\sum_{l=0}^{\infty}(v^{l+1}(t_1,t_2) - v^l(t_1,t_2))\right|$$
$$\leq \sum_{l=0}^{\infty}|v^{l+1}(t_1,t_2) - v^l(t_1,t_2)|$$

$$
\begin{aligned}
&\le \sum_{l=0}^{\infty} M^{l+1} h_{l+1}(t_1,t_1^0)h_{l+1}(t_2,t_2^0)\\
&\le \sum_{l=0}^{\infty} M^{l+1} \frac{(\sigma_1(t_1)-t_1^0)^{l+1}(\sigma_2(t_2)-t_2^0)^{l+1}}{((l+1)!)^2}\\
&\le \sum_{l=0}^{\infty} M^{l+1} \frac{P^{l+1}}{((l+1)!)^2}\\
&\le \sum_{l=0}^{\infty} M^{l+1} \frac{P^{l+1}}{(l+1)!}\\
&\le e^{MP}\\
&< \infty,
\end{aligned}
$$

$(t_1,t_2) \in D$, $t_1 \ge t_1^0$, $t_2 \ge t_2^0$. Therefore $v^l(t_1,t_2)$ converges uniformly to a solution $v(t_1,t_2)$, $(t_1,t_2) \in D$, $t_1 \ge t_1^0$, $t_2 \ge t_2^0$, of the problem (7.1), (7.2). Suppose that $v_1(t_1,t_2)$ and $v_2(t_1,t_2)$ are two solutions of the problem (7.1), (7.2), $(t_1,t_2) \in D$, $t_1 \ge t_1^0$, $t_2 \ge t_2^0$. Then

$$
\begin{aligned}
|v_1(t_1,t_2) - v_2(t_1,t_2)| &= \Bigg|1 + \int_{t_1^0}^{t_1}\int_{t_2^0}^{t_2} b(s_1,s_2)v_1(s_1,s_2)\Delta_2 s_2 \Delta_1 s_1\\
&\quad - 1 - \int_{t_1^0}^{t_1}\int_{t_2^0}^{t_2} b(s_1,s_2)v_2(s_1,s_2)\Delta_2 s_2 \Delta_1 s_1\Bigg|\\
&= \left|\int_{t_1^0}^{t_1}\int_{t_2^0}^{t_2} b(s_1,s_2)(v_1(s_1,s_2) - v_2(s_1,s_2))\Delta_2 s_2 \Delta_1 s_1\right|\\
&\le \int_{t_1^0}^{t_1}\int_{t_2^0}^{t_2} |b(s_1,s_2)||v_1(s_1,s_2) - v_2(s_1,s_2)|\Delta_2 s_2 \Delta_1 s_1,
\end{aligned}
$$

$(t_1,t_2) \in D$, $t_1 \ge t_1^0$, $t_2 \ge t_2^0$. Hence, by Lemma 7.1.2, we get that

$$
v_1(t_1,t_2) - v_2(t_1,t_2) = 0, \quad (t_1,t_2) \in D,\ t_1 \ge t_1^0,\ t_2 \ge t_2^0.
$$

This completes the proof. □

As in above one can prove the following theorem.

Theorem 7.1.4. *Let $b \in \mathcal{C}(\overline{D})$. Then the problem*

$$
v_{t_1 t_2}^{\Delta_1 \Delta_2} - b(t_1,t_2)v^{\sigma_1\sigma_2} = 0, \quad (t_1,t_2) \in D,\ t_1 \ge t_1^0,\ t_2 \ge t_2^0, \tag{7.6}
$$

$$
v^{\sigma_1}(t_1,t_2^0) = 1, \quad v(t_1^0,t_2) = 1, \quad (t_1,t_2) \in D,\ t_1 \ge t_1^0,\ t_2 \ge t_2^0, \tag{7.7}
$$

has a unique solution $v \in C^2_{1,2}(D)$ *which can be represented in the form*

$$v(t_1, t_2) = \int_{t_1^0}^{t_1} \int_{t_2^0}^{t_2} b(s_1, s_2)\Delta_2 s_2 \Delta_1 s_1 + 1.$$

7.2 Snow's inequalities

Theorem 7.2.1 (Snow's Inequality). *Let* u*,* a*, and* b *be nonnegative continuous functions on* D*. For any* (t_1^0, t_2^0) *by* $v(t_1, t_2)$ *we denote the nonnegative solution of the problem (7.1), (7.2) in* D*. If, for* $(t_1^0, t_2^0) \in D$*,*

$$u(t_1^0, t_2^0) \le a(t_1^0, t_2^0) + \int_{t_1^0}^{t_1} \int_{t_2^0}^{t_2} b(s_1, s_2) u^{\sigma_1\sigma_2}(s_1, s_2)\Delta_2 s_2 \Delta_1 s_1,$$

$(t_1, t_2) \in D$*,* $t_1 \ge t_1^0$*,* $t_2 \ge t_2^0$*, then*

$$u(t_1^0, t_2^0) \le a(t_1^0, t_2^0) + \int_{t_1^0}^{t_1} \int_{t_2^0}^{t_2} v(s_1, s_2) a^{\sigma_1\sigma_2}(s_1, s_2) b(s_1, s_2)\Delta_2 s_2 \Delta_1 s_1,$$

$(t_1, t_2) \in D$*,* $t_1 \ge t_1^0$*,* $t_2 \ge t_2^0$*.*

Proof. Let

$$z(t_1^0, t_2^0) = \int_{t_1}^{t_1^0} \int_{t_2}^{t_2^0} b(s_1, s_2) u^{\sigma_1\sigma_2}(s_1, s_2)\Delta_2 s_2 \Delta_1 s_1,$$

$(t_1, t_2) \in D$, $t_1 \ge t_1^0$, $t_2 \ge t_2^0$. Then

$$\begin{aligned}
u(t_1^0, t_2^0) &\le a(t_1^0, t_2^0) + z(t_1^0, t_2^0),\\
z(t_1, t_2^0) &= 0,\\
z^{\Delta_1}_{t_1^0}(t_1^0, t_2^0) &= \int_{t_2}^{t_2^0} b(t_1^0, s_2) u^{\sigma_1\sigma_2}(t_1^0, s_2)\Delta_2 s_2,\\
z^{\Delta_1}_{t_1^0}(t_1^0, t_2) &= 0,\\
z^{\Delta_1\Delta_2}_{t_1^0 t_2^0}(t_1^0, t_2^0) &= b(t_1^0, t_2^0) u^{\sigma_1\sigma_2}(t_1^0, t_2^0)\\
&\le a^{\sigma_1\sigma_2}(t_1^0, t_2^0) b(t_1^0, t_2^0) + b(t_1^0, t_2^0) z^{\sigma_1\sigma_2}(t_1^0, t_2^0),
\end{aligned}$$

whereupon

$$z^{\Delta_1\Delta_2}_{t_1^0 t_2^0}(t_1^0, t_2^0) - b(t_1^0, t_2^0) z^{\sigma_1\sigma_2}(t_1^0, t_2^0) \le a^{\sigma_1\sigma_2}(t_1^0, t_2^0) b(t_1^0, t_2^0).$$

We fix $(t_1, t_2) \in D$, $t_1 \geq t_1^0$, $t_2 \geq t_2^0$. Let

$$R = \{(s_1, s_2) \in D : t_1 \geq s_1 \geq t_1^0,\ t_2 \geq s_2 \geq t_2^0\}$$

and ∂R be its boundary. For $(s_1, s_2) \in R$ and $x \in \mathcal{C}^2_{1,2}(\overline{D})$, we define the operators

$$\begin{aligned} L(x)(s_1, s_2) &= x^{\Delta_1\Delta_2}_{s_1 s_2}(s_1, s_2) - b(s_1, s_2)x^{\sigma_1\sigma_2}(s_1, s_2), \\ M(x)(s_1, s_2) &= x^{\Delta_1\Delta_2}_{s_1 s_2}(s_1, s_2) - b(s_1, s_2)x(s_1, s_2). \end{aligned}$$

Then, for $x, y \in \mathcal{C}^2_{1,2}(\overline{D})$ and $(s_1, s_2) \in D$, we have

$$\begin{aligned} & y(s_1, s_2)L(x)(s_1, s_2) - x^{\sigma_1\sigma_2}(s_1, s_2)M(y)(s_1, s_2) \\ &\quad = y(s_1, s_2)(x^{\Delta_1\Delta_2}_{s_1 s_2}(s_1, s_2) - b(s_1, s_2)x^{\sigma_1\sigma_2}(s_1, s_2)) \\ &\qquad - x^{\sigma_1\sigma_2}(s_1, s_2)(y^{\Delta_1\Delta_2}_{s_1 s_2}(s_1, s_2) - b(s_1, s_2)y(s_1, s_2)) \\ &\quad = y(s_1, s_2)x^{\Delta_1\Delta_2}_{s_1 s_2}(s_1, s_2) - x^{\sigma_1\sigma_2}(s_1, s_2)y^{\Delta_1\Delta_2}_{s_1 s_2}(s_1, s_2), \end{aligned}$$

and

$$\begin{aligned} & (yx^{\Delta_1}_{s_1})^{\Delta_2}_{s_2}(s_1, s_2) - (x^{\sigma_2}y^{\Delta_2}_{s_2})^{\Delta_1}_{s_1}(s_1, s_2) \\ &\quad = y(s_1, s_2)x^{\Delta_1\Delta_2}_{s_1 s_2}(s_1, s_2) + y^{\Delta_2}_{s_2}(s_1, s_2)x^{\Delta_1\sigma_2}_{s_1}(s_1, s_2) \\ &\qquad - x^{\sigma_1\sigma_2}(s_1, s_2)y^{\Delta_1\Delta_2}_{s_1 s_2}(s_1, s_2) - x^{\Delta_1\sigma_2}_{s_1}(s_1, s_2)y^{\Delta_2}_{s_2}(s_1, s_2) \\ &\quad = y(s_1, s_2)x^{\Delta_1\Delta_2}_{s_1 s_2}(s_1, s_2) - x^{\sigma_1\sigma_2}(s_1, s_2)y^{\Delta_1\Delta_2}_{s_1 s_2}(s_1, s_2), \end{aligned}$$

i. e.,

$$\begin{aligned} & y(s_1, s_2)L(x)(s_1, s_2) - x^{\sigma_1\sigma_2}(s_1, s_2)M(y)(s_1, s_2) \\ &\quad = (yx^{\Delta_1}_{s_1})^{\Delta_2}_{s_2}(s_1, s_2) - (x^{\sigma_2}y^{\Delta_2}_{s_2})^{\Delta_1}_{s_1}(s_1, s_2). \end{aligned}$$

Applying Green's formula, we get

$$\begin{aligned} & \iint_R v(s_1, s_2)L(z)(s_1, s_2)\Delta_2 s_2 \Delta_1 s_1 \\ &\quad = \iint_R (v(s_1, s_2)L(z)(s_1, s_2) - z^{\sigma_1\sigma_2}(s_1, s_2)M(v)(s_1, s_2))\Delta_2 s_2 \Delta_1 s_1 \\ &\quad = \iint_R ((vz^{\Delta_1}_{s_1})^{\Delta_2}_{s_2}(s_1, s_2) \\ &\qquad - (z^{\sigma_2}v^{\Delta_2}_{s_2})^{\Delta_1}_{s_1}(s_1, s_2))\Delta_2 s_2 \Delta_1 s_1 \\ &\quad = -\int_{\partial R} z^{\sigma_2}(s_1, s_2)v^{\Delta_2}_{s_2}(s_1, s_2)\Delta_2 s_2 + v(s_1, s_2)z^{\Delta_1}_{s_1}(s_1, s_2)\Delta_1 s_1 \end{aligned}$$

$$
\begin{aligned}
&= -\int_{t_2^0}^{t_2} z^{\sigma_2}(t_1, s_2) v_{s_2}^{\Delta_2}(t_1, s_2)\Delta_2 s_2 \\
&\quad + \int_{t_2^0}^{t_2} z^{\sigma_2}(t_1^0, s_2) v_{s_2}^{\Delta_2}(t_1^0, s_2)\Delta_2 s_2 \\
&\quad - \int_{t_1^0}^{t_1} v(s_1, t_2^0) z_{s_1}^{\Delta_1}(s_1, t_2^0)\Delta_1 s_1 \\
&\quad + \int_{t_1^0}^{t_1} v(s_1, t_2) z_{s_1}^{\Delta_1}(s_1, t_2)\Delta_1 s_1 \\
&= -\int_{t_1^0}^{t_1} v(s_1, t_2^0) z_{s_1}^{\Delta_1}(s_1, t_2^0)\Delta_1 s_1 \\
&= -\int_{t_1^0}^{t_1} z_{s_1}^{\Delta_1}(s_1, t_2^0)\Delta_1 s_1 \\
&= -z(t_1, t_2^0) + z(t_1^0, t_2^0) \\
&= z(t_1^0, t_2^0),
\end{aligned}
$$

i. e.,

$$
\begin{aligned}
z(t_1^0, t_2^0) &= \iint_R v(s_1, s_2) L(z)(s_1, s_2)\Delta_2 s_2 \Delta_1 s_1 \\
&\le \int_{t_1}^{t_1^0}\int_{t_2}^{t_2^0} v(s_1, s_2) a^{\sigma_1\sigma_2}(s_1, s_2) b(s_1, s_2)\Delta_2 s_2 \Delta_1 s_1.
\end{aligned}
$$

Therefore

$$
\begin{aligned}
u(t_1^0, t_2^0) &\le a(t_1^0, t_2^0) + z(t_1^0, t_2^0) \\
&\le a(t_1^0, t_2^0) \\
&\quad + \int_{t_1}^{t_1^0}\int_{t_2}^{t_2^0} v(s_1, s_2) a^{\sigma_1\sigma_2}(s_1, s_2) b(s_1, s_2)\Delta_2 s_2 \Delta_1 s_1.
\end{aligned}
$$

This completes the proof. □

Theorem 7.2.2 (Snow's Inequality). *Let u, a, b, c, and f be nonnegative continuous functions on D. For $(t_1^0, t_2^0) \in D$, by v we denote a nonnegative solution of the problem*

$$
v_{t_1 t_2}^{\Delta_1 \Delta_2} - (b(t_1, t_2) + f(t_1, t_2)) v^{\sigma_1\sigma_2} = 0,
$$

$(t_1,t_2) \in D$, $t_1 \ge t_1^0$, $t_2 \ge t_2^0$,

$$v(t_1,t_2^0) = 1, \quad v_{t_2}^{\Delta_2}(t_1^0,t_2) = 0,$$

$(t_1,t_2) \in D$, $t_1 \ge t_1^0$, $t_2 \ge t_2^0$. *If*

$$u(t_1^0,t_2^0) \le a(t_1^0,t_2^0) + \int_{t_1^0}^{t_1}\int_{t_2^0}^{t_2} b(s_1,s_2)u(s_1,s_2)\Delta_2 s_2\Delta_1 s_1$$
$$+ \int_{t_1^0}^{t_1}\int_{t_2^0}^{t_2} b(s_1,s_2)\left(c(s_1,s_2) + \int_{s_1}^{t_1}\int_{s_2}^{t_2} f(\xi_1,\xi_2)u(\xi_1,\xi_2)\Delta_2\xi_2\Delta_1\xi_1 \right)\Delta_2 s_2\Delta_1 s_1,$$

$(t_1,t_2) \in D$, $t_1 \ge t_1^0$, $t_2 \ge t_2^0$, *then*

$$u(t_1^0,t_2^0) \le a(t_1^0,t_2^0)$$
$$+ \int_{t_1}^{t_1^0}\int_{t_2}^{t_2^0} v(s_1,s_2)(b(s_1,s_2)(a(s_1,s_2) + c(s_1,s_2))$$
$$+ f(s_1,s_2)a(s_1,s_2))\Delta_2 s_2\Delta_1 s_1,$$

$(t_1,t_2) \in D$, $t_1 \ge t_1^0$, $t_2 \ge t_2^0$.

Proof. Let

$$\phi(t_1^0,t_2^0) = \int_{t_1^0}^{t_1}\int_{t_2^0}^{t_2} b(s_1,s_2)u(s_1,s_2)\Delta_2 s_2\Delta_1 s_1$$
$$+ \int_{t_1^0}^{t_1}\int_{t_2^0}^{t_2} b(s_1,s_2)\left(c(s_1,s_2) + \int_{s_1}^{t_1}\int_{s_2}^{t_2} f(\xi_1,\xi_2)u(\xi_1,\xi_2)\Delta_2\xi_2\Delta_1\xi_1 \right)\Delta_2 s_2\Delta_1 s_1,$$

$(t_1,t_2) \in D$, $t_1 \ge t_1^0$, $t_2 \ge t_2^0$. Then

$$u(t_1^0,t_2^0) \le a(t_1^0,t_2^0) + \phi(t_1^0,t_2^0),$$
$$\phi_{t_1^0}^{\Delta_1}(t_1^0,t_2^0) = \int_{t_2}^{t_2^0} b(t_1^0,s_2)u(t_1^0,s_2)\Delta_2 s_2$$
$$+ \int_{t_2}^{t_2^0} b(t_1^0,s_2)\left(c(t_1^0,s_2) + \int_{t_1^0}^{t_1}\int_{s_2}^{t_2} f(\xi_1,\xi_2)u(\xi_1,\xi_2)\Delta_2\xi_2\Delta_1\xi_1 \right)\Delta_2 s_2,$$
$$\phi_{t_1^0 t_2^0}^{\Delta_1\Delta_2}(t_1^0,t_2^0) = b(t_1^0,t_2^0)u(t_1^0,t_2^0)$$

$$
\begin{aligned}
&+ b(t_1^0, t_2^0)\Bigg(c(t_1^0, t_2^0) \\
&+ \int_{t_1^0}^{t_1} \int_{t_2^0}^{t_2} f(s_1, s_2) u(s_1, s_2) \Delta_2 s_2 \Delta_1 s_1 \Bigg) \\
&\leq b(t_1^0, t_2^0)(a(t_1^0, t_2^0) + \phi(t_1^0, t_2^0)) \\
&+ b(t_1^0, t_2^0)\Bigg(c(t_1^0, t_2^0) + \int_{t_1^0}^{t_1} \int_{t_2^0}^{t_2} f(s_1, s_2)(a(s_1, s_2) \\
&+ \phi(s_1, s_2)) \Delta_2 s_2 \Delta_1 s_1 \Bigg) \\
&= b(t_1^0, t_2^0)\Bigg(a(t_1^0, t_2^0) + c(t_1^0, t_2^0) + \phi(t_1^0, t_2^0) \\
&+ \int_{t_1^0}^{t_1} \int_{t_2^0}^{t_2} f(s_1, s_2)(a(s_1, s_2) + \phi(s_1, s_2)) \Delta_2 s_2 \Delta_1 s_1 \Bigg),
\end{aligned}
$$

$$
\begin{aligned}
\phi(t_1^0, t_2) &= 0, \\
\phi(t_1, t_2^0) &= 0,
\end{aligned}
$$

$(t_1, t_2) \in D$, $t_1 \geq t_1^0$, $t_2 \geq t_2^0$. Let

$$
z(t_1^0, t_2^0) = \phi(t_1^0, t_2^0) + \int_{t_1^0}^{t_1} \int_{t_2^0}^{t_2} f(s_1, s_2)(a(s_1, s_2) + \phi(s_1, s_2)) \Delta_2 s_2 \Delta_1 s_1,
$$

$(t_1, t_2) \in D$, $t_1 \geq t_1^0$, $t_2 \geq t_2^0$. Then

$$
\begin{aligned}
z(t_1^0, t_2) &= \phi(t_1^0, t_2) \\
&= 0, \\
z(t_1, t_2^0) &= \phi(t_1, t_2^0) \\
&= 0, \\
z_{t_1^0}^{\Delta_1}(t_1^0, t_2^0) &= \phi_{t_1^0}^{\Delta_1}(t_1^0, t_2^0) \\
&+ \int_{t_2}^{t_2^0} f(t_1^0, s_2)(a(t_1^0, s_2) + \phi(t_1^0, s_2)) \Delta_2 s_2, \\
z_{t_1^0 t_2^0}^{\Delta_1 \Delta_2}(t_1^0, t_2^0) &= \phi_{t_1^0 t_2^0}^{\Delta_1 \Delta_2}(t_1^0, t_2^0) + f(t_1^0, t_2^0)(a(t_1^0, t_2^0) + \phi(t_1^0, t_2^0)),
\end{aligned}
$$

$(t_1, t_2) \in D$, $t_1 \geq t_1^0$, $t_2 \geq t_2^0$. Note that $\phi(t_1^0, t_2^0)$ is a nonnegative nonincreasing function for $(t_1^0, t_2^0) \in D$. Hence, $z(t_1^0, t_2^0)$ is a nonnegative nonincreasing function for $(t_1^0, t_2^0) \in D$,

and

$$\phi^{\Delta_1\Delta_2}_{t_1^0 t_2^0}(t_1^0,t_2^0) \le b(t_1^0,t_2^0)(a(t_1^0,t_2^0)+c(t_1^0,t_2^0)+z(t_1^0,t_2^0)),$$
$$\phi(t_1^0,t_2^0) \le z(t_1^0,t_2^0),$$

$(t_1^0,t_2^0) \in D$. Then

$$\begin{aligned} z^{\Delta_1\Delta_2}_{t_1^0 t_2^0}(t_1^0,t_2^0) &\le b(t_1^0,t_2^0)(a(t_1^0,t_2^0)+c(t_1^0,t_2^0)+z(t_1^0,t_2^0)) \\ &\quad + f(t_1^0,t_2^0)(a(t_1^0,t_2^0)+z(t_1^0,t_2^0)) \\ &= (b(t_1^0,t_2^0)+f(t_1^0,t_2^0))z(t_1^0,t_2^0) \\ &\quad + b(t_1^0,t_2^0)(a(t_1^0,t_2^0)+c(t_1^0,t_2^0)) \\ &\quad + f(t_1^0,t_2^0)a(t_1^0,t_2^0), \end{aligned}$$

$(t_1^0,t_2^0) \in D$. We set

$$\begin{aligned} g(t_1^0,t_2^0) &= b(t_1^0,t_2^0)+f(t_1^0,t_2^0), \\ h(t_1^0,t_2^0) &= b(t_1^0,t_2^0)(a(t_1^0,t_2^0)+c(t_1^0,t_2^0)) \\ &\quad + f(t_1^0,t_2^0)a(t_1^0,t_2^0), \end{aligned}$$

$(t_1^0,t_2^0) \in D$. Therefore

$$z^{\Delta_1\Delta_2}_{t_1^0 t_2^0}(t_1^0,t_2^0) \le g(t_1^0,t_2^0)z(t_1^0,t_2^0)+h(t_1^0,t_2^0),$$

or

$$z^{\Delta_1\Delta_2}_{t_1^0 t_2^0}(t_1^0,t_2^0) - g(t_1^0,t_2^0)z(t_1^0,t_2^0) \le h(t_1^0,t_2^0),$$

$(t_1^0,t_2^0) \in D$. We fix $(t_1,t_2) \in D$, $t_1 \ge t_1^0$, $t_2 \ge t_2^0$ and let R be as in the proof of Theorem 7.2.1. For $(s_1,s_2) \in R$ and $x \in \mathcal{C}^2_{1,2}(\overline{D})$, we define the operators

$$\begin{aligned} L(x)(s_1,s_2) &= x^{\Delta_1\Delta_2}_{s_1 s_2}(s_1,s_2) - g(s_1,s_2)x(s_1,s_2), \\ M(x)(s_1,s_2) &= x^{\Delta_1\Delta_2}_{s_1 s_2}(s_1,s_2) - g(s_1,s_2)x^{\sigma_1\sigma_2}(s_1,s_2). \end{aligned}$$

Then, for $x,y \in \mathcal{C}^2_{1,2}(\overline{D})$ and $(s_1,s_2) \in D$, we have

$$\begin{aligned} & y^{\sigma_1\sigma_2}(s_1,s_2)L(x)(s_1,s_2) - x(s_1,s_2)M(y)(s_1,s_2) \\ &\quad = y^{\sigma_1\sigma_2}(s_1,s_2)(x^{\Delta_1\Delta_2}_{s_1 s_2}(s_1,s_2) - g(s_1,s_2)x(s_1,s_2)) \\ &\quad\quad - x(s_1,s_2)(y^{\Delta_1\Delta_2}_{s_1 s_2}(s_1,s_2) - g(s_1,s_2)y^{\sigma_1\sigma_2}(s_1,s_2)) \\ &\quad = y^{\sigma_1\sigma_2}(s_1,s_2)x^{\Delta_1\Delta_2}_{s_1 s_2}(s_1,s_2) - x(s_1,s_2)y^{\Delta_1\Delta_2}_{s_1 s_2}(s_1,s_2), \end{aligned}$$

and

$$\begin{aligned}
(y^{\sigma_1}x_{s_1}^{\Delta_1})_{s_2}^{\Delta_2}(s_1,s_2) - (xy_{s_2}^{\Delta_2})_{s_1}^{\Delta_1}(s_1,s_2)
&= y^{\sigma_1\sigma_2}(s_1,s_2)x_{s_1s_2}^{\Delta_1\Delta_2}(s_1,s_2) + y_{s_2}^{\Delta_2\sigma_1}(s_1,s_2)x_{s_1}^{\Delta_1}(s_1,s_2)\\
&\quad - x(s_1,s_2)y_{s_1s_2}^{\Delta_1\Delta_2}(s_1,s_2) - x_{s_1}^{\Delta_1}(s_1,s_2)y_{s_2}^{\Delta_2\sigma_1}(s_1,s_2)\\
&= y^{\sigma_1\sigma_2}(s_1,s_2)x_{s_1s_2}^{\Delta_1\Delta_2}(s_1,s_2) - x(s_1,s_2)y_{s_1s_2}^{\Delta_1\Delta_2}(s_1,s_2),
\end{aligned}$$

i. e.,

$$\begin{aligned}
&y^{\sigma_1\sigma_2}(s_1,s_2)L(x)(s_1,s_2) - x(s_1,s_2)M(y)(s_1,s_2)\\
&\quad = (y^{\sigma_1}x_{s_1}^{\Delta_1})_{s_2}^{\Delta_2}(s_1,s_2) - (xy_{s_2}^{\Delta_2})_{s_1}^{\Delta_1}(s_1,s_2).
\end{aligned}$$

Applying Green's formula, we get

$$\begin{aligned}
&\iint\limits_R v^{\sigma_1\sigma_2}(s_1,s_2)L(z)(s_1,s_2)\Delta_2 s_2\Delta_1 s_1\\
&\quad = \iint\limits_R (v^{\sigma_1\sigma_2}(s_1,s_2)L(z)(s_1,s_2) - z(s_1,s_2)M(v)(s_1,s_2))\Delta_2 s_2\Delta_1 s_1\\
&\quad = \iint\limits_R ((v^{\sigma_1}z_{s_1}^{\Delta_1})_{s_2}^{\Delta_2}(s_1,s_2)\\
&\qquad - (zv_{s_2}^{\Delta_2})_{s_1}^{\Delta_1}(s_1,s_2))\Delta_2 s_2(s_1,s_2)\Delta_1 s_1\\
&\quad = -\int\limits_{\partial R} z(s_1,s_2)v_{s_2}^{\Delta_2}(s_1,s_2)\Delta_2 s_2 + v^{\sigma_1}(s_1,s_2)z_{s_1}^{\Delta_1}(s_1,s_2)\Delta_1 s_1\\
&\quad = -\int_{t_2^0}^{t_2} z(t_1,s_2)v_{s_2}^{\Delta_2}(t_1,s_2)\Delta_2 s_2\\
&\qquad + \int_{t_2^0}^{t_2} z(t_1^0,s_2)v_{s_2}^{\Delta_2}(t_1^0,s_2)\Delta_2 s_2\\
&\qquad - \int_{t_1^0}^{t_1} v^{\sigma_1}(s_1,t_2^0)z_{s_1}^{\Delta_1}(s_1,t_2^0)\Delta_1 s_1\\
&\qquad + \int_{t_1^0}^{t_1} v^{\sigma_1}(s_1,t_2)z_{s_1}^{\Delta_1}(s_1,t_2)\Delta_1 s_1\\
&\quad = -\int_{t_1^0}^{t_1} v^{\sigma_1}(s_1,t_2^0)z_{s_1}^{\Delta_1}(s_1,t_2^0)\Delta_1 s_1
\end{aligned}$$

$$
\begin{aligned}
&= -\int_{t_1^0}^{t_1} z_{s_1}^{\Delta_1}(s_1, t_2^0)\Delta_1 s_1 \\
&= -z(t_1, t_2^0) + z(t_1^0, t_2^0) \\
&= z(t_1^0, t_2^0),
\end{aligned}
$$

i. e.,

$$
\begin{aligned}
z(t_1^0, t_2^0) &= \iint_R v(s_1, s_2)L(z)(s_1, s_2)\Delta_2 s_2 \Delta_1 s_1 \\
&\le \int_{t_1}^{t_1^0}\int_{t_2}^{t_2^0} v(s_1, s_2)h(s_1, s_2)\Delta_2 s_2 \Delta_1 s_1 \\
&\le \int_{t_1}^{t_1^0}\int_{t_2}^{t_2^0} v(s_1, s_2)(b(s_1, s_2)(a(s_1, s_2) + c(s_1, s_2)) \\
&\quad + f(s_1, s_2)a(s_1, s_2))\Delta_2 s_2 \Delta_1 s_1, \\
\phi(t_1^0, t_2^0) &\le z(t_1^0, t_2^0) \\
&\le \int_{t_1}^{t_1^0}\int_{t_2}^{t_2^0} v(s_1, s_2)(b(s_1, s_2)(a(s_1, s_2) + c(s_1, s_2)) \\
&\quad + f(s_1, s_2)a(s_1, s_2))\Delta_2 s_2 \Delta_1 s_1, \\
u(t_1^0, t_2^0) &\le a(t_1^0, t_2^0) + \phi(t_1^0, t_2^0) \\
&\le a(t_1^0, t_2^0) \\
&\quad + \int_{t_1}^{t_1^0}\int_{t_2}^{t_2^0} v(s_1, s_2)(b(s_1, s_2)(a(s_1, s_2) + c(s_1, s_2)) \\
&\quad + f(s_1, s_2)a(s_1, s_2))\Delta_2 s_2 \Delta_1 s_1,
\end{aligned}
$$

$(t_1, t_2) \in D$, $t_1 \ge t_1^0$, $t_2 \ge t_2^0$. This completes the proof. □

Theorem 7.2.3 (Snow's Inequality). *Let a, b, c, p, q, and u be nonnegative continuous functions on D. Let, for $(t_1^0, t_2^0) \in D$, v and w be the nonnegative solutions of the problems*

$$
v_{t_1 t_2}^{\Delta_1 \Delta_2} - (p(t_1, t_2) + b(t_1, t_2)(c(t_1, t_2) + q(t_1, t_2)))v = 0,
$$

$(t_1, t_2) \in D$, $t_1 \ge t_1^0$, $t_2 \ge t_2^0$,

$$
v^{\sigma_1}(t_1, t_2^0) = 1, \quad v_{t_2}^{\Delta_2}(t_1^0, t_2) = 0,
$$

$(t_1, t_2) \in D$, $t_1 \ge t_1^0$, $t_2 \ge t_2^0$,

$$
w_{t_1 t_2}^{\Delta_1 \Delta_2} - b(t_1, t_2)c(t_1, t_2)w = 0,
$$

$(t_1, t_2) \in D$, $t_1 \geq t_1^0$, $t_2 \geq t_2^0$,

$$w^{\sigma_1}(t_1, t_2^0) = 1, \quad w_{t_2}^{\Delta_2}(t_1^0, t_2) = 0,$$

$(t_1, t_2) \in D$, $t_1 \geq t_1^0$, $t_2 \geq t_2^0$, respectively. If

$$u(t_1^0, t_2^0) \leq a(t_1^0, t_2^0) + b(t_1^0, t_2^0)\Bigg(\int_{t_1}^{t_1^0}\int_{t_2}^{t_2^0} c(s_1, s_2)u(s_1, s_2)\Delta_2 s_2 \Delta_1 s_1$$
$$+ \int_{t_1}^{t_1^0}\int_{t_2}^{t_2^0} p(s_1, s_2)\Bigg(\int_{t_1}^{s_1}\int_{t_2}^{s_2} q(\xi_1, \xi_2)u(\xi_1, \xi_2)\Delta_2\xi_2\Delta_1\xi_1\Bigg)\Delta_2 s_2 \Delta_1 s_1\Bigg),$$

$(t_1, t_2) \in D$, $t_1 \geq t_1^0$, $t_2 \geq t_2^0$, then

$$u(t_1^0, t_2^0) \leq a(t_1^0, t_2^0) + b(t_1^0, t_2^0)\int_{t_1}^{t_1^0}\int_{t_2}^{t_2^0} p(s_1, s_2)w(s_1, s_2)$$
$$\times \Bigg(\int_{t_1}^{s_1}\int_{t_2}^{s_2} v(\xi_1, \xi_2)a(\xi_1, \xi_2)(c(\xi_1, \xi_2) + q(\xi_1, \xi_2))\Delta_2\xi_2\Delta_1\xi_1 + a(s_1, s_2)c(s_1, s_2)\Bigg)\Delta_2 s_2 \Delta_1 s_1,$$

$(t_1, t_2) \in D$, $t_1 \geq t_1^0$, $t_2 \geq t_2^0$.

Proof. Let

$$\phi(t_1^0, t_2^0) = \int_{t_1}^{t_1^0}\int_{t_2}^{t_2^0} c(s_1, s_2)u(s_1, s_2)\Delta_2 s_2 \Delta_1 s_1$$
$$+ \int_{t_1}^{t_1^0}\int_{t_2}^{t_2^0} p(s_1, s_2)\Bigg(\int_{t_1}^{s_1}\int_{t_2}^{s_2} q(\xi_1, \xi_2)u(\xi_1, \xi_2)\Delta_2\xi_2\Delta_1\xi_1\Bigg)\Delta_2 s_2 \Delta_1 s_1,$$

$(t_1, t_2) \in D$, $t_1 \geq t_1^0$, $t_2 \geq t_2^0$. Then

$$u(t_1^0, t_2^0) \leq a(t_1^0, t_2^0) + b(t_1^0, t_2^0)\phi(t_1^0, t_2^0),$$
$$\phi(t_1, t_2^0) = 0,$$
$$\phi(t_1^0, t_2) = 0,$$
$$\phi_{t_1^0}^{\Delta_1}(t_1^0, t_2^0) = \int_{t_2}^{t_2^0} c(t_1^0, s_2)u(t_1^0, s_2)\Delta_2 s_2$$

$$
\begin{aligned}
&+\int_{t_2}^{t_2^0} p(t_1^0,s_2)\left(\int_{t_1}^{t_1^0}\int_{t_2}^{s_2} q(\xi_1,\xi_2)\right.\\
&\qquad \left.\times u(\xi_1,\xi_2)\Delta_2\xi_2\Delta_1\xi_1\right)\Delta_2 s_2,\\
\phi_{t_1^0 t_2^0}^{\Delta_1\Delta_2}(t_1^0,t_2^0) &= c(t_1^0,t_2^0)u(t_1^0,t_2^0)\\
&\quad + p(t_1^0,t_2^0)\int_{t_1}^{t_1^0}\int_{t_2}^{t_2^0} q(s_1,s_2)u(s_1,s_2)\Delta_2 s_2\Delta_1 s_1\\
&\le c(t_1^0,t_2^0)(a(t_1^0,t_2^0)+b(t_1^0,t_2^0)\phi(t_1^0,t_2^0))\\
&\quad + p(t_1^0,t_2^0)\int_{t_1}^{t_1^0}\int_{t_2}^{t_2^0} q(s_1,s_2)(a(s_1,s_2)\\
&\quad + b(s_1,s_2)\phi(s_1,s_2))\Delta_2 s_2\Delta_1 s_1\\
&\le c(t_1^0,t_2^0)(a(t_1^0,t_2^0)+b(t_1^0,t_2^0)\phi(t_1^0,t_2^0))\\
&\quad + p(t_1^0,t_2^0)\phi(t_1^0,t_2^0)+\int_{t_1}^{t_1^0}\int_{t_2}^{t_2^0} q(s_1,s_2)\\
&\quad \times (a(s_1,s_2)+b(s_1,s_2)\phi(s_1,s_2))\Delta_2 s_2\Delta_1 s_1,
\end{aligned}
$$

$(t_1,t_2)\in D$, $t_1\ge t_1^0$, $t_2\ge t_2^0$. Let

$$
\begin{aligned}
z(t_1^0,t_2^0) &= \phi(t_1^0,t_2^0)\\
&\quad + \int_{t_1}^{t_1^0}\int_{t_2}^{t_2^0} q(s_1,s_2)(a(s_1,s_2)+b(s_1,s_2)\phi(s_1,s_2))\Delta_2 s_2\Delta_1 s_1,
\end{aligned}
$$

$(t_1,t_2)\in D$, $t_1\ge t_1^0$, $t_2\ge t_2^0$. Then

$$
\begin{aligned}
\phi(t_1^0,t_2^0) &\le z(t_1^0,t_2^0),\\
\phi_{t_1^0 t_2^0}^{\Delta_1\Delta_2}(t_1^0,t_2^0) &\le c(t_1^0,t_2^0)(a(t_1^0,t_2^0)+b(t_1^0,t_2^0)\phi(t_1^0,t_2^0))\\
&\quad + p(t_1^0,t_2^0)z(t_1^0,t_2^0)\\
&\le c(t_1^0,t_2^0)(a(t_1^0,t_2^0)+b(t_1^0,t_2^0)z(t_1^0,t_2^0))\\
&\quad + p(t_1^0,t_2^0)z(t_1^0,t_2^0)\\
&= a(t_1^0,t_2^0)c(t_1^0,t_2^0)\\
&\quad + z(t_1^0,t_2^0)(p(t_1^0,t_2^0)+b(t_1^0,t_2^0)c(t_1^0,t_2^0)),\\
z(t_1,t_2^0) &= \phi(t_1,t_2^0)\\
&= 0,
\end{aligned}
$$

$$
\begin{aligned}
z(t_1^0, t_2) &= \phi(t_1^0, t_2) \\
&= 0, \\
z_{t_1^0}^{\Delta_1}(t_1^0, t_2^0) &= \phi_{t_1^0}^{\Delta_1}(t_1^0, t_2^0) \\
&\quad + \int_{t_2}^{t_2^0} q(t_1^0, s_2)(a(t_1^0, s_2) + b(t_1^0, s_2)\phi(t_1^0, s_2))\Delta_2 s_2, \\
z_{t_1^0 t_2^0}^{\Delta_1\Delta_2}(t_1^0, t_2^0) &= \phi_{t_1^0 t_2^0}^{\Delta_1\Delta_2}(t_1^0, t_2^0) \\
&\quad + q(t_1^0, t_2^0)(a(t_1^0, t_2^0) + b(t_1^0, t_2^0)\phi(t_1^0, t_2^0)) \\
&\leq a(t_1^0, t_2^0)c(t_1^0, t_2^0) \\
&\quad + z(t_1^0, t_2^0)(p(t_1^0, t_2^0) + b(t_1^0, t_2^0)c(t_1^0, t_2^0)) \\
&\quad + q(t_1^0, t_2^0)(a(t_1^0, t_2^0) + b(t_1^0, t_2^0)z(t_1^0, t_2^0)) \\
&= a(t_1^0, t_2^0)(c(t_1^0, t_2^0) + q(t_1^0, t_2^0)) \\
&\quad + z(t_1^0, t_2^0)(p(t_1^0, t_2^0) + b(t_1^0, t_2^0) \\
&\quad \times (c(t_1^0, t_2^0) + q(t_1^0, t_2^0))).
\end{aligned}
$$

Let

$$
\begin{aligned}
g_1(t_1^0, t_2^0) &= a(t_1^0, t_2^0)(c(t_1^0, t_2^0) + q(t_1^0, t_2^0)), \\
h_1(t_1^0, t_2^0) &= p(t_1^0, t_2^0) + b(t_1^0, t_2^0) \\
&\quad \times (c(t_1^0, t_2^0) + q(t_1^0, t_2^0)).
\end{aligned}
$$

Then

$$
z_{t_1^0 t_2^0}^{\Delta_1\Delta_2}(t_1^0, t_2^0) \leq g_1(t_1^0, t_2^0) + h_1(t_1^0, t_2^0)z(t_1^0, t_2^0),
$$

or

$$
z_{t_1^0 t_2^0}^{\Delta_1\Delta_2}(t_1^0, t_2^0) - h_1(t_1^0, t_2^0)z(t_1^0, t_2^0) \leq g_1(t_1^0, t_2^0).
$$

We fix $(t_1, t_2) \in D$, $t_1 \geq t_1^0$, $t_2 \geq t_2^0$. Let R and ∂R be as in the proof of Theorem 7.2.1. For $(s_1, s_2) \in R$ and $x \in \mathcal{C}_{1,2}^2(\overline{D})$, we define the operators

$$
\begin{aligned}
L(x)(s_1, s_2) &= x_{s_1 s_2}^{\Delta_1\Delta_2}(s_1, s_2) - h_1(s_1, s_2)x(s_1, s_2), \\
M(x)(s_1, s_2) &= x_{s_1 s_2}^{\Delta_1\Delta_2}(s_1, s_2) - h_1(s_1, s_2)x^{\sigma_1\sigma_2}(s_1, s_2).
\end{aligned}
$$

Then, for $x, y \in \mathcal{C}_{1,2}^2(\overline{D})$ and $(s_1, s_2) \in D$, we have

$$
\begin{aligned}
&y^{\sigma_1\sigma_2}(s_1, s_2)L(x)(s_1, s_2) - x(s_1, s_2)M(y)(s_1, s_2) \\
&\quad = y^{\sigma_1\sigma_2}(s_1, s_2)(x_{s_1 s_2}^{\Delta_1\Delta_2}(s_1, s_2) - h_1(s_1, s_2)x(s_1, s_2))
\end{aligned}
$$

$$\begin{aligned}
&- x(s_1, s_2)(y^{\Delta_1\Delta_2}_{s_1 s_2}(s_1, s_2) - h_1(s_1, s_2)y^{\sigma_1\sigma_2}(s_1, s_2)) \\
&= y^{\sigma_1\sigma_2}(s_1, s_2)x^{\Delta_1\Delta_2}_{s_1 s_2}(s_1, s_2) - x(s_1, s_2)y^{\Delta_1\Delta_2}_{s_1 s_2}(s_1, s_2),
\end{aligned}$$

and

$$\begin{aligned}
&(y^{\sigma_1}(s_1, s_2)x^{\Delta_1}_{s_1}(s_1, s_2))^{\Delta_2}_{s_2}(s_1, s_2) - (x(s_1, s_2)y^{\Delta_2}_{s_2}(s_1, s_2))^{\Delta_1}_{s_1}(s_1, s_2) \\
&\quad = y^{\sigma_1\sigma_2}(s_1, s_2)x^{\Delta_1\Delta_2}_{s_1 s_2}(s_1, s_2) + y^{\Delta_2\sigma_1}_{s_2}(s_1, s_2)x^{\Delta_1}_{s_1}(s_1, s_2) \\
&\qquad - x(s_1, s_2)y^{\Delta_1\Delta_2}_{s_1 s_2}(s_1, s_2) - x^{\Delta_1}_{s_1}(s_1, s_2)y^{\Delta_2\sigma_1}_{s_2}(s_1, s_2) \\
&\quad = y^{\sigma_1\sigma_2}(s_1, s_2)x^{\Delta_1\Delta_2}_{s_1 s_2}(s_1, s_2) - x(s_1, s_2)y^{\Delta_1\Delta_2}_{s_1 s_2}(s_1, s_2),
\end{aligned} \tag{7.8}$$

i. e.,

$$\begin{aligned}
&y^{\sigma_1\sigma_2}(s_1, s_2)L(x)(s_1, s_2) - x(s_1, s_2)M(y)(s_1, s_2) \\
&\quad = (y^{\sigma_1}x^{\Delta_1}_{s_1})^{\Delta_2}_{s_2}(s_1, s_2) - (xy^{\Delta_2}_{s_2})^{\Delta_1}_{s_1}(s_1, s_2).
\end{aligned}$$

Applying Green's formula, we get

$$\begin{aligned}
&\iint_R v^{\sigma_1\sigma_2}(s_1, s_2)L(z)(s_1, s_2)\Delta_2 s_2\Delta_1 s_1 \\
&\quad = \iint_R (v^{\sigma_1\sigma_2}(s_1, s_2)L(z)(s_1, s_2) - z(s_1, s_2)M(v)(s_1, s_2))\Delta_2 s_2\Delta_1 s_1 \\
&\quad = \iint_R ((v^{\sigma_1}z^{\Delta_1}_{s_1})^{\Delta_2}_{s_2}(s_1, s_2) \\
&\qquad - (zv^{\Delta_2}_{s_2})^{\Delta_1}_{s_1}(s_1, s_2))\Delta_2 s_2\Delta_1 s_1 \\
&\quad = -\int_{\partial R} z(s_1, s_2)v^{\Delta_2}_{s_2}(s_1, s_2)\Delta_2 s_2 + v^{\sigma_1}(s_1, s_2)z^{\Delta_1}_{s_1}(s_1, s_2)\Delta_1 s_1 \\
&\quad = -\int_{t_2^0}^{t_2} z(t_1, s_2)v^{\Delta_2}_{s_2}(t_1, s_2)\Delta_2 s_2 \\
&\qquad + \int_{t_2^0}^{t_2} z(t_1^0, s_2)v^{\Delta_2}_{s_2}(t_1^0, s_2)\Delta_2 s_2 \\
&\qquad - \int_{t_1^0}^{t_1} v^{\sigma_1}(s_1, t_2^0)z^{\Delta_1}_{s_1}(s_1, t_2^0)\Delta_1 s_1 \\
&\qquad + \int_{t_1^0}^{t_1} v^{\sigma_1}(s_1, t_2)z^{\Delta_1}_{s_1}(s_1, t_2)\Delta_1 s_1
\end{aligned}$$

$$\begin{aligned}
&= -\int_{t_1^0}^{t_1} v^{\sigma_1}(s_1, t_2^0) z_{s_1}^{\Delta_1}(s_1, t_2^0)\Delta_1 s_1 \\
&= -\int_{t_1^0}^{t_1} z_{s_1}^{\Delta_1}(s_1, t_2^0)\Delta_1 s_1 \\
&= -z(t_1, t_2^0) + z(t_1^0, t_2^0) \\
&= z(t_1^0, t_2^0),
\end{aligned}$$

i. e.,

$$\begin{aligned}
z(t_1^0, t_2^0) &= \iint_R v(s_1, s_2) L(z)(s_1, s_2)\Delta_2 s_2 \Delta_1 s_1 \\
&\leq \int_{t_1}^{t_1^0}\int_{t_2}^{t_2^0} v(s_1, s_2) g_1(s_1, s_2)\Delta_2 s_2 \Delta_1 s_1,
\end{aligned}$$

and from here

$$\begin{aligned}
\phi_{t_1^0 t_2^0}^{\Delta_1\Delta_2}(t_1^0, t_2^0) &\leq c(t_1^0, t_2^0)(a(t_1^0, t_2^0) + b(t_1^0, t_2^0)\phi(t_1^0, t_2^0)) \\
&\quad + p(t_1^0, t_2^0) z(t_1^0, t_2^0) \\
&\leq b(t_1^0, t_2^0) c(t_1^0, t_2^0)\phi(t_1^0, t_2^0) \\
&\quad + a(t_1^0, t_2^0) c(t_1^0, t_2^0) \\
&\quad + p(t_1^0, t_2^0) \int_{t_1}^{t_1^0}\int_{t_2}^{t_2^0} v(s_1, s_2) g_1(s_1, s_2)\Delta_2 s_2 \Delta_1 s_1.
\end{aligned}$$

Let

$$\begin{aligned}
g_2(t_1^0, t_2^0) &= b(t_1^0, t_2^0) c(t_1^0, t_2^0), \\
h_2(t_1^0, t_2^0) &= p(t_1^0, t_2^0) \int_{t_1}^{t_1^0}\int_{t_2}^{t_2^0} v(s_1, s_2) g_1(s_1, s_2)\Delta_2 s_2 \Delta_1 s_1 \\
&\quad + a(t_1^0, t_2^0) c(t_1^0, t_2^0).
\end{aligned}$$

Then

$$\phi_{t_1^0 t_2^0}^{\Delta_1\Delta_2}(t_1^0, t_2^0) \leq g_2(t_1^0, t_2^0) z(t_1^0, t_2^0) + h_2(t_1^0, t_2^0),$$

or

$$\phi_{t_1^0 t_2^0}^{\Delta_1\Delta_2}(t_1^0, t_2^0) - g_2(t_1^0, t_2^0) z(t_1^0, t_2^0) \leq h_2(t_1^0, t_2^0).$$

We fix $(t_1, t_2) \in D$, $t_1 \geq t_1^0$, $t_2 \geq t_2^0$. Let R and ∂R be as in the proof of Theorem 7.2.1. For $(s_1, s_2) \in R$ and $x \in \mathcal{C}^2_{1,2}(\overline{D})$, we define the operators

$$L_1(x)(s_1, s_2) = x^{\Delta_1\Delta_2}_{s_1 s_2}(s_1, s_2) - g_2(s_1, s_2)x(s_1, s_2),$$
$$M_1(x)(s_1, s_2) = x^{\Delta_1\Delta_2}_{s_1 s_2}(s_1, s_2) - g_2(s_1, s_2)x^{\sigma_1\sigma_2}(s_1, s_2).$$

Then, for $x, y \in \mathcal{C}^2_{1,2}(\overline{D})$ and $(s_1, s_2) \in D$, we have

$$\begin{aligned}
&y^{\sigma_1\sigma_2}(s_1, s_2)L_1(x)(s_1, s_2) - x(s_1, s_2)M_1(y)(s_1, s_2)\\
&\quad = y^{\sigma_1\sigma_2}(s_1, s_2)(x^{\Delta_1\Delta_2}_{s_1 s_2}(s_1, s_2) - g_2(s_1, s_2)x(s_1, s_2))\\
&\qquad - x(s_1, s_2)(y^{\Delta_1\Delta_2}_{s_1 s_2}(s_1, s_2) - g_2(s_1, s_2)y^{\sigma_1\sigma_2}(s_1, s_2))\\
&\quad = y^{\sigma_1\sigma_2}(s_1, s_2)x^{\Delta_1\Delta_2}_{s_1 s_2}(s_1, s_2) - x(s_1, s_2)y^{\Delta_1\Delta_2}_{s_1 s_2}(s_1, s_2),
\end{aligned}$$

and using (7.8), we get

$$\begin{aligned}
&y^{\sigma_1\sigma_2}(s_1, s_2)L_1(x)(s_1, s_2) - x(s_1, s_2)M_1(y)(s_1, s_2)\\
&\quad = (y^{\sigma_1}x^{\Delta_1}_{s_1})^{\Delta_2}_{s_2}(s_1, s_2) - (xy^{\Delta_2}_{s_2})^{\Delta_1}_{s_1}(s_1, s_2).
\end{aligned}$$

Applying Green's formula, we get

$$\begin{aligned}
&\iint_R w^{\sigma_1\sigma_2}(s_1, s_2)L_1(\phi)(s_1, s_2)\Delta_2 s_2\Delta_1 s_1\\
&\quad = \iint_R (w^{\sigma_1\sigma_2}(s_1, s_2)L_1(\phi)(s_1, s_2) - \phi(s_1, s_2)M_1(w)(s_1, s_2))\Delta_2 s_2\Delta_1 s_1\\
&\quad = \iint_R ((w^{\sigma_1}\phi^{\Delta_1}_{s_1})^{\Delta_2}_{s_2}(s_1, s_2)\\
&\qquad - (\phi w^{\Delta_2}_{s_2})^{\Delta_1}_{s_1}(s_1, s_2))\Delta_2 s_2\Delta_1 s_1\\
&\quad = -\int_{\partial R} \phi(s_1, s_2)w^{\Delta_2}_{s_2}(s_1, s_2)\Delta_2 s_2 + w^{\sigma_1}(s_1, s_2)\phi^{\Delta_1}_{s_1}(s_1, s_2)\Delta_1 s_1\\
&\quad = -\int_{t_2^0}^{t_2} \phi(t_1, s_2)w^{\Delta_2}_{s_2}(t_1, s_2)\Delta_2 s_2\\
&\qquad + \int_{t_2^0}^{t_2} \phi(t_1^0, s_2)w^{\Delta_2}_{s_2}(t_1^0, s_2)\Delta_2 s_2\\
&\qquad - \int_{t_1^0}^{t_1} w^{\sigma_1}(s_1, t_2^0)\phi^{\Delta_1}_{s_1}(s_1, t_2^0)\Delta_1 s_1
\end{aligned}$$

$$
\begin{aligned}
&+\int_{t_1^0}^{t_1} w^{\sigma_1}(s_1,t_2)\phi_{s_1}^{\Delta_1}(s_1,t_2)\Delta_1 s_1\\
&=-\int_{t_1^0}^{t_1} w^{\sigma_1}(s_1,t_2^0)\phi_{s_1}^{\Delta_1}(s_1,t_2^0)\Delta_1 s_1\\
&=-\int_{t_1^0}^{t_1} \phi_{s_1}^{\Delta_1}(s_1,t_2^0)\Delta_1 s_1\\
&=-\phi(t_1,t_2^0)+\phi(t_1^0,t_2^0)\\
&=\phi(t_1^0,t_2^0),
\end{aligned}
$$

i. e.,

$$
\begin{aligned}
\phi(t_1^0,t_2^0) &= \iint_R w(s_1,s_2)L_1(\phi)(s_1,s_2)\Delta_2 s_2\Delta_1 s_1\\
&\le \int_{t_1}^{t_1^0}\int_{t_2}^{t_2^0} w(s_1,s_2)h_2(s_1,s_2)\Delta_2 s_2\Delta_1 s_1,
\end{aligned}
$$

$(t_1,t_2)\in D$, $t_1\ge t_1^0$, $t_2\ge t_2^0$. Then

$$
\begin{aligned}
u(t_1^0,t_2^0) &\le a(t_1^0,t_2^0)+b(t_1^0,t_2^0)\phi(t_1^0,t_2^0)\\
&\le a(t_1^0,t_2^0)\\
&\quad + b(t_1^0,t_2^0)\int_{t_1}^{t_1^0}\int_{t_2}^{t_2^0} p(s_1,s_2)w(s_1,s_2)\\
&\quad \times\Bigg(\int_{t_1}^{s_1}\int_{t_2}^{s_2} v(\xi_1,\xi_2)a(\xi_1,\xi_2)(c(\xi_1,\xi_2)\\
&\quad + q(\xi_1,\xi_2))\Delta_2\xi_2\Delta_1\xi_1 + a(s_1,s_2)c(s_1,s_2)\Bigg)\Delta_2 s_2\Delta_1 s_1,
\end{aligned}
$$

$(t_1,t_2)\in D$, $t_1\ge t_1^0$, $t_2\ge t_2^0$. This completes the proof. □

Theorem 7.2.4 (Snow's Inequality). *Let a, b, c, h, p, q, and u be nonnegative continuous functions on D. Let also, for $(t_1^0,t_2^0)\in D$, v_1, v_2, and v_3 be the nonnegative solutions of the problems*

$$
v_{1t_1t_2}^{\Delta_1\Delta_2} - g_1(t_1,t_2)v_1 = 0,
$$

$(t_1,t_2)\in D$, $t_1\ge t_1^0$, $t_2\ge t_2^0$,

$$
v_1(t_1,t_2^0)=1,\quad v_{1t_2}^{\Delta_2}(t_1^0,t_2)=0,
$$

$(t_1, t_2) \in D$, $t_1 \geq t_1^0$, $t_2 \geq t_2^0$,

$$v_{2t_1t_2}^{\Delta_1\Delta_2} - g_2(t_1, t_2)v_2 = 0,$$

$(t_1, t_2) \in D$, $t_1 \geq t_1^0$, $t_2 \geq t_2^0$,

$$v_2(t_1, t_2^0) = 1, \quad v_{2t_2}^{\Delta_2}(t_1^0, t_2) = 0,$$

$(t_1, t_2) \in D$, $t_1 \geq t_1^0$, $t_2 \geq t_2^0$,

$$v_{3t_1t_2}^{\Delta_1\Delta_2} - b(t_1, t_2)v_3 = 0,$$

$(t_1, t_2) \in D$, $t_1 \geq t_1^0$, $t_2 \geq t_2^0$,

$$v_3(t_1, t_2^0) = 1, \quad v_{3t_2}^{\Delta_2}(t_1^0, t_2) = 0,$$

$(t_1, t_2) \in D$, $t_1 \geq t_1^0$, $t_2 \geq t_2^0$, *respectively. If*

$$\begin{aligned}
u(t_1^0, t_2^0) \leq a(t_1^0, t_2^0) &+ \int_{t_1}^{t_1^0}\int_{t_2}^{t_2^0} b(s_1, s_2)u(s_1, s_2)\Delta_2 s_2 \Delta_1 s_1 \\
&+ \int_{t_1}^{t_1^0}\int_{t_2}^{t_2^0} c(s_1, s_2)\left(\int_{t_1}^{s_1}\int_{t_2}^{s_2} h(\xi_1, \xi_2)u(\xi_1, \xi_2)\Delta_2\xi_2\Delta_1\xi_1\right)\Delta_2 s_2 \Delta_1 s_1 \\
&+ \int_{t_1}^{t_1^0}\int_{t_2}^{t_2^0} c(s_1, s_2)\left(\int_{t_1}^{s_1}\int_{t_2}^{s_2} p(\xi_1, \xi_2)\right. \\
&\times \left.\left(\int_{t_1}^{\xi_1}\int_{t_2}^{\xi_2} q(\eta_1, \eta_2)u(\eta_1, \eta_2)\Delta_2\eta_2\Delta_1\eta_1\right)\Delta_2\xi_2\Delta_1\xi_1\right)\Delta_2 s_2 \Delta_1 s_1,
\end{aligned}$$

$(t_1, t_2) \in D$, $t_1 \geq t_1^0$, $t_2 \geq t_2^0$, *then*

$$u(t_1^0, t_2^0) \leq a(t_1^0, t_2^0) + \int_{t_1}^{t_1^0}\int_{t_2}^{t_2^0} v_3(s_1, s_2)h_3(s_1, s_2)\Delta_2 s_2 \Delta_1 s_1,$$

$(t_1, t_2) \in D$, $t_1 \geq t_1^0$, $t_2 \geq t_2^0$, *where*

$$\begin{aligned}
h_1(t_1^0, t_2^0) &= a(t_1^0, t_2^0)b(t_1^0, t_2^0) + h(t_1^0, t_2^0)a(t_1^0, t_2^0) \\
&\quad + q(t_1^0, t_2^0)a(t_1^0, t_2^0), \\
g_1(t_1^0, t_2^0) &= b(t_1^0, t_2^0) + c(t_1^0, t_2^0) + h(t_1^0, t_2^0) + p(t_1^0, t_2^0) \\
&\quad + q(t_1^0, t_2^0),
\end{aligned}$$

$$
\begin{aligned}
h_2(t_1^0,t_2^0) &= a(t_1^0,t_2^0)(b(t_1^0,t_2^0)+h(t_1^0,t_2^0)) \\
&\quad + p(t_1^0,t_2^0) \\
&\quad \times \int_{t_1}^{t_1^0}\int_{t_2}^{t_2^0} v_1(s_1,s_2)h_1(s_1,s_2)\Delta_2 s_2\Delta_1 s_1, \\
g_2(t_1^0,t_2^0) &= b(t_1^0,t_2^0)+c(t_1^0,t_2^0)+h(t_1^0,t_2^0), \\
h_3(t_1^0,t_2^0) &= a(t_1^0,t_2^0)b(t_1^0,t_2^0) \\
&\quad + c(t_1^0,t_2^0) \\
&\quad \times \int_{t_1}^{t_1^0}\int_{t_2}^{t_2^0} v_2(s_1,s_2)h_2(s_1,s_2)\Delta_2 s_2\Delta_1 s_1,
\end{aligned}
$$

$(t_1,t_2)\in D$, $t_1\ge t_1^0$, $t_2\ge t_2^0$.

Proof. Let

$$
\begin{aligned}
\phi(t_1^0,t_2^0) &= \int_{t_1}^{t_1^0}\int_{t_2}^{t_2^0} b(s_1,s_2)u(s_1,s_2)\Delta_2 s_2\Delta_1 s_1 \\
&\quad + \int_{t_1}^{t_1^0}\int_{t_2}^{t_2^0} c(s_1,s_2)\left(\int_{t_1}^{s_1}\int_{t_2}^{s_2} h(\xi_1,\xi_2)u(\xi_1,\xi_2)\Delta_2\xi_2\Delta_1\xi_1\right)\Delta_2 s_2\Delta_1 s_1 \\
&\quad + \int_{t_1}^{t_1^0}\int_{t_2}^{t_2^0} c(s_1,s_2)\left(\int_{t_1}^{s_1}\int_{t_2}^{s_2} p(\xi_1,\xi_2)\right. \\
&\quad \left.\times\left(\int_{t_1}^{\xi_1}\int_{t_2}^{\xi_2} q(\eta_1,\eta_2)u(\eta_1,\eta_2)\Delta_2\eta_2\Delta_1\eta_1\right)\Delta_2\xi_2\Delta_1\xi_1\right)\Delta_2 s_2\Delta_1 s_1,
\end{aligned}
$$

$(t_1,t_2)\in D$, $t_1\ge t_1^0$, $t_2\ge t_2^0$. Then

$$
\begin{aligned}
u(t_1^0,t_2^0) &\le a(t_1^0,t_2^0)+\phi(t_1^0,t_2^0), \\
\phi(t_1,t_2^0) &= 0, \\
\phi(t_1^0,t_2) &= 0, \\
\phi^{\Delta_1\Delta_2}_{t_1^0t_2^0}(t_1^0,t_2^0) &= b(t_1^0,t_2^0)u(t_1^0,t_2^0) \\
&\quad + c(t_1^0,t_2^0)\left(\int_{t_1}^{t_1^0}\int_{t_2}^{t_2^0} h(s_1,s_2)u(s_1,s_2)\Delta_2 s_2\Delta_1 s_1\right)
\end{aligned}
$$

$$
\begin{aligned}
&+ c(t_1^0, t_2^0) \int_{t_1}^{t_1^0} \int_{t_2}^{t_2^0} p(s_1, s_2) \left(\int_{t_1}^{s_1} \int_{t_2}^{s_2} q(\xi_1, \xi_2) \right. \\
&\left. \times u(\xi_1, \xi_2) \Delta_2 \xi_2 \Delta_1 \xi_1 \right) \Delta_2 s_2 \Delta_1 s_1 \\
&\leq b(t_1^0, t_2^0)(a(t_1^0, t_2^0) + \phi(t_1^0, t_2^0)) \\
&+ c(t_1^0, t_2^0) \left(\phi(t_1^0, t_2^0) \right. \\
&+ \int_{t_1}^{t_1^0} \int_{t_2}^{t_2^0} h(s_1, s_2)(a(s_1, s_2) + \phi(s_1, s_2)) \Delta_2 s_2 \Delta_1 s_1 \\
&+ \int_{t_1}^{t_1^0} \int_{t_2}^{t_2^0} p(s_1, s_2) \left(\int_{t_1}^{s_1} \int_{t_2}^{s_2} q(\xi_1, \xi_2) \right. \\
&\left. \left. \times (a(\xi_1, \xi_2) + \phi(\xi_1, \xi_2)) \Delta_2 \xi_2 \Delta_1 \xi_1 \right) \Delta_2 s_2 \Delta_1 s_1 \right),
\end{aligned}
$$

$(t_1, t_2) \in D$, $t_1 \geq t_1^0$, $t_2 \geq t_2^0$. Let

$$
\begin{aligned}
\psi(t_1^0, t_2^0) &= \phi(t_1^0, t_2^0) \\
&+ \int_{t_1}^{t_1^0} \int_{t_2}^{t_2^0} h(s_1, s_2)(a(s_1, s_2) + \phi(s_1, s_2)) \Delta_2 s_2 \Delta_1 s_1 \\
&+ \int_{t_1}^{t_1^0} \int_{t_2}^{t_2^0} p(s_1, s_2) \left(\int_{t_1}^{s_1} \int_{t_2}^{s_2} q(\xi_1, \xi_2) \right. \\
&\left. \times (a(\xi_1, \xi_2) + \phi(\xi_1, \xi_2)) \Delta_2 \xi_2 \Delta_1 \xi_1 \right) \Delta_2 s_2 \Delta_1 s_1,
\end{aligned}
$$

$(t_1, t_2) \in D$, $t_1 \geq t_1^0$, $t_2 \geq t_2^0$. Then

$$
\begin{aligned}
\psi(t_1, t_2^0) &= \phi(t_1, t_2^0) \\
&= 0, \\
\psi(t_1^0, t_2) &= \phi(t_1^0, t_2) \\
&= 0, \\
\phi(t_1^0, t_2^0) &\leq \psi(t_1^0, t_2^0), \\
\phi_{t_1^0 t_2^0}^{\Delta_1 \Delta_2}(t_1^0, t_2^0) &\leq b(t_1^0, t_2^0)(a(t_1^0, t_2^0) + \psi(t_1^0, t_2^0)) \\
&+ c(t_1^0, t_2^0) \psi(t_1^0, t_2^0)
\end{aligned}
$$

$$
\begin{aligned}
&= a(t_1^0,t_2^0)b(t_1^0,t_2^0)\\
&\quad + (b(t_1^0,t_2^0)+c(t_1^0,t_2^0))\psi(t_1^0,t_2^0),\\
\psi^{\Delta_1\Delta_2}_{t_1^0t_2^0}(t_1^0,t_2^0) &= \phi^{\Delta_1\Delta_2}_{t_1^0t_2^0}(t_1^0,t_2^0)\\
&\quad + h(t_1^0,t_2^0)(a(t_1^0,t_2^0)+\phi(t_1^0,t_2^0))\\
&\quad + p(t_1^0,t_2^0)\int_{t_1}^{t_1^0}\int_{t_2}^{t_2^0} q(s_1,s_2)\\
&\quad \times (a(s_1,s_2)+\phi(s_1,s_2))\Delta_2 s_2\Delta_1 s_1\\
&\le a(t_1^0,t_2^0)b(t_1^0,t_2^0)\\
&\quad + (b(t_1^0,t_2^0)+c(t_1^0,t_2^0))\psi(t_1^0,t_2^0)\\
&\quad + h(t_1^0,t_2^0)(a(t_1^0,t_2^0)+\psi(t_1^0,t_2^0))\\
&\quad + p(t_1^0,t_2^0)\Bigg(\psi(t_1^0,t_2^0)\\
&\quad + \int_{t_1}^{t_1^0}\int_{t_2}^{t_2^0} q(s_1,s_2)(a(s_1,s_2)+\psi(s_1,s_2))\Delta_2 s_2\Delta_1 s_1\Bigg),
\end{aligned}
$$

$(t_1,t_2)\in D$, $t_1\ge t_1^0$, $t_2\ge t_2^0$. Let

$$
\begin{aligned}
z(t_1^0,t_2^0) &= \psi(t_1^0,t_2^0)\\
&\quad + \int_{t_1}^{t_1^0}\int_{t_2}^{t_2^0} q(s_1,s_2)(a(s_1,s_2)+\psi(s_1,s_2))\Delta_2 s_2\Delta_1 s_1,
\end{aligned}
$$

$(t_1,t_2)\in D$, $t_1\ge t_1^0$, $t_2\ge t_2^0$. Then

$$
\begin{aligned}
\psi(t_1^0,t_2^0) &\le z(t_1^0,t_2^0),\\
z(t_1,t_2^0) &= \psi(t_1,t_2^0)\\
&= 0,\\
z(t_1^0,t_2) &= \psi(t_1^0,t_2)\\
&= 0,\\
\psi^{\Delta_1\Delta_2}_{t_1^0t_2^0}(t_1^0,t_2^0) &\le a(t_1^0,t_2^0)b(t_1^0,t_2^0)\\
&\quad + (b(t_1^0,t_2^0)+c(t_1^0,t_2^0))z(t_1^0,t_2^0)\\
&\quad + h(t_1^0,t_2^0)(a(t_1^0,t_2^0)+z(t_1^0,t_2^0))\\
&\quad + p(t_1^0,t_2^0)z(t_1^0,t_2^0),\\
z^{\Delta_1\Delta_2}_{t_1^0t_2^0}(t_1^0,t_2^0) &= \psi^{\Delta_1\Delta_2}_{t_1^0t_2^0}(t_1^0,t_2^0)\\
&\quad + q(t_1^0,t_2^0)(a(t_1^0,t_2^0)+\psi(t_1^0,t_2^0))
\end{aligned}
$$

$$
\begin{aligned}
&\leq a(t_1^0,t_2^0)b(t_1^0,t_2^0)+h(t_1^0,t_2^0)a(t_1^0,t_2^0)\\
&\quad+q(t_1^0,t_2^0)a(t_1^0,t_2^0)\\
&\quad+(b(t_1^0,t_2^0)+c(t_1^0,t_2^0)+h(t_1^0,t_2^0)\\
&\quad+p(t_1^0,t_2^0)+q(t_1^0,t_2^0))z(t_1^0,t_2^0)\\
&=g_1(t_1^0,t_2^0)z(t_1^0,t_2^0)+h_1(t_1^0,t_2^0),
\end{aligned}
$$

or

$$z^{\Delta_1\Delta_2}_{t_1^0t_2^0}(t_1^0,t_2^0)-g_1(t_1^0,t_2^0)z(t_1^0,t_2^0)\leq h_1(t_1^0,t_2^0).$$

We fix $(t_1,t_2)\in D$, $t_1\geq t_1^0$, $t_2\geq t_2^0$. Let R and ∂R be as in the proof of Theorem 7.2.1. For $(s_1,s_2)\in R$ and $x\in C^2_{1,2}(\overline{D})$, we define the operators

$$
\begin{aligned}
L(x)(s_1,s_2)&=x^{\Delta_1\Delta_2}_{s_1s_2}(s_1,s_2)-g_1(s_1,s_2)x(s_1,s_2),\\
M(x)(s_1,s_2)&=x^{\Delta_1\Delta_2}_{s_1s_2}(s_1,s_2)-g_1(s_1,s_2)x^{\sigma_1\sigma_2}(s_1,s_2).
\end{aligned}
$$

Then, for $x,y\in C^2_{1,2}(\overline{D})$ and $(s_1,s_2)\in D$, we have

$$
\begin{aligned}
&y^{\sigma_1\sigma_2}(s_1,s_2)L(x)(s_1,s_2)-x(s_1,s_2)M(y)(s_1,s_2)\\
&\quad=y^{\sigma_1\sigma_2}(s_1,s_2)(x^{\Delta_1\Delta_2}_{s_1s_2}(s_1,s_2)-g_1(s_1,s_2)x(s_1,s_2))\\
&\qquad-x(s_1,s_2)(y^{\Delta_1\Delta_2}_{s_1s_2}(s_1,s_2)-g_1(s_1,s_2)y^{\sigma_1\sigma_2}(s_1,s_2))\\
&\quad=y^{\sigma_1\sigma_2}(s_1,s_2)x^{\Delta_1\Delta_2}_{s_1s_2}(s_1,s_2)-x(s_1,s_2)y^{\Delta_1\Delta_2}_{s_1s_2}(s_1,s_2),
\end{aligned}
$$

and

$$
\begin{aligned}
&(y^{\sigma_1}x^{\Delta_1}_{s_1})^{\Delta_2}_{s_2}(s_1,s_2)-(xy^{\Delta_2}_{s_2})^{\Delta_1}_{s_1}(s_1,s_2)\\
&\quad=y^{\sigma_1\sigma_2}(s_1,s_2)x^{\Delta_1\Delta_2}_{s_1s_2}(s_1,s_2)+y^{\Delta_2\sigma_1}_{s_2}(s_1,s_2)x^{\Delta_1}_{s_1}(s_1,s_2)\\
&\qquad-x(s_1,s_2)y^{\Delta_1\Delta_2}_{s_1s_2}(s_1,s_2)-x^{\Delta_1}_{s_1}(s_1,s_2)y^{\Delta_2\sigma_1}_{s_2}(s_1,s_2)\\
&\quad=y^{\sigma_1\sigma_2}(s_1,s_2)x^{\Delta_1\Delta_2}_{s_1s_2}(s_1,s_2)-x(s_1,s_2)y^{\Delta_1\Delta_2}_{s_1s_2}(s_1,s_2),
\end{aligned}
\tag{7.9}
$$

i. e.,

$$
\begin{aligned}
&y^{\sigma_1\sigma_2}(s_1,s_2)L(x)(s_1,s_2)-x(s_1,s_2)M(y)(s_1,s_2)\\
&\quad=(y^{\sigma_1}x^{\Delta_1}_{s_1})^{\Delta_2}_{s_2}(s_1,s_2)-(xy^{\Delta_2}_{s_2})^{\Delta_1}_{s_1}(s_1,s_2).
\end{aligned}
$$

Applying Green's formula, we get

$$
\begin{aligned}
&\iint_R v_1^{\sigma_1\sigma_2}(s_1,s_2)L(z)(s_1,s_2)\Delta_2s_2\Delta_1s_1\\
&\quad=\iint_R(v_1^{\sigma_1\sigma_2}(s_1,s_2)L(z)(s_1,s_2)-z(s_1,s_2)M(v_1)(s_1,s_2))\Delta_2s_2\Delta_1s_1
\end{aligned}
$$

$$
\begin{aligned}
&= \iint\limits_R \left((v_1^{\sigma_1} z_{s_1}^{\Delta_1})_{s_2}^{\Delta_2}(s_1,s_2) \right. \\
&\qquad \left. - (z v_{1s_2}^{\Delta_2})_{s_1}^{\Delta_1}(s_1,s_2) \right) \Delta_2 s_2 \Delta_1 s_1 \\
&= -\int\limits_{\partial R} z(s_1,s_2) v_{s_2}^{\Delta_2}(s_1,s_2)\Delta_2 s_2 + v_1^{\sigma_1}(s_1,s_2) z_{s_1}^{\Delta_1}(s_1,s_2)\Delta_1 s_1 \\
&= -\int_{t_2^0}^{t_2} z(t_1,s_2) v_{1s_2}^{\Delta_2}(t_1,s_2)\Delta_2 s_2 \\
&\qquad + \int_{t_2^0}^{t_2} z(t_1^0,s_2) v_{1s_2}^{\Delta_2}(t_1^0,s_2)\Delta_2 s_2 \\
&\qquad - \int_{t_1^0}^{t_1} v_1^{\sigma_1}(s_1,t_2^0) z_{s_1}^{\Delta_1}(s_1,t_2^0)\Delta_1 s_1 \\
&\qquad + \int_{t_1^0}^{t_1} v_1^{\sigma_1}(s_1,t_2) z_{s_1}^{\Delta_1}(s_1,t_2)\Delta_1 s_1 \\
&= -\int_{t_1^0}^{t_1} v_1^{\sigma_1}(s_1,t_2^0) z_{s_1}^{\Delta_1}(s_1,t_2^0)\Delta_1 s_1 \\
&= -\int_{t_1^0}^{t_1} z_{s_1}^{\Delta_1}(s_1,t_2^0)\Delta_1 s_1 \\
&= -z(t_1,t_2^0) + z(t_1^0,t_2^0) \\
&= z(t_1^0,t_2^0),
\end{aligned}
$$

i. e.,

$$
\begin{aligned}
z(t_1^0,t_2^0) &= \iint\limits_R v_1(s_1,s_2) L(z)(s_1,s_2)\Delta_2 s_2 \Delta_1 s_1 \\
&\le \int_{t_1}^{t_1^0}\int_{t_2}^{t_2^0} v_1(s_1,s_2) h_1(s_1,s_2)\Delta_2 s_2 \Delta_1 s_1.
\end{aligned}
$$

Hence,

$$
\begin{aligned}
\psi_{t_1^0 t_2^0}^{\Delta_1\Delta_2}(t_1^0,t_2^0) &\le a(t_1^0,t_2^0)(b(t_1^0,t_2^0) + h(t_1^0,t_2^0)) \\
&\quad + p(t_1^0,t_2^0) z(t_1^0,t_2^0) \\
&\quad + (b(t_1^0,t_2^0) + c(t_1^0,t_2^0) \\
&\quad + h(t_1^0,t_2^0))\psi(t_1^0,t_2^0)
\end{aligned}
$$

$$\begin{aligned}
&\le a(t_1^0,t_2^0)(b(t_1^0,t_2^0)+h(t_1^0,t_2^0))\\
&\quad + p(t_1^0,t_2^0)\\
&\quad \times \int_{t_1}^{t_1^0}\int_{t_2}^{t_2^0} v_1(s_1,s_2)h_1(s_1,s_2)\Delta_2 s_2\Delta_1 s_1\\
&\quad + (b(t_1^0,t_2^0)+c(t_1^0,t_2^0)\\
&\quad + h(t_1^0,t_2^0))\psi(t_1^0,t_2^0)\\
&= g_2(t_1^0,t_2^0)\psi(t_1^0,t_2^0)+h_2(t_1^0,t_2^0),
\end{aligned}$$

or

$$\psi^{\Delta_1\Delta_2}_{t_1^0 t_2^0}(t_1^0,t_2^0) - g_2(t_1^0,t_2^0)\psi(t_1^0,t_2^0) \le h_2(t_1^0,t_2^0).$$

We fix $(t_1,t_2)\in D$, $t_1\ge t_1^0$, $t_2\ge t_2^0$. Let R and ∂R be as in the proof of Theorem 7.2.1. For $(s_1,s_2)\in R$ and $x\in \mathcal{C}^2_{1,2}(\overline{D})$, we define the operators

$$\begin{aligned}
L_1(x)(s_1,s_2) &= x^{\Delta_1\Delta_2}_{s_1 s_2}(s_1,s_2) - g_2(s_1,s_2)x(s_1,s_2),\\
M_1(x)(s_1,s_2) &= x^{\Delta_1\Delta_2}_{s_1 s_2}(s_1,s_2) - g_2(s_1,s_2)x^{\sigma_1\sigma_2}(s_1,s_2).
\end{aligned}$$

Then, for $x,y\in \mathcal{C}^2_{1,2}(\overline{D})$ and $(s_1,s_2)\in D$, we have

$$\begin{aligned}
&y^{\sigma_1\sigma_2}(s_1,s_2)L_1(x)(s_1,s_2) - x(s_1,s_2)M_1(y)(s_1,s_2)\\
&\quad = y^{\sigma_1\sigma_2}(s_1,s_2)(x^{\Delta_1\Delta_2}_{s_1 s_2}(s_1,s_2) - g_2(s_1,s_2)x(s_1,s_2))\\
&\qquad - x(s_1,s_2)(y^{\Delta_1\Delta_2}_{s_1 s_2}(s_1,s_2) - g_2(s_1,s_2)y^{\sigma_1\sigma_2}(s_1,s_2))\\
&\quad = y^{\sigma_1\sigma_2}(s_1,s_2)x^{\Delta_1\Delta_2}_{s_1 s_2}(s_1,s_2) - x(s_1,s_2)y^{\Delta_1\Delta_2}_{s_1 s_2}(s_1,s_2),
\end{aligned}$$

and, using (7.9), get

$$\begin{aligned}
&y^{\sigma_1\sigma_2}(s_1,s_2)L_1(x)(s_1,s_2) - x(s_1,s_2)M_1(y)(s_1,s_2)\\
&\quad = (y^{\sigma_1}x^{\Delta_1}_{s_1})^{\Delta_2}_{s_2}(s_1,s_2) - (xy^{\Delta_2}_{s_2})^{\Delta_1}_{s_1}(s_1,s_2).
\end{aligned}$$

Applying Green's formula, we get

$$\begin{aligned}
&\iint_R v_2^{\sigma_1\sigma_2}(s_1,s_2)L_1(\psi)(s_1,s_2)\Delta_2 s_2\Delta_1 s_1\\
&\quad = \iint_R (v_2^{\sigma_1\sigma_2}(s_1,s_2)L_1(\psi)(s_1,s_2) - \psi(s_1,s_2)M_1(v_2)(s_1,s_2))\Delta_2 s_2\Delta_1 s_1\\
&\quad = \iint_R ((v_2^{\sigma_1}\psi^{\Delta_1}_{s_1})^{\Delta_2}_{s_2}(s_1,s_2)\\
&\qquad - (\psi v^{\Delta_2}_{2s_2})^{\Delta_1}_{s_1}(s_1,s_2))\Delta_2 s_2\Delta_1 s_1
\end{aligned}$$

$$
\begin{aligned}
&= -\int_{\partial R} \psi(s_1,s_2)v_{2s_2}^{\Delta_2}(s_1,s_2)\Delta_2 s_2 + v_2^{\sigma_1}(s_1,s_2)\psi_{s_1}^{\Delta_1}(s_1,s_2)\Delta_1 s_1 \\
&= -\int_{t_2^0}^{t_2} \psi(t_1,s_2)v_{2s_2}^{\Delta_2}(t_1,s_2)\Delta_2 s_2 \\
&\quad + \int_{t_2^0}^{t_2} \psi(t_1^0,s_2)v_{2s_2}^{\Delta_2}(t_1^0,s_2)\Delta_2 s_2 \\
&\quad - \int_{t_1^0}^{t_1} v_2^{\sigma_1}(s_1,t_2^0)\psi_{s_1}^{\Delta_1}(s_1,t_2^0)\Delta_1 s_1 \\
&\quad + \int_{t_1^0}^{t_1} v_2^{\sigma_1}(s_1,t_2)\psi_{s_1}^{\Delta_1}(s_1,t_2)\Delta_1 s_1 \\
&= -\int_{t_1^0}^{t_1} v_2^{\sigma_1}(s_1,t_2^0)\psi_{s_1}^{\Delta_1}(s_1,t_2^0)\Delta_1 s_1 \\
&= -\int_{t_1^0}^{t_1} \psi_{s_1}^{\Delta_1}(s_1,t_2^0)\Delta_1 s_1 \\
&= -\psi(t_1,t_2^0) + \psi(t_1^0,t_2^0) \\
&= \psi(t_1^0,t_2^0),
\end{aligned}
$$

i. e.,

$$
\begin{aligned}
\psi(t_1^0,t_2^0) &= \iint_R v_2(s_1,s_2)L_1(\psi)(s_1,s_2)\Delta_2 s_2 \Delta_1 s_1 \\
&\le \int_{t_1^0}^{t_1}\int_{t_2^0}^{t_2} v_2(s_1,s_2)h_2(s_1,s_2)\Delta_2 s_2 \Delta_1 s_1.
\end{aligned}
$$

From here,

$$
\begin{aligned}
\phi_{t_1^0 t_2^0}^{\Delta_1\Delta_2}(t_1^0,t_2^0) &\le b(t_1^0,t_2^0)(a(t_1^0,t_2^0) + \phi(t_1^0,t_2^0)) \\
&\quad + c(t_1^0,t_2^0)\psi(t_1^0,t_2^0) \\
&\le a(t_1^0,t_2^0)b(t_1^0,t_2^0) \\
&\quad + c(t_1^0,t_2^0) \\
&\quad \times \int_{t_1^0}^{t_1}\int_{t_2^0}^{t_2} v_2(s_1,s_2)h_2(s_1,s_2)\Delta_2 s_2 \Delta_1 s_1 \\
&\quad + b(t_1^0,t_2^0)\phi(t_1^0,t_2^0)
\end{aligned}
$$

$$= b(t_1^0, t_2^0)\phi(t_1^0, t_2^0) + h_3(t_1^0, t_2^0),$$

or

$$\phi^{\Delta_1\Delta_2}_{t_1^0 t_2^0}(t_1^0, t_2^0) - b(t_1^0, t_2^0)\phi(t_1^0, t_2^0) \le h_3(t_1^0, t_2^0).$$

We fix $(t_1, t_2) \in D$, $t_1 \ge t_1^0$, $t_2 \ge t_2^0$. Let R and ∂R be as in the proof of Theorem 7.2.1. For $(s_1, s_2) \in R$ and $x \in \mathcal{C}^2_{1,2}(\overline{D})$, we define the operators

$$\begin{aligned}
L_2(x)(s_1, s_2) &= x^{\Delta_1\Delta_2}_{s_1 s_2}(s_1, s_2) - b(s_1, s_2)x(s_1, s_2),\\
M_2(x)(s_1, s_2) &= x^{\Delta_1\Delta_2}_{s_1 s_2}(s_1, s_2) - b(s_1, s_2)x^{\sigma_1\sigma_2}(s_1, s_2).
\end{aligned}$$

Then, for $x, y \in \mathcal{C}^2_{1,2}(\overline{D})$ and $(s_1, s_2) \in D$, we have

$$\begin{aligned}
&y^{\sigma_1\sigma_2}(s_1, s_2)L_2(x)(s_1, s_2) - x(s_1, s_2)M_2(y)(s_1, s_2)\\
&\quad = y^{\sigma_1\sigma_2}(s_1, s_2)(x^{\Delta_1\Delta_2}_{s_1 s_2}(s_1, s_2) - b(s_1, s_2)x(s_1, s_2))\\
&\qquad - x(s_1, s_2)(y^{\Delta_1\Delta_2}_{s_1 s_2}(s_1, s_2) - b(s_1, s_2)y^{\sigma_1\sigma_2}(s_1, s_2))\\
&\quad = y^{\sigma_1\sigma_2}(s_1, s_2)x^{\Delta_1\Delta_2}_{s_1 s_2}(s_1, s_2) - x(s_1, s_2)y^{\Delta_1\Delta_2}_{s_1 s_2}(s_1, s_2),
\end{aligned}$$

and, using (7.9), get

$$\begin{aligned}
&y^{\sigma_1\sigma_2}(s_1, s_2)L_2(x)(s_1, s_2) - x(s_1, s_2)M_2(y)(s_1, s_2)\\
&\quad = (y^{\sigma_1}x^{\Delta_1}_{s_1})^{\Delta_2}_{s_2}(s_1, s_2) - (xy^{\Delta_2}_{s_2})^{\Delta_1}_{s_1}(s_1, s_2).
\end{aligned}$$

Applying Green's formula, we get

$$\begin{aligned}
&\iint_R v_3^{\sigma_1\sigma_2}(s_1, s_2)L_2(\phi)(s_1, s_2)\Delta_2 s_2\Delta_1 s_1\\
&\quad = \iint_R (v_3^{\sigma_1\sigma_2}(s_1, s_2)L_2(\phi)(s_1, s_2) - \phi(s_1, s_2)M_2(v_3)(s_1, s_2))\Delta_2 s_2\Delta_1 s_1\\
&\quad = \iint_R ((v_3^{\sigma_1}\phi^{\Delta_1}_{s_1})^{\Delta_2}_{s_2}(s_1, s_2)\\
&\qquad - (\phi v^{\Delta_2}_{3s_2})^{\Delta_1}_{s_1}(s_1, s_2))\Delta_2 s_2\Delta_1 s_1\\
&\quad = -\int_{\partial R} \phi^{\sigma_2}(s_1, s_2)v^{\Delta_2}_{3s_2}(s_1, s_2)\Delta_2 s_2 + v_3(s_1, s_2)\phi^{\Delta_1}_{s_1}(s_1, s_2)\Delta_1 s_1\\
&\quad = -\int_{t_2^0}^{t_2} \phi(t_1, s_2)v^{\Delta_2}_{3s_2}(t_1, s_2)\Delta_2 s_2\\
&\qquad + \int_{t_2^0}^{t_2} \phi(t_1^0, s_2)v^{\Delta_2}_{3s_2}(t_1^0, s_2)\Delta_2 s_2
\end{aligned}$$

$$
\begin{aligned}
&- \int_{t_1^0}^{t_1} v_3^{\sigma_1}(s_1, t_2^0)\phi_{s_1}^{\Delta_1}(s_1, t_2^0)\Delta_1 s_1 \\
&+ \int_{t_1^0}^{t_1} v_3^{\sigma_1}(s_1, t_2)\phi_{s_1}^{\Delta_1}(s_1, t_2)\Delta_1 s_1 \\
&= - \int_{t_1^0}^{t_1} v_3^{\sigma_1}(s_1, t_2^0)\phi_{s_1}^{\Delta_1}(s_1, t_2^0)\Delta_1 s_1 \\
&= - \int_{t_1^0}^{t_1} \phi_{s_1}^{\Delta_1}(s_1, t_2^0)\Delta_1 s_1 \\
&= -\phi(t_1, t_2^0) + \phi(t_1^0, t_2^0) \\
&= \phi(t_1^0, t_2^0),
\end{aligned}
$$

i. e.,

$$
\begin{aligned}
\phi(t_1^0, t_2^0) &= \iint_R v_3(s_1, s_2)L_2(\phi)(s_1, s_2)\Delta_2 s_2 \Delta_1 s_1 \\
&\le \int_{t_1^0}^{t_1}\int_{t_2^0}^{t_2} v_3(s_1, s_2)h_3(s_1, s_2)\Delta_2 s_2 \Delta_1 s_1,
\end{aligned}
$$

$(t_1, t_2) \in D$, $t_1 \ge t_1^0$, $t_2 \ge t_2^0$. Therefore

$$
\begin{aligned}
u(t_1^0, t_2^0) &\le a(t_1^0, t_2^0) + \phi(t_1^0, t_2^0) \\
&\le a(t_1^0, t_2^0) \\
&\quad + \int_{t_1^0}^{t_1}\int_{t_2^0}^{t_2} v_3(s_1, s_2)h_3(s_1, s_2)\Delta_2 s_2 \Delta_1 s_1,
\end{aligned}
$$

$(t_1, t_2) \in D$, $t_1 \ge t_1^0$, $t_2 \ge t_2^0$. This completes the proof. □

8 Two-dimensional linear integro-dynamic inequalities

In this chapter are investigated some two dimensional Pachpatte type linear integro-dynamic inequalities and some of their modifications. The material in this chapter is based on some results in [24] and [25].

Let $\mathbb{T}_1$ and $\mathbb{T}_2$ be time scales with forward jump operators and delta differentiation operators σ_1, σ_2 and Δ_1, Δ_2, respectively. Suppose that $0 \in \mathbb{T}_1$, $0 \in \mathbb{T}_2$.

8.1 Pachpatte's inequalities

Theorem 8.1.1. *Let $u(t_1, t_2)$ be a nonnegative twice continuously-differentiable function for $(t_1, t_2) \in (\mathbb{R}_+ \cap \mathbb{T}_1) \times (\mathbb{R}_+ \cap \mathbb{T}_2)$, $u_{t_1 t_2}^{\Delta_1 \Delta_2}(t_1, t_2)$ be a nonnegative function for $(t_1, t_2) \in (\mathbb{R}_+ \cap \mathbb{T}_1) \times (\mathbb{R}_+ \cap \mathbb{T}_2)$, $a(t_1, t_2)$ be a positive nondecreasing function in each variable t_1, t_2, $(t_1, t_2) \in (\mathbb{R}_+ \cap \mathbb{T}_1) \times (\mathbb{R}_+ \cap \mathbb{T}_2)$, and $b(t_1, t_2)$ be a nonnnegative continuous function for $(t_1, t_2) \in (\mathbb{R}_+ \cap \mathbb{T}_1) \times (\mathbb{R}_+ \cap \mathbb{T}_2)$. If*

$$u_{t_1 t_2}^{\Delta_1 \Delta_2}(t_1, t_2) \le a(t_1, t_2) + \int_0^{t_1} \int_0^{t_2} b(s_1, s_2) u_{t_1 t_2}^{\Delta_1 \Delta_2}(s_1, s_2) \Delta_2 s_2 \Delta_1 s_1,$$

$(t_1, t_2) \in (\mathbb{R}_+ \cap \mathbb{T}_1) \times (\mathbb{R}_+ \cap \mathbb{T}_2)$, then

$$u_{t_1 t_2}^{\Delta_1 \Delta_2}(t_1, t_2) \le a(t_1, t_2) e_f(t_1, 0), \quad (t_1, t_2) \in (\mathbb{R}_+ \cap \mathbb{T}_1) \times (\mathbb{R}_+ \cap \mathbb{T}_2),$$

where

$$f(t_1, t_2) = \int_0^{t_2} b(t_1, s_2) \Delta_2 s_2, \quad (t_1, t_2) \in (\mathbb{R}_+ \cap \mathbb{T}_1) \times (\mathbb{R}_+ \cap \mathbb{T}_2).$$

Proof. Let

$$z(t_1, t_2) = u_{t_1 t_2}^{\Delta_1 \Delta_2}(t_1, t_2), \quad (t_1, t_2) \in (\mathbb{R}_+ \cap \mathbb{T}_1) \times (\mathbb{R}_+ \cap \mathbb{T}_2).$$

Then we can rewrite the given inequality in the following way:

$$z(t_1, t_2) \le a(t_1, t_2) + \int_0^{t_1} \int_0^{t_2} b(s_1, s_2) z(s_1, s_2) \Delta_2 s_2 \Delta_1 s_1, \quad (t_1, t_2) \in (\mathbb{R}_+ \cap \mathbb{T}_1) \times (\mathbb{R}_+ \cap \mathbb{T}_2).$$

Hence,

$$\frac{z(t_1, t_2)}{a(t_1, t_2)} \le 1 + \int_0^{t_1} \int_0^{t_2} b(s_1, s_2) \frac{z(s_1, s_2)}{u(t_1, t_2)} \Delta_2 s_2 \Delta_1 s_1$$

https://doi.org/10.1515/9783110705553-008

$$\leq 1 + \int_0^{t_1}\int_0^{t_2} b(s_1, s_2)\frac{z(s_1, s_2)}{a(s_1, s_2)}\Delta_2 s_2 \Delta_1 s_1,$$

$(t_1, t_2) \in (\mathbb{R}_+ \cap \mathbb{T}_1) \times (\mathbb{R}_+ \cap \mathbb{T}_2)$. Let

$$v(t_1, t_2) = \frac{z(t_1, t_2)}{a(t_1, t_2)}, \quad (t_1, t_2) \in (\mathbb{R}_+ \cap \mathbb{T}_1) \times (\mathbb{R}_+ \cap \mathbb{T}_2).$$

Then

$$v(t_1, t_2) \leq 1 + \int_0^{t_1}\int_0^{t_2} b(s_1, s_2)v(s_1, s_2)\Delta_2 s_2 \Delta_1 s_1, \quad (t_1, t_2) \in (\mathbb{R}_+ \cap \mathbb{T}_1) \times (\mathbb{R}_+ \cap \mathbb{T}_2).$$

Now we apply Wendroff's inequality, Theorem 6.1.1, and get

$$v(t_1, t_2) \leq e_f(t_1, 0), \quad (t_1, t_2) \in (\mathbb{R}_+ \cap \mathbb{T}_1) \times (\mathbb{R}_+ \cap \mathbb{T}_2),$$

whereupon

$$z(t_1, t_2) \leq a(t_1, t_2)e_f(t_1, 0), \quad (t_1, t_2) \in (\mathbb{R}_+ \cap \mathbb{T}_1) \times (\mathbb{R}_+ \cap \mathbb{T}_2).$$

This completes the proof. □

Lemma 8.1.2. *Let $u(t_1, t_2)$ be a nonnegative twice continuously-differentiable function for $(t_1, t_2) \in (\mathbb{R}_+ \cap \mathbb{T}_1) \times (\mathbb{R}_+ \cap \mathbb{T}_2)$ such that $u(t_1, t_2)$ is a nondecreasing function with respect to t_2, $u_{t_2}^{\Delta_2}(t_1, t_2)$ be a nonnegative function for $(t_1, t_2) \in (\mathbb{R}_+ \cap \mathbb{T}_1) \times (\mathbb{R}_+ \cap \mathbb{T}_2)$,*

$$\begin{aligned} u_{t_1}^{\Delta_1}(t_1, 0) &= a(t_1), \quad u(t_1, 0) = b(t_1), \quad t \in \mathbb{R}_+ \cap \mathbb{T}_1, \\ u(0, t_2) &= c(t_2), \quad t_2 \in \mathbb{R}_+ \cap \mathbb{T}_2, \end{aligned}$$

where $a(t_1)$ is a nonnegative continuous function for $t_1 \in \mathbb{R}_+ \cap \mathbb{T}_1$, $b(t_1)$ is a positive continuous function for $t_1 \in \mathbb{R}_+ \cap \mathbb{T}_1$, $c(t_2)$ is a nonnegative continuous function for $t_2 \in \mathbb{R}_+ \cap \mathbb{T}_2$, and $b(0) = c(0)$. Let also $f(t_1, t_2)$ be a nonnegative continuous function for $(t_1, t_2) \in (\mathbb{R}_+ \cap \mathbb{T}_1) \times (\mathbb{R}_+ \cap \mathbb{T}_2)$. If

$$u_{t_1 t_2}^{\Delta_1 \Delta_2}(t_1, t_2) \leq f(t_1, t_2)u(t_1, t_2), \quad (t_1, t_2) \in (\mathbb{R}_+ \cap \mathbb{T}_1) \times (\mathbb{R}_+ \cap \mathbb{T}_2),$$

then

$$u(t_1, t_2) \leq c(t_2)e_g(t_1, 0), \quad (t_1, t_2) \in (\mathbb{R}_+ \cap \mathbb{T}_1) \times (\mathbb{R}_+ \cap \mathbb{T}_2),$$

where

$$g(t_1, t_2) = \frac{a(t_1)}{b(t_1)} + \int_0^{t_2} f(t_1, s_2)\Delta_2 s_2, \quad (t_1, t_2) \in (\mathbb{R}_+ \cap \mathbb{T}_1) \times (\mathbb{R}_+ \cap \mathbb{T}_2).$$

Proof. Since $u(t_1,t_2)$ is a nondecreasing function in t_2, $t_2 \in \mathbb{R}_+ \cap \mathbb{T}_2$, we have

$$u(t_1,t_2) \le u(t_1,\sigma_2(t_2)), \quad (t_1,t_2) \in (\mathbb{R}_+ \cap \mathbb{T}_1) \times (\mathbb{R}_+ \cap \mathbb{T}_2).$$

Hence,

$$\begin{aligned}\frac{u_{t_1t_2}^{\Delta_1\Delta_2}(t_1,t_2)u(t_1,t_2)}{u(t_1,t_2)u(t_1,\sigma_2(t_2))} &\le \frac{u_{t_1t_2}^{\Delta_1\Delta_2}(t_1,t_2)}{u(t_1,t_2)}\\ &\le f(t_1,t_2)\\ &\le f(t_1,t_2)\\ &\quad + \frac{u_{t_2}^{\Delta_2}(t_1,t_2)}{u(t_1,t_2)u(t_1,\sigma_2(t_2))},\end{aligned}$$

$(t_1,t_2) \in (\mathbb{R}_+ \cap \mathbb{T}_1) \times (\mathbb{R}_+ \cap \mathbb{T}_2)$. Therefore

$$\begin{aligned}\left(\frac{u_{t_1}^{\Delta_1}(t_1,t_2)}{u(t_1,t_2)}\right)_{t_2}^{\Delta_2} &\le f(t_1,t_2),\\ \frac{u_{t_1}^{\Delta_1}(t_1,t_2)}{u(t_1,t_2)} - \frac{u_{t_1}^{\Delta_1}(t_1,0)}{u(t_1,0)} &\le \int_0^{t_2} f(t_1,s_2)\Delta_2 s_2,\\ \frac{u_{t_1}^{\Delta_1}(t_1,t_2)}{u(t_1,t_2)} &\le \frac{a(t_1)}{b(t_1)} + \int_0^{t_2} f(t_1,s_2)\Delta_2 s_2\\ &= g(t_1,t_2),\\ u_{t_1}^{\Delta_1}(t_1,t_2) &\le g(t_1,t_2)u(t_1,t_2),\end{aligned}$$

$(t_1,t_2) \in (\mathbb{R}_+ \cap \mathbb{T}_1) \times (\mathbb{R}_+ \cap \mathbb{T}_2)$. From the last inequality and Lemma 2.1.1, we obtain

$$\begin{aligned}u(t_1,t_2) &\le u(0,t_2)e_g(t_1,0)\\ &= c(t_2)e_g(t_1,0), \quad (t_1,t_2) \in (\mathbb{R}_+ \cap \mathbb{T}_1) \times (\mathbb{R}_+ \cap \mathbb{T}_2).\end{aligned}$$

This completes the proof. □

Theorem 8.1.3 (Pachpatte's Inequality). *Let $u(t_1,t_2)$ be a nonnegative twice continuously-differentiable function for $(t_1,t_2) \in (\mathbb{R}_+ \cap \mathbb{T}_1) \times (\mathbb{R}_+ \cap \mathbb{T}_2)$, $u_{t_1t_2}^{\Delta_1\Delta_2}(t_1,t_2)$, $a(t_1,t_2)$, $b(t_1,t_2)$, and $c(t_1,t_2)$ be nonnegative continuous functions,*

$$\begin{aligned}&u(t_1,0) = f(t_1), \quad t_1 \in \mathbb{R}_+ \cap \mathbb{T}_1, \quad u(0,t_2) = g(t_2), \quad t_2 \in \mathbb{R}_+ \cap \mathbb{T}_2,\\ &f(0) = g(0),\end{aligned}$$

where $f(t_1)$ is a nonnegative continuously-differentiable function for $t_1 \in \mathbb{R}_+ \cap \mathbb{T}_1$, $g(t_2)$ is a nonnegative continuously-differentiable function for $t_2 \in \mathbb{R}_+ \cap \mathbb{T}_2$. If

$$\begin{aligned}u_{t_1t_2}^{\Delta_1\Delta_2}(t_1,t_2) &\le a(t_1,t_2)\\ &\quad + b(t_1,t_2)\int_0^{t_1}\int_0^{t_2} c(s_1,s_2)(u(s_1,s_2) + u_{t_1t_2}^{\Delta_1\Delta_2}(s_1,s_2))\Delta_2 s_2\Delta_1 s_1,\end{aligned}$$

$(t_1,t_2) \in (\mathbb{R}_+ \cap \mathbb{T}_1) \times (\mathbb{R}_+ \cap \mathbb{T}_2)$, *then*

$$u^{\Delta_1\Delta_2}_{t_1t_2}(t_1,t_2) \le a(t_1,t_2) + b(t_1,t_2)\int_0^{t_1} h_3(s_1,s_2)e_{\ominus h_4}(\sigma_1(s_1),t_1)\Delta_1 s_1,$$

$(t_1,t_2) \in (\mathbb{R}_+ \cap \mathbb{T}_1) \times (\mathbb{R}_+ \cap \mathbb{T}_2)$, *where*

$$h_1(t_1,t_2) = f(t_1) + g(t_2) - g(0) + \int_0^{t_1}\int_0^{t_2} a(s_1,s_2)\Delta_2 s_2\Delta_1 s_1,$$

$$h_2(t_1,t_2) = \int_0^{t_1}\int_0^{t_2} b(s_1,s_2)\Delta_2 s_2\Delta_1 s_1,$$

$$h_3(t_1,t_2) = \int_0^{t_2} c(t_1,s_2)(h_1(t_1,s_2) + a(t_1,s_2))\Delta_2 s_2,$$

$$h_4(t_1,t_2) = \int_0^{t_2} c(t_1,s_2)(h_2(t_1,s_2) + b(t_1,s_2))\Delta_2 s_2,$$

$(t_1,t_2) \in (\mathbb{R}_+ \cap \mathbb{T}_1) \times (\mathbb{R}_+ \cap \mathbb{T}_2)$.

Proof. Let

$$z(t_1,t_2) = \int_0^{t_1}\int_0^{t_2} c(s_1,s_2)(u(s_1,s_2) + u^{\Delta_1\Delta_2}_{t_1t_2}(s_1,s_2))\Delta_2 s_2\Delta_1 s_1,$$

$(t_1,t_2) \in (\mathbb{R}_+ \cap \mathbb{T}_1) \times (\mathbb{R}_+ \cap \mathbb{T}_2)$. Then

$$u^{\Delta_1\Delta_2}_{t_1t_2}(t_1,t_2) \le a(t_1,t_2) + b(t_1,t_2)z(t_1,t_2), \quad (t_1,t_2) \in (\mathbb{R}_+ \cap \mathbb{T}_1) \times (\mathbb{R}_+ \cap \mathbb{T}_2).$$

Hence,

$$u^{\Delta_1}_{t_1}(t_1,t_2) - u^{\Delta_1}_{t_1}(t_1,0) \le \int_0^{t_2} (a(t_1,s_2) + b(t_1,s_2)z(t_1,s_2))\Delta_2 s_2,$$

$$u(t_1,t_2) - u(0,t_2) - u(t_1,0) + u(0,0) \le \int_0^{t_1}\int_0^{t_2} (a(s_1,s_2) + b(s_1,s_2)z(s_1,s_2))\Delta_2 s_2\Delta_1 s_1,$$

$$\begin{aligned} u(t_1,t_2) &\le f(t_1) + g(t_2) - g(0) \\ &\quad + \int_0^{t_1}\int_0^{t_2} (a(s_1,s_2) + b(s_1,s_2)z(s_1,s_2))\Delta_2 s_2\Delta_1 s_1, \end{aligned}$$

$(t_1, t_2) \in (\mathbb{R}_+ \cap \mathbb{T}_1) \times (\mathbb{R}_+ \cap \mathbb{T}_2)$. Note that $z(t_1, t_2)$ is a nondecreasing function with respect to t_1 and t_2, $(t_1, t_2) \in (\mathbb{R}_+ \cap \mathbb{T}_1) \times (\mathbb{R}_+ \cap \mathbb{T}_2)$. Therefore

$$\begin{aligned} u(t_1, t_2) &\le f(t_1) + g(t_2) - g(0) + \int_0^{t_1}\int_0^{t_2} a(s_1, s_2)\Delta_2 s_2 \Delta_1 s_1 \\ &\quad + \left(\int_0^{t_1}\int_0^{t_2} b(s_1, s_2)\Delta_2 s_2 \Delta_1 s_1\right) z(t_1, t_2) \\ &= h_1(t_1, t_2) + h_2(t_1, t_2) z(t_1, t_2), \end{aligned}$$

$(t_1, t_2) \in (\mathbb{R}_+ \cap \mathbb{T}_1) \times (\mathbb{R}_+ \cap \mathbb{T}_2)$. We have

$$\begin{aligned} z_{t_1}^{\Delta_1}(t_1, t_2) &= \int_0^{t_2} c(t_1, s_2)(u(t_1, s_2) + u_{t_1 t_2}^{\Delta_1 \Delta_2}(t_1, s_2))\Delta_2 s_2 \\ &\le \int_0^{t_2} c(t_1, s_2)(h_1(t_1, s_2) + h_2(t_1, s_2) z(t_1, s_2))\Delta_2 s_2 \\ &\quad + \int_0^{t_2} c(t_1, s_2)(a(t_1, s_2) + b(t_1, s_2) z(t_1, s_2))\Delta_2 s_2 \\ &= \int_0^{t_2} c(t_1, s_2)(h_1(t_1, s_2) + a(t_1, s_2))\Delta_2 s_2 \\ &\quad + \int_0^{t_2} c(t_1, s_2)(h_2(t_1, s_2) + b(t_1, s_2)) z(t_1, s_2)\Delta_2 s_2 \\ &\le h_3(t_1, t_2) \\ &\quad + \left(\int_0^{t_2} c(t_1, s_2)(h_2(t_1, s_2) + b(t_1, s_2))\Delta_2 s_2\right) z(t_1, t_2) \\ &= h_3(t_1, t_2) + h_4(t_1, t_2) z(t_1, t_2), \end{aligned}$$

$(t_1, t_2) \in (\mathbb{R}_+ \cap \mathbb{T}_1) \times (\mathbb{R}_+ \cap \mathbb{T}_2)$. By Lemma 2.1.1, we obtain

$$\begin{aligned} z(t_1, t_2) &\le z(0, t_2) e_{h_4}(t_1, 0) \\ &\quad + \int_0^{t_1} h_3(s_1, t_2) e_{\ominus h_4}(\sigma_1(s_1), t_1)\Delta_1 s_1 \\ &= \int_0^{t_1} h_3(s_1, t_2) e_{\ominus h_4}(\sigma_1(s_1), t_1)\Delta_1 s_1, \end{aligned}$$

$(t_1, t_2) \in (\mathbb{R}_+ \cap \mathbb{T}_1) \times (\mathbb{R}_+ \cap \mathbb{T}_2)$. Therefore

$$u_{t_1 t_2}^{\Delta_1 \Delta_2}(t_1, t_2) \leq a(t_1, t_2) + b(t_1, t_2) \int_0^{t_1} h_3(s_1, t_2) e_{\ominus h_4}(\sigma_1(s_1), t_1) \Delta_1 s_1,$$

$(t_1, t_2) \in (\mathbb{R}_+ \cap \mathbb{T}_1) \times (\mathbb{R}_+ \cap \mathbb{T}_2)$. This completes the proof. □

Theorem 8.1.4 (Pachpatte's Inequality). *Let $u(t_1, t_2)$ be a positive twice continuously-differentiable function for $(t_1, t_2) \in (\mathbb{R}_+ \cap \mathbb{T}_1) \times (\mathbb{R}_+ \cap \mathbb{T}_2)$,*

$$\begin{aligned} u_{t_1}^{\Delta_1}(t_1, t_2) \geq 0, \quad & u_{t_2}^{\Delta_2}(t_1, t_2) \geq 0, \\ u(t_1, t_2) \geq u_0, \quad & u(t_1, 0) = p(t_1), \\ u_{t_1}^{\Delta_1}(t_1, 0) = q(t_1), \quad & u(0, t_2) = r(t_2), \quad p(0) = r(0), \end{aligned}$$

where u_0 is a positive constant, $p(t_1)$ and $q(t_1)$ are positive continuous functions for $t_1 \in \mathbb{R}_+ \cap \mathbb{T}_1$, $r(t_2)$ is a positive continuous function for $t_2 \in \mathbb{R}_+ \cap \mathbb{T}_2$. Let also $a(t_1, t_2)$, $b(t_1, t_2)$, and $c(t_1, t_2)$ be nonnegative continuous functions for $(t_1, t_2) \in (\mathbb{R}_+ \cap \mathbb{T}_1) \times (\mathbb{R}_+ \cap \mathbb{T}_2)$. If

$$\begin{aligned} 0 &\leq u_{t_1 t_2}^{\Delta_1 \Delta_2}(t_1, t_2) \\ &\leq a(t_1, t_2) + b(t_1, t_2) \Bigg(u(t_1, t_2) \\ &\quad + \int_0^{t_1} \int_0^{t_2} c(s_1, s_2)(u(s_1, s_2) + u_{t_1 t_2}^{\Delta_1 \Delta_2}(s_1, s_2)) \Delta_2 s_2 \Delta_1 s_1 \Bigg), \end{aligned}$$

$(t_1, t_2) \in (\mathbb{R}_+ \cap \mathbb{T}_1) \times (\mathbb{R}_+ \cap \mathbb{T}_2)$, then

$$u(t_1, t_2) \leq a(t_1, t_2) + b(t_1, t_2) e_g(t_1, 0),$$

$(t_1, t_2) \in (\mathbb{R}_+ \cap \mathbb{T}_1) \times (\mathbb{R}_+ \cap \mathbb{T}_2)$, where

$$\begin{aligned} f(t_1, t_2) &= \frac{a(t_1, t_2)(1 + c(t_1, t_2))}{u_0} + b(t_1, t_2) + c(t_1, t_2) + b(t_1, t_2) c(t_1, t_2), \\ g(t_1, t_2) &= \frac{q(t_1)}{p(t_1)} + \int_0^{t_2} f(t_1, s_2) \Delta_2 s_2, \end{aligned}$$

$(t_1, t_2) \in (\mathbb{R}_+ \cap \mathbb{T}_1) \times (\mathbb{R}_+ \cap \mathbb{T}_2)$.

Proof. Let

$$z(t_1, t_2) = u(t_1, t_2) + \int_0^{t_1} \int_0^{t_2} c(s_1, s_2)(u(s_1, s_2) + u_{t_1 t_2}^{\Delta_1 \Delta_2}(s_1, s_2)) \Delta_2 s_2 \Delta_1 s_1,$$

$(t_1, t_2) \in (\mathbb{R}_+ \cap \mathbb{T}_1) \times (\mathbb{R}_+ \cap \mathbb{T}_2)$. Then

$$u(t_1, t_2) \leq z(t_1, t_2),$$
$$u^{\Delta_1\Delta_2}_{t_1t_2}(t_1, t_2) \leq a(t_1, t_2) + b(t_1, t_2)z(t_1, t_2),$$

$(t_1, t_2) \in (\mathbb{R}_+ \cap \mathbb{T}_1) \times (\mathbb{R}_+ \cap \mathbb{T}_2)$. Hence,

$$z^{\Delta_1}_{t_1}(t_1, t_2) = u^{\Delta_1}_{t_1}(t_1, t_2) + \int_0^{t_2} c(t_1, s_2)(u(t_1, s_2) + u^{\Delta_1\Delta_2}_{t_1t_2}(t_1, s_2))\Delta_2 s_2,$$

$$\begin{aligned} z^{\Delta_1\Delta_2}_{t_1t_2}(t_1, t_2) &= u^{\Delta_1\Delta_2}_{t_1t_2}(t_1, t_2) + c(t_1, t_2)(u(t_1, t_2) + u^{\Delta_1\Delta_2}_{t_1t_2}(t_1, t_2)) \\ &\leq a(t_1, t_2) + b(t_1, t_2)z(t_1, t_2) + c(t_1, t_2)z(t_1, t_2) \\ &\quad + a(t_1, t_2)c(t_1, t_2) + b(t_1, t_2)c(t_1, t_2)z(t_1, t_2) \\ &= a(t_1, t_2)(1 + c(t_1, t_2)) \\ &\quad + (b(t_1, t_2) + c(t_1, t_2) + b(t_1, t_2)c(t_1, t_2))z(t_1, t_2), \end{aligned}$$

$(t_1, t_2) \in (\mathbb{R}_+ \cap \mathbb{T}_1) \times (\mathbb{R}_+ \cap \mathbb{T}_2)$. Then

$$\begin{aligned} \frac{z^{\Delta_1\Delta_2}_{t_1t_2}(t_1, t_2)}{z(t_1, t_2)} &\leq \frac{a(t_1, t_2)(1 + c(t_1, t_2))}{z(t_1, t_2)} \\ &\quad + b(t_1, t_2) + c(t_1, t_2) + b(t_1, t_2)c(t_1, t_2) \\ &\leq \frac{a(t_1, t_2)(1 + c(t_1, t_2))}{u_0} \\ &\quad + b(t_1, t_2) + c(t_1, t_2) + b(t_1, t_2)c(t_1, t_2) \\ &= f(t_1, t_2), \end{aligned} \tag{8.1}$$

$(t_1, t_2) \in (\mathbb{R}_+ \cap \mathbb{T}_1) \times (\mathbb{R}_+ \cap \mathbb{T}_2)$. By the definition of the function $z(t_1, t_2)$, using that $u(t_1, t_2)$ is a nondecreasing function with respect to t_1 and t_2, $(t_1, t_2) \in (\mathbb{R}_+ \cap \mathbb{T}_1) \times (\mathbb{R}_+ \cap \mathbb{T}_2)$, it follows that $z(t_1, t_2)$ is a nondecreasing function with respect to t_1 and t_2, $(t_1, t_2) \in (\mathbb{R}_+ \cap \mathbb{T}_1) \times (\mathbb{R}_+ \cap \mathbb{T}_2)$. Therefore

$$z(t_1, t_2) \leq z(t_1, \sigma_2(t_2)), \quad (t_1, t_2) \in (\mathbb{R}_+ \cap \mathbb{T}_1) \times (\mathbb{R}_+ \cap \mathbb{T}_2).$$

From here and (8.1), we get

$$\begin{aligned} \frac{z^{\Delta_1\Delta_2}_{t_1t_2}(t_1, t_2)z(t_1, t_2)}{z(t_1, t_2)z(t_1, \sigma_2(t_2))} &\leq \frac{z^{\Delta_1\Delta_2}_{t_1t_2}(t_1, t_2)}{z(t_1, t_2)} \\ &\leq f(t_1, t_2), \end{aligned} \tag{8.2}$$

$(t_1, t_2) \in (\mathbb{R}_+ \cap \mathbb{T}_1) \times (\mathbb{R}_+ \cap \mathbb{T}_2)$. Note that

$$\begin{aligned} z^{\Delta_2}_{t_2}(t_1, t_2) &= u^{\Delta_2}_{t_2}(t_1, t_2) + \int_0^{t_1} c(s_1, t_2)(u(s_1, t_2) + u^{\Delta_1\Delta_2}_{t_1t_2}(s_1, t_2))\Delta_1 s_1 \\ &\geq 0, \quad (t_1, t_2) \in (\mathbb{R}_+ \cap \mathbb{T}_1) \times (\mathbb{R}_+ \cap \mathbb{T}_2). \end{aligned}$$

Also,

$$z^{\Delta_1}_{t_1}(t_1,t_2) \geq 0, \quad (t_1,t_2) \in (\mathbb{R}_+ \cap \mathbb{T}_1) \times (\mathbb{R}_+ \cap \mathbb{T}_2).$$

Hence, by (8.2), we obtain

$$\begin{aligned}\frac{z^{\Delta_1\Delta_2}_{t_1t_2}(t_1,t_2)z(t_1,t_2)}{z(t_1,t_2)z(t_1,\sigma_2(t_2))} &\leq f(t_1,t_2)\\ &\leq f(t_1,t_2)\\ &\quad + \frac{z^{\Delta_1}_{t_1}(t_1,t_2)z^{\Delta_2}_{t_2}(t_1,t_2)}{z(t_1,t_2)z(t_1,\sigma_2(t_2))},\end{aligned}$$

$(t_1,t_2) \in (\mathbb{R}_+ \cap \mathbb{T}_1) \times (\mathbb{R}_+ \cap \mathbb{T}_2)$. Consequently,

$$\begin{aligned}\left(\frac{z^{\Delta_1}_{t_1}(t_1,t_2)}{z(t_1,t_2)}\right)^{\Delta_2}_{t_2} &\leq f(t_1,t_2),\\ \frac{z^{\Delta_1}_{t_1}(t_1,t_2)}{z(t_1,t_2)} - \frac{z^{\Delta_1}_{t_1}(t_1,0)}{z(t_1,0)} &\leq \int_0^{t_2} f(t_1,s_2)\Delta_2 s_2,\\ \frac{z^{\Delta_1}_{t_1}(t_1,t_2)}{z(t_1,t_2)} - \frac{u^{\Delta_1}_{t_1}(t_1,0)}{u(t_1,0)} &\leq \int_0^{t_2} f(t_1,s_2)\Delta_2 s_2,\\ \frac{z^{\Delta_1}_{t_1}(t_1,t_2)}{z(t_1,t_2)} &\leq \frac{q(t_1)}{p(t_1)} + \int_0^{t_2} f(t_1,s_2)\Delta_2 s_2\\ &= g(t_1,t_2),\\ z^{\Delta_1}_{t_1}(t_1,t_2) &\leq g(t_1,t_2)z(t_1,t_2),\end{aligned}$$

$(t_1,t_2) \in (\mathbb{R}_+ \cap \mathbb{T}_1) \times (\mathbb{R}_+ \cap \mathbb{T}_2)$. From the last inequality and Lemma 2.1.1, we obtain

$$\begin{aligned}z(t_1,t_2) &\leq z(0,t_2)e_g(t_1,0)\\ &= u(0,t_2)e_g(t_1,0)\\ &= r(t_2)e_g(t_1,0),\end{aligned}$$

$(t_1,t_2) \in (\mathbb{R}_+ \cap \mathbb{T}_1) \times (\mathbb{R}_+ \cap \mathbb{T}_2)$. Therefore

$$\begin{aligned}u(t_1,t_2) &\leq a(t_1,t_2) + b(t_1,t_2)z(t_1,t_2)\\ &\leq a(t_1,t_2) + b(t_1,t_2)r(t_2)e_g(t_1,0),\end{aligned}$$

$(t_1,t_2) \in (\mathbb{R}_+ \cap \mathbb{T}_1) \times (\mathbb{R}_+ \cap \mathbb{T}_2)$. This completes the proof. □

Theorem 8.1.5 (Pachpatte's Inequality). *Let $u(t_1,t_2)$ be a nonnegative twice continuously-differentiable function for $(t_1,t_2) \in (\mathbb{R}_+ \cap \mathbb{T}_1) \times (\mathbb{R}_+ \cap \mathbb{T}_2)$, $u^{\Delta_1\Delta_2}_{t_1t_2}(t_1,t_2)$ be a* *nonnegative function for* $(t_1,t_2) \in (\mathbb{R}_+ \cap \mathbb{T}_1) \times (\mathbb{R}_+ \cap \mathbb{T}_2)$, $a(t_1,t_2)$, $b(t_1,t_2)$ *and* $c(t_1,t_2)$ *be nonnegative continuous functions for* $(t_1,t_2) \in (\mathbb{R}_+ \cap \mathbb{T}_1) \times (\mathbb{R}_+ \cap \mathbb{T}_2)$, u_0 *be a positive constant. If*

$$u_{t_1t_2}^{\Delta_1\Delta_2}(t_1,t_2) \le u_0 + \int_0^{t_1}\int_0^{t_2} a(s_1,s_2)u_{t_1t_2}^{\Delta_1\Delta_2}(s_1,s_2)\Delta_2 s_2\Delta_1 s_1$$
$$+ \int_0^{t_1}\int_0^{t_2} b(s_1,s_2)\left(\int_0^{s_1}\int_0^{s_2} c(\tau_1,\tau_2)u_{t_1t_2}^{\Delta_1\Delta_2}(\tau_1,\tau_2)\Delta_2\tau_2\Delta_1\tau_1\right)\Delta_2 s_2\Delta_1 s_1,$$

$(t_1,t_2) \in (\mathbb{R}_+ \cap \mathbb{T}_1) \times (\mathbb{R}_+ \cap \mathbb{T}_2)$, *then*

$$u_{t_1t_2}^{\Delta_1\Delta_2}(t_1,t_2) \le u_0 e_g(t_1,0),$$

$(t_1,t_2) \in (\mathbb{R}_+ \cap \mathbb{T}_1) \times (\mathbb{R}_+ \cap \mathbb{T}_2)$, *where*

$$f(t_1,t_2) = a(t_1,t_2) + b(t_1,t_2)\int_0^{t_1}\int_0^{t_2} c(\tau_1,\tau_2)\Delta_2\tau_2\Delta_1\tau_1,$$

$$g(t_1,t_2) = \int_0^{t_2} f(t_1,s_2)\Delta_2 s_2,$$

$(t_1,t_2) \in (\mathbb{R}_+ \cap \mathbb{T}_1) \times (\mathbb{R}_+ \cap \mathbb{T}_2)$.

Proof. Let

$$z(t_1,t_2) = u_0 + \int_0^{t_1}\int_0^{t_2} a(s_1,s_2)u_{t_1t_2}^{\Delta_1\Delta_2}(s_1,s_2)\Delta_2 s_2\Delta_1 s_1$$
$$+ \int_0^{t_1}\int_0^{t_2} b(s_1,s_2)\left(\int_0^{s_1}\int_0^{s_2} c(\tau_1,\tau_2)u_{t_1t_2}^{\Delta_1\Delta_2}(\tau_1,\tau_2)\Delta_2\tau_2\Delta_1\tau_1\right)\Delta_2 s_2\Delta_1 s_1,$$

$(t_1,t_2) \in (\mathbb{R}_+ \cap \mathbb{T}_1) \times (\mathbb{R}_+ \cap \mathbb{T}_2)$. Note that $z(t_1,t_2)$ is a positive continuous nondecreasing function in the variables t_1 and t_2, $(t_1,t_2) \in (\mathbb{R}_+ \cap \mathbb{T}_1) \times (\mathbb{R}_+ \cap \mathbb{T}_2)$. Then

$$u_{t_1t_2}^{\Delta_1\Delta_2}(t_1,t_2) \le z(t_1,t_2),$$

$(t_1,t_2) \in (\mathbb{R}_+ \cap \mathbb{T}_1) \times (\mathbb{R}_+ \cap \mathbb{T}_2)$. We have

$$z_{t_1}^{\Delta_1}(t_1,t_2) = \int_0^{t_2} a(t_1,s_2)u_{t_1t_2}^{\Delta_1\Delta_2}(t_1,s_2)\Delta_2 s_2$$
$$+ \int_0^{t_2} b(t_1,s_2)\left(\int_0^{t_1}\int_0^{s_2} c(\tau_1,\tau_2)u_{t_1t_2}^{\Delta_1\Delta_2}(\tau_1,\tau_2)\Delta_2\tau_2\Delta_1\tau_1\right)\Delta_2 s_2\Delta_1 s_1,$$

$$\begin{aligned}
z^{\Delta_1\Delta_2}_{t_1t_2}(t_1,t_2) &= a(t_1,t_2)u^{\Delta_1\Delta_2}_{t_1t_2}(t_1,t_2) \\
&\quad + b(t_1,t_2)\int_0^{t_1}\int_0^{t_2} c(\tau_1,\tau_2)u^{\Delta_1\Delta_2}_{t_1t_2}(\tau_1,\tau_2)\Delta_2\tau_2\Delta_1\tau_1 \\
&\le a(t_1,t_2)z(t_1,t_2) \\
&\quad + b(t_1,t_2)\int_0^{t_1}\int_0^{t_2} c(\tau_1,\tau_2)z(\tau_1,\tau_2)\Delta_2\tau_2\Delta_1\tau_1 \\
&\le a(t_1,t_2)z(t_1,t_2) \\
&\quad + b(t_1,t_2)\left(\int_0^{t_1}\int_0^{t_2} c(\tau_1,\tau_2)\Delta_2\tau_2\Delta_1\tau_1\right)z(t_1,t_2) \\
&= \left(a(t_1,t_2) + b(t_1,t_2)\left(\int_0^{t_1}\int_0^{t_2} c(\tau_1,\tau_2)\Delta_2\tau_2\Delta_1\tau_1\right)\right) \\
&\quad \times z(t_1,t_2) \\
&= f(t_1,t_2)z(t_1,t_2),
\end{aligned}$$

$(t_1,t_2) \in (\mathbb{R}_+ \cap \mathbb{T}_1) \times (\mathbb{R}_+ \cap \mathbb{T}_2)$, i. e.,

$$z^{\Delta_1\Delta_2}_{t_1t_2}(t_1,t_2) \le f(t_1,t_2)z(t_1,t_2), \tag{8.3}$$

$(t_1,t_2) \in (\mathbb{R}_+ \cap \mathbb{T}_1) \times (\mathbb{R}_+ \cap \mathbb{T}_2)$. Note that

$$\begin{aligned}
z^{\Delta_2}_{t_2}(t_1,t_2) &= \int_0^{t_1} a(s_1,t_2)u^{\Delta_1\Delta_2}_{t_1t_2}(s_1,t_2)\Delta_1 s_1 \\
&\quad + \int_0^{t_1} b(s_1,t_2)\left(\int_0^{s_1}\int_0^{t_2} c(\tau_1,\tau_2)u^{\Delta_1\Delta_2}_{t_1t_2}(\tau_1,\tau_2)\Delta_2\tau_2\Delta_1\tau_1\right)\Delta_1 s_1, \\
z^{\Delta_1}_{t_1}(t_1,0) &= 0, \\
z(t_1,0) &= u_0, \\
z(0,t_2) &= u_0,
\end{aligned}$$

$(t_1,t_2) \in (\mathbb{R}_+ \cap \mathbb{T}_1) \times (\mathbb{R}_+ \cap \mathbb{T}_2)$. Hence, using (8.3) and Lemma 8.1.2, we get

$$z(t_1,t_2) \le u_0 e_g(t_1,0),$$

$(t_1,t_2) \in (\mathbb{R}_+ \cap \mathbb{T}_1) \times (\mathbb{R}_+ \cap \mathbb{T}_2)$. Consequently,

$$u^{\Delta_1\Delta_2}_{t_1t_2}(t_1,t_2) \le u_0 e_g(t_1,0),$$

$(t_1,t_2) \in (\mathbb{R}_+ \cap \mathbb{T}_1) \times (\mathbb{R}_+ \cap \mathbb{T}_2)$. This completes the proof. □

Theorem 8.1.6. *Let $u(t_1, t_2)$ be nonnegative twice continuously-differentiable function for $(t_1, t_2) \in (\mathbb{R}_+ \cap \mathbb{T}_1) \times (\mathbb{R}_+ \cap \mathbb{T}_2)$,*

$$u(0, t_2) = p(t_2), \quad u(t_1, 0) = q(t_1),$$
$$p(0) = q(0),$$

where $p(t_2)$ is a nonnegative continuous function for $t_2 \in \mathbb{R}_+ \cap \mathbb{T}_2$, $q(t_1)$ is a nonnegative continuous function for $t_1 \in \mathbb{R}_+ \cap \mathbb{T}_1$. Let $a(t_1, t_2)$, $b(t_1, t_2)$ and $c(t_1, t_2)$ be nonnegative continuous functions for $(t_1, t_2) \in (\mathbb{R}_+ \cap \mathbb{T}_1) \times (\mathbb{R}_+ \cap \mathbb{T}_2)$, u_0 a positive constant. If

$$\begin{aligned} u^{\Delta_1\Delta_2}_{t_1 t_2}(t_1, t_2) \le u_0 &+ \int_0^{t_1}\int_0^{t_2} a(s_1, s_2)(u(s_1, s_2) + u^{\Delta_1\Delta_2}_{t_1 t_2}(s_1, s_2))\Delta_2 s_2 \Delta_1 s_1 \\ &+ \int_0^{t_1}\int_0^{t_2} b(s_1, s_2)\left(\int_0^{s_1}\int_0^{s_2} c(\tau_1, \tau_2) u^{\Delta_1\Delta_2}_{t_1 t_2}(\tau_1, \tau_2)\Delta_2\tau_2\Delta_1\tau_1\right)\Delta_2 s_2 \Delta_1 s_1, \end{aligned}$$

$(t_1, t_2) \in (\mathbb{R}_+ \cap \mathbb{T}_1) \times (\mathbb{R}_+ \cap \mathbb{T}_2)$, then

$$u(t_1, t_2) \le u_0 e_g(t_1, 0) + \int_0^{t_1} f(s_1, t_2) e_{\ominus g}(\tau_1(s_1), t_1)\Delta_1 s_1,$$

$(t_1, t_2) \in (\mathbb{R}_+ \cap \mathbb{T}_1) \times (\mathbb{R}_+ \cap \mathbb{T}_2)$, where

$$\begin{aligned} f(t_1, t_2) &= \int_0^{t_2} a(t_1, s_2)(p(s_2) + q(t_1) - q(0))\Delta_2 s_2, \\ g(t_1, t_2) &= \int_0^{t_2} a(t_1, s_2)(t_1 s_2 + 1)\Delta_2 s_2 \\ &\quad + \int_0^{t_2} b(t_1, s_2)\left(\int_0^{t_1}\int_0^{s_2} c(\tau_1, \tau_2)\Delta_2\tau_2\Delta_1\tau_1\right)\Delta_2 s_2, \end{aligned}$$

$(t_1, t_2) \in (\mathbb{R}_+ \cap \mathbb{T}_1) \times (\mathbb{R}_+ \cap \mathbb{T}_2)$.

Proof. Let

$$\begin{aligned} z(t_1, t_2) = u_0 &+ \int_0^{t_1}\int_0^{t_2} a(s_1, s_2)(u(s_1, s_2) + u^{\Delta_1\Delta_2}_{t_1 t_2}(s_1, s_2))\Delta_2 s_2 \Delta_1 s_1 \\ &+ \int_0^{t_1}\int_0^{t_2} b(s_1, s_2)\left(\int_0^{s_1}\int_0^{s_2} c(\tau_1, \tau_2) u^{\Delta_1\Delta_2}_{t_1 t_2}(\tau_1, \tau_2)\Delta_2\tau_2\Delta_1\tau_1\right)\Delta_2 s_2 \Delta_1 s_1, \end{aligned}$$

$(t_1, t_2) \in (\mathbb{R}_+ \cap \mathbb{T}_1) \times (\mathbb{R}_+ \cap \mathbb{T}_2)$. Note that $z(t_1, t_2)$ is a positive nondecreasing function with respect to the variables t_1 and t_2, $(t_1, t_2) \in (\mathbb{R}_+ \cap \mathbb{T}_1) \times (\mathbb{R}_+ \cap \mathbb{T}_2)$. We have

$$
\begin{aligned}
u^{\Delta_1\Delta_2}_{t_1 t_2}(t_1, t_2) &\le z(t_1, t_2),\\
u^{\Delta_1}_{t_1}(t_1, t_2) &\le u^{\Delta_1}_{t_1}(t_1, 0) + \int_0^{t_2} z(t_1, s_2)\Delta_2 s_2,\\
u(t_1, t_2) &\le u(0, t_2) + u(t_1, 0) - u(0, 0)\\
&\quad + \int_0^{t_1}\int_0^{t_2} z(s_1, s_2)\Delta_2 s_2 \Delta_1 s_1\\
&\le (p(t_2) + q(t_1) - q(0)) + t_1 t_2 z(t_1, t_2),\\
z^{\Delta_1}_{t_1}(t_1, t_2) &= \int_0^{t_2} a(t_1, s_2)(u(t_1, s_2) + u^{\Delta_1\Delta_2}_{t_1 t_2}(t_1, s_2))\Delta_2 s_2\\
&\quad + \int_0^{t_2} b(t_1, s_2)\left(\int_0^{t_1}\int_0^{s_2} c(\tau_1, \tau_2) u^{\Delta_1\Delta_2}_{t_1 t_2}(\tau_1, \tau_2)\Delta_2\tau_2\Delta_1\tau_1\right)\Delta_2 s_2\\
&\le \int_0^{t_2} a(t_1, s_2)(p(s_2) + q(t_1) - q(0) + t_1 s_2 z(t_1, s_2) + z(t_1, s_2))\Delta_2 s_2\\
&\quad + \int_0^{t_2} b(t_1, s_2)\left(\int_0^{t_1}\int_0^{s_2} c(\tau_1, \tau_2) z(\tau_1, \tau_2)\Delta_2\tau_2\Delta_1\tau_1\right)\Delta_2 s_2\\
&\le \int_0^{t_2} a(t_1, s_2)(p(s_2) + q(t_1) - q(0))\Delta_2 s_2\\
&\quad \left(\int_0^{t_2} a(t_1, s_2)(t_1 s_2 + 1)\Delta_2 s_2\right) z(t_1, t_2)\\
&\quad + \left(\int_0^{t_2} b(t_1, s_2)\left(\int_0^{t_1}\int_0^{s_2} c(\tau_1, \tau_2)\Delta_2\tau_2\Delta_1\tau_1\right)\Delta_2 s_2\right)\\
&\quad \times z(t_1, t_2)\\
&= f(t_1, t_2) + g(t_1, t_2) z(t_1, t_2),
\end{aligned}
$$

$(t_1, t_2) \in (\mathbb{R}_+ \cap \mathbb{T}_1) \times (\mathbb{R}_+ \cap \mathbb{T}_2)$. Hence, by Lemma 2.1.1, we get

$$
z(t_1, t_2) \le z(0, t_2) e_g(t_1, 0) + \int_0^{t_1} f(s_1, t_2) e_{\ominus g}(\tau_1(s_1), t_1)\Delta_1 s_1
$$

$$= u_0 e_g(t_1, 0) + \int_0^{t_1} f(s_1, t_2) e_{\ominus g}(\sigma_1(s_1), t_1)\Delta_1 s_1,$$

$(t_1, t_2) \in (\mathbb{R}_+ \cap \mathbb{T}_1) \times (\mathbb{R}_+ \cap \mathbb{T}_2)$. Consequently,

$$\begin{aligned} u_{t_1 t_2}^{\Delta_1 \Delta_2}(t_1, t_2) &\le z(t_1, t_2) \\ &\le u_0 e_g(t_1, 0) + \int_0^{t_1} f(s_1, t_2) e_{\ominus g}(\sigma_1(s_1), t_1)\Delta_1 s_1, \end{aligned}$$

$(t_1, t_2) \in (\mathbb{R}_+ \cap \mathbb{T}_1) \times (\mathbb{R}_+ \cap \mathbb{T}_2)$. This completes the proof. □

Theorem 8.1.7 (Pachpatte's Inequality). *Let $u(t_1, t_2)$ be a nonnegative twice continuously-differentiable function for $(t_1, t_2) \in (\mathbb{R}_+ \cap \mathbb{T}_1) \times (\mathbb{R}_+ \cap \mathbb{T}_2)$, $u_{t_1 t_2}^{\Delta_1 \Delta_2}(t_1, t_2)$ be a nonnegative function for $(t_1, t_2) \in (\mathbb{R}_+ \cap \mathbb{T}_1) \times (\mathbb{R}_+ \cap \mathbb{T}_2)$,*

$$u_{t_1}^{\Delta_1}(t_1, 0) = f(t_1), \quad u(0, t_2) = g(t_2),$$

$f(t_1)$ be a nonnegative continuous function for $t_1 \in \mathbb{R}_+ \cap \mathbb{T}_1$, $g(t_2)$ be a nonnegative continuous function for $t_2 \in \mathbb{R}_+ \cap \mathbb{T}_2$, $a(t_1, t_2)$ and $b(t_1, t_2)$ be nonnegative continuous functions for $(t_1, t_2) \in (\mathbb{R}_+ \cap \mathbb{T}_1) \times (\mathbb{R}_+ \cap \mathbb{T}_2)$, u_0 be a nonnegative constant. If

$$\begin{aligned} u_{t_1 t_2}^{\Delta_1 \Delta_2}(t_1, t_2) &\le u_0 + \int_0^{t_1}\int_0^{t_2} a(s_1, s_2)\big(u(s_1, s_2) + u_{t_1 t_2}^{\Delta_1 \Delta_2}(s_1, s_2)\big)\Delta_2 s_2 \Delta_1 s_1 \\ &\quad + \int_0^{t_1}\int_0^{t_2} a(s_1, s_2)\left(\int_0^{s_1}\int_0^{s_2} b(\tau_1, \tau_2)\big(u(\tau_1, \tau_2) + u_{t_1 t_2}^{\Delta_1 \Delta_2}(\tau_1, \tau_2)\big)\Delta_2 \tau_2 \Delta_1 \tau_1\right)\Delta_2 s_2 \Delta_1 s_1, \end{aligned}$$

$(t_1, t_2) \in (\mathbb{R}_+ \cap \mathbb{T}_1) \times (\mathbb{R}_+ \cap \mathbb{T}_2)$, then

$$u_{t_1 t_2}^{\Delta_1 \Delta_2}(t_1, t_2) \le u_0 e_{h_2}(t_1, 0) + \int_0^{t_1} h_1(s_1, t_2) e_{\ominus h_2}(\sigma_1(s_1), t_2)\Delta_1 s_1,$$

$(t_1, t_2) \in (\mathbb{R}_+ \cap \mathbb{T}_1) \times (\mathbb{R}_+ \cap \mathbb{T}_2)$, and

$$\begin{aligned} u(t_1, t_2) &\le g(t_2) + \int_0^{t_1} f(s_1)\Delta_1 s_1 \\ &\quad + \int_0^{t_1}\int_0^{t_2}\left(u_0 e_{h_2}(s_1, 0) + \int_0^{s_1} h_1(y_1, s_2) e_{\ominus h_2}(\sigma_1(y_1), s_2)\Delta_1 y_1\right)\Delta_2 s_2 \Delta_1 s_1, \end{aligned}$$

$(t_1, t_2) \in (\mathbb{R}_+ \cap \mathbb{T}_1) \times (\mathbb{R}_+ \cap \mathbb{T}_2)$, *where*

$$h_1(t_1, t_2) = \int_0^{t_2} a(t_1, s_2)\left(g(s_2) + \int_0^{t_1} f(s_1)\Delta_1 s_1\right)\Delta_2 s_2 + \int_0^{t_2} a(t_1, s_2)\left(\int_0^{t_1}\int_0^{s_2} b(\tau_1, \tau_2)\left(g(\tau_2) + \int_0^{\tau_1} f(y_1)\Delta_1 y_1\right)\Delta_2\tau_2\Delta_1\tau_1\right)\Delta_2 s_2,$$

$(t_1, t_2) \in (\mathbb{R}_+ \cap \mathbb{T}_1) \times (\mathbb{R}_+ \cap \mathbb{T}_2)$, *and*

$$h_2(t_1, t_2) = \int_0^{t_2} a(t_1, s_2)(t_1 s_2 + 1)\Delta_2 s_2 + \int_0^{t_2} a(t_1, s_2)\left(\int_0^{t_1}\int_0^{s_2}(\tau_1\tau_2 + 1)b(\tau_1, \tau_2)\Delta_2\tau_2\Delta_1\tau_1\right)\Delta_2 s_2,$$

$(t_1, t_2) \in (\mathbb{R}_+ \cap \mathbb{T}_1) \times (\mathbb{R}_+ \cap \mathbb{T}_2)$.

Proof. Let

$$z(t_1, t_2) = u_0 + \int_0^{t_1}\int_0^{t_2} a(s_1, s_2)(u(s_1, s_2) + u_{t_1 t_2}^{\Delta_1\Delta_2}(s_1, s_2))\Delta_2 s_2\Delta_1 s_2 + \int_0^{t_1}\int_0^{t_2} a(s_1, s_2)\left(\int_0^{s_1}\int_0^{s_2} b(\tau_1, \tau_2)(u(\tau_1, \tau_2) + u_{t_1 t_2}^{\Delta_1\Delta_2}(\tau_1, \tau_2))\Delta_2\tau_2\Delta_1\tau_1\right)\Delta_2 s_2\Delta_1 s_1,$$

$(t_1, t_2) \in (\mathbb{R}_+ \cap \mathbb{T}_1) \times (\mathbb{R}_+ \cap \mathbb{T}_2)$. Then

$$u_{t_1 t_2}^{\Delta_1\Delta_2}(t_1, t_2) \le z(t_1, t_2),$$
$$z(t_1, 0) = u_0,$$
$$z(0, t_2) = u_0,$$

and

$$u_{t_1}^{\Delta_1}(t_1, t_2) - u_{t_1}^{\Delta_1}(t_1, 0) \le \int_0^{t_2} z(t_1, s_2)\Delta_2 s_2,$$

$$u_{t_1}^{\Delta_1}(t_1, t_2) \le f(t_1) + \int_0^{t_2} z(t_1, s_2)\Delta_2 s_2,$$

$$u(t_1, t_2) - u(0, t_2) \le \int_0^{t_1} f(s_1)\Delta_1 s_1 + \int_0^{t_1}\int_0^{t_2} z(s_1, s_2)\Delta_2 s_2\Delta_1 s_1,$$

$$u(t_1,t_2) \le g(t_2) + \int_0^{t_1} f(s_1)\Delta_1 s_1 + \int_0^{t_1}\int_0^{t_2} z(s_1,s_2)\Delta_2 s_2 \Delta_1 s_1,$$

$(t_1,t_2) \in (\mathbb{R}_+ \cap \mathbb{T}_1)\times(\mathbb{R}_+ \cap \mathbb{T}_2)$. Note that $z(t_1,t_2)$ is a nonnegative nondecreasing function in each variable t_1, t_2, $(t_1,t_2) \in (\mathbb{R}_+ \cap \mathbb{T}_1)\times(\mathbb{R}_+ \cap \mathbb{T}_2)$. Next,

$$\begin{aligned}
z_{t_1}^{\Delta_1}(t_1,t_2) &= \int_0^{t_2} a(t_1,s_2)(u(t_1,s_2) + u_{t_1t_2}^{\Delta_1\Delta_2}(t_1,s_2))\Delta_2 s_2 \\
&\quad + \int_0^{t_2} a(t_1,s_2)\left(\int_0^{t_1}\int_0^{s_2} b(\tau_1,\tau_2)(u(\tau_1,\tau_2) + u_{t_1t_2}^{\Delta_1\Delta_2}(\tau_1,\tau_2))\Delta_2\tau_2\Delta_1\tau_1\right)\Delta_2 s_2\Delta_1 s_1 \\
&\le \int_0^{t_2} a(t_1,s_2)u(t_1,s_2)\Delta_2 s_2 \\
&\quad + \int_0^{t_2} a(t_1,s_2)z(t_1,s_2)\Delta_2 s_2 \\
&\quad + \int_0^{t_2} a(t_1,s_2)\left(\int_0^{t_1}\int_0^{s_2} b(\tau_1,\tau_2)u(\tau_1,\tau_2)\Delta_2\tau_2\Delta_1\tau_1\right)\Delta_2 s_2\Delta_1 s_1 \\
&\quad + \int_0^{t_2} a(t_1,s_2)\left(\int_0^{t_1}\int_0^{s_2} b(\tau_1,\tau_2)z(\tau_1,\tau_2)\Delta_2\tau_2\Delta_1\tau_1\right)\Delta_2 s_2\Delta_1 s_1 \\
&\le \int_0^{t_2} a(t_1,s_2)\left(g(s_2) + \int_0^{t_1} f(s_1)\Delta_1 s_1 + \int_0^{t_1}\int_0^{s_2} z(s_1,\tau_2)\Delta_2\tau_2\Delta_1 s_1\right)\Delta_2 s_2 \\
&\quad + \left(\int_0^{t_2} a(t_1,s_2)\Delta_2 s_2\right)z(t_1,t_2) \\
&\quad + \int_0^{t_2} a(t_1,s_2)\left(\int_0^{t_1}\int_0^{s_2} b(\tau_1,\tau_2)\left(g(\tau_2) + \int_0^{\tau_1} f(y_1)\Delta_1 y_1\right.\right. \\
&\quad \left.\left. + \int_0^{\tau_1}\int_0^{\tau_2} z(y_1,y_2)\Delta_2 y_2\Delta_1 y_1\right)\Delta_2\tau_2\Delta_1\tau_1\right)\Delta_2 s_2 \\
&\quad + \left(\int_0^{t_2} a(t_1,s_2)\left(\int_0^{t_1}\int_0^{s_2} b(\tau_1,\tau_2)\Delta_2\tau_2\Delta_1\tau_1\right)\Delta_2 s_2\right)z(t_1,t_2)
\end{aligned}$$

$$\begin{aligned}
&\leq \int_0^{t_2} a(t_1,s_2)\left(g(s_2)+\int_0^{t_1} f(s_1)\Delta_1 s_1\right)\Delta_2 s_2\\
&\quad+\left(\int_0^{t_2} a(t_1,s_2)t_1 s_2\Delta_2 s_2\right)z(t_1,t_2)\\
&\quad+\left(\int_0^{t_2} a(t_1,s_2)\Delta_2 s_2\right)z(t_1,t_2)\\
&\quad+\int_0^{t_2} a(t_1,s_2)\left(\int_0^{t_1}\int_0^{s_2} b(\tau_1,\tau_2)\left(g(\tau_2)+\int_0^{\tau_1} f(y_1)\Delta_1 y_1\right)\Delta_2\tau_2\Delta_1\tau_1\right)\Delta_2 s_2\\
&\quad+\left(\int_0^{t_2} a(t_1,s_2)\left(\int_0^{t_1}\int_0^{s_2}\tau_1\tau_2 b(\tau_1,\tau_2)\Delta_2\tau_2\Delta_1\tau_1\right)\Delta_2 s_2\right)z(t_1,t_2)\\
&\quad+\left(\int_0^{t_2} a(t_1,s_2)\left(\int_0^{t_1}\int_0^{s_2} b(\tau_1,\tau_2)\Delta_2\tau_2\Delta_1\tau_1\right)\Delta_2 s_2\right)z(t_1,t_2)\\
&= h_1(t_1,t_2)+h_2(t_1,t_2)z(t_1,t_2),
\end{aligned}$$

$(t_1,t_2)\in(\mathbb{R}_+\cap\mathbb{T}_1)\times(\mathbb{R}_+\cap\mathbb{T}_2)$, i. e.,

$$z_{t_1}^{\Delta_1}(t_1,t_2)\leq h_1(t_1,t_2)+h_2(t_1,t_2)z(t_1,t_2),$$

$(t_1,t_2)\in(\mathbb{R}_+\cap\mathbb{T}_1)\times(\mathbb{R}_+\cap\mathbb{T}_2)$. From the last inequality and Lemma 2.1.1, we get

$$\begin{aligned}
z(t_1,t_2)&\leq z(0,t_2)e_{h_2}(t_1,0)+\int_0^{t_1} h_1(s_1,t_2)e_{\ominus h_2}(\sigma_1(s_1),t_1)\Delta_1 s_1\\
&= u_0 e_{h_2}(t_1,0)+\int_0^{t_1} h_1(s_1,t_2)e_{\ominus h_2}(\sigma_1(s_1),t_1)\Delta_1 s_1,
\end{aligned}$$

$(t_1,t_2)\in(\mathbb{R}_+\cap\mathbb{T}_1)\times(\mathbb{R}_+\cap\mathbb{T}_2)$. Therefore

$$\begin{aligned}
u_{t_1t_2}^{\Delta_1\Delta_2}(t_1,t_2)&\leq z(t_1,t_2)\\
&\leq u_0 e_{h_2}(t_1,0)+\int_0^{t_1} h_1(s_1,t_2)e_{\ominus h_2}(\sigma_1(s_1),t_1)\Delta_1 s_1,
\end{aligned}$$

$(t_1,t_2)\in(\mathbb{R}_+\cap\mathbb{T}_1)\times(\mathbb{R}_+\cap\mathbb{T}_2)$, and

$$u_{t_1}^{\Delta_1}(t_1,t_2)-u_{t_1}^{\Delta_1}(t_1,0)\leq\int_0^{t_2}\left(u_0 e_{h_2}(t_1,0)+\int_0^{t_1} h_1(s_1,s_2)e_{\ominus h_2}(\sigma_1(s_1),s_1)\Delta_1 s_1\right)\Delta_2 s_2,$$

$(t_1, t_2) \in (\mathbb{R}_+ \cap \mathbb{T}_1) \times (\mathbb{R}_+ \cap \mathbb{T}_2)$, or

$$u_{t_1}^{\Delta_1}(t_1, t_2) \leq f(t_1) + \int_0^{t_2} \left(u_0 e_{h_2}(t_1, 0) + \int_0^{t_1} h_1(s_1, s_2) e_{\ominus h_2}(\sigma_1(s_1), s_1) \Delta_1 s_1 \right) \Delta_2 s_2,$$

$(t_1, t_2) \in (\mathbb{R}_+ \cap \mathbb{T}_1) \times (\mathbb{R}_+ \cap \mathbb{T}_2)$, and

$$u(t_1, t_2) - u(0, t_2) \leq \int_0^{t_1} f(s_1) \Delta_1 s_1 + \int_0^{t_1} \int_0^{t_2} \left(u_0 e_{h_2}(s_1, 0) + \int_0^{s_1} h_1(y_1, s_2) e_{\ominus h_2}(\sigma_1(y_1), s_1) \Delta_1 y_1 \right) \Delta_2 s_2 \Delta_1 s_1,$$

$(t_1, t_2) \in (\mathbb{R}_+ \cap \mathbb{T}_1) \times (\mathbb{R}_+ \cap \mathbb{T}_2)$, or

$$u(t_1, t_2) \leq g(t_2) + \int_0^{t_1} f(s_1) \Delta_1 s_1 + \int_0^{t_1} \int_0^{t_2} \left(u_0 e_{h_2}(s_1, 0) + \int_0^{s_1} h_1(y_1, s_2) e_{\ominus h_2}(\sigma_1(y_1), s_1) \Delta_1 y_1 \right) \Delta_2 s_2 \Delta_1 s_1,$$

$(t_1, t_2) \in (\mathbb{R}_+ \cap \mathbb{T}_1) \times (\mathbb{R}_+ \cap \mathbb{T}_2)$. This completes the proof. □

8.2 Modifications of Pachpatte's inequalities

Theorem 8.2.1. *Let $u(t_1, t_2)$ be a nonnegative twice continuously-differentiable function for $(t_1, t_2) \in (\mathbb{R}_+ \cap \mathbb{T}_1) \times (\mathbb{R}_+ \cap \mathbb{T}_2)$,*

$$u_{t_1}^{\Delta_1}(t_1, t_2) \geq 0, \quad u_{t_2}^{\Delta_2}(t_1, t_2) \geq 0, \quad u_{t_1 t_2}^{\Delta_1 \Delta_2}(t_1, t_2) \geq 0,$$

$(t_1, t_2) \in (\mathbb{R}_+ \cap \mathbb{T}_1) \times (\mathbb{R}_+ \cap \mathbb{T}_2)$,

$$u_{t_1}^{\Delta_1}(t_1, 0) = f(t_1), \quad u(0, t_2) = g(t_2),$$

where $f(t_1)$ is a nonnegative continuous function for $t_1 \in \mathbb{R}_+ \cap \mathbb{T}_1$ and $g(t_2)$ is a nonnegative continuous function for $t_2 \in \mathbb{R}_+ \cap \mathbb{T}_2$. Let $a(t_1, t_2)$ and $b(t_1, t_2)$ be nonnegative continuous functions for $(t_1, t_2) \in (\mathbb{R}_+ \cap \mathbb{T}_1) \times (\mathbb{R}_+ \cap \mathbb{T}_2)$. If

$$u_{t_1 t_2}^{\Delta_1 \Delta_2}(t_1, t_2) \leq a(t_1, t_2) + \int_0^{t_1} \int_0^{t_2} b(s_1, s_2) u_{t_1}^{\Delta_1}(s_1, s_2) \Delta_2 s_2 \Delta_1 s_1,$$

$(t_1, t_2) \in (\mathbb{R}_+ \cap \mathbb{T}_1) \times (\mathbb{R}_+ \cap \mathbb{T}_2)$, *then*

$$
\begin{aligned}
u_{t_1 t_2}^{\Delta_1 \Delta_2}(t_1, t_2) &\leq h_3(t_1, t_2), \\
u_{t_1}^{\Delta_1}(t_1, t_2) &\leq f(t_1) + \int_0^{t_2} h_3(t_1, s_2)\Delta_2 s_2, \\
u(t_1, t_2) &\leq g(t_2) + \int_0^{t_1} \left(f(s_1) + \int_0^{t_2} h_3(s_1, s_2)\Delta_2 s_2 \right)\Delta_1 s_1,
\end{aligned}
$$

$(t_1, t_2) \in (\mathbb{R}_+ \cap \mathbb{T}_1) \times (\mathbb{R}_+ \cap \mathbb{T}_2)$, *where*

$$
\begin{aligned}
f_1(t_1, t_2) &= f(t_1) + \int_0^{t_2} a(t_1, s_2)\Delta_2 s_2, \\
h_1(t_1, t_2) &= \int_0^{t_2} f_1(t_1, s_2) b(t_1, s_2)\Delta_2 s_2, \\
h_2(t_1, t_2) &= \int_0^{t_2} b(t_1, s_2) s_2 \Delta_2 s_2, \\
h_3(t_1, t_2) &= a(t_1, t_2) \\
&\quad + \int_0^{t_1} h_1(s_1, t_2) e_{\ominus h_2}(\sigma_1(s_1), t_1)\Delta_1 s_1,
\end{aligned}
$$

$(t_1, t_2) \in (\mathbb{R}_+ \cap \mathbb{T}_1) \times (\mathbb{R}_+ \cap \mathbb{T}_2)$.

Proof. Let

$$
z(t_1, t_2) = \int_0^{t_1} \int_0^{t_2} b(s_1, s_2) u_{t_1}^{\Delta_1}(s_1, s_2)\Delta_2 s_2 \Delta_1 s_1,
$$

$(t_1, t_2) \in (\mathbb{R}_+ \cap \mathbb{T}_1) \times (\mathbb{R}_+ \cap \mathbb{T}_2)$. Note that $z(t_1, t_2)$ is a nonnegative continuous nondecreasing function in each variable t_1 and t_2, $(t_1, t_2) \in (\mathbb{R}_+ \cap \mathbb{T}_1) \times (\mathbb{R}_+ \cap \mathbb{T}_2)$. Then

$$
\begin{aligned}
u_{t_1 t_2}^{\Delta_1 \Delta_2}(t_1, t_2) &\leq z(t_1, t_2) + a(t_1, t_2), \\
u_{t_1}^{\Delta_1}(t_1, t_2) - u_{t_1}^{\Delta_1}(t_1, 0) &\leq \int_0^{t_2} z(t_1, s_2)\Delta_2 s_2 + \int_0^{t_2} a(t_1, s_2)\Delta_2 s_2, \\
u_{t_1}^{\Delta_1}(t_1, t_2) &\leq f(t_1) + \int_0^{t_2} z(t_1, s_2)\Delta_2 s_2 + \int_0^{t_2} a(t_1, s_2)\Delta_2 s_2 \\
&\leq f_1(t_1, t_2) + t_2 z(t_1, t_2),
\end{aligned}
$$

$(t_1,t_2) \in (\mathbb{R}_+ \cap \mathbb{T}_1) \times (\mathbb{R}_+ \cap \mathbb{T}_2)$. Next,

$$\begin{aligned}
z_{t_1}^{\Delta_1}(t_1,t_2) &= \int_0^{t_2} b(t_1,s_2) u_{t_1}^{\Delta_1}(t_1,s_2)\Delta_2 s_2 \\
&\le \int_0^{t_2} b(t_1,s_2)(f_1(t_1,s_2) + s_2 z(t_1,s_2))\Delta_2 s_2 \\
&= \int_0^{t_2} b(t_1,s_2) f_1(t_1,s_2)\Delta_2 s_2 \\
&\quad + \int_0^{t_2} b(t_1,s_2) s_2 z(t_1,s_2)\Delta_2 s_2 \\
&\le \int_0^{t_2} f_1(t_1,s_2) b(t_1,s_2)\Delta_2 s_2 \\
&\quad + \left(\int_0^{t_2} b(t_1,s_2) s_2 \Delta_2 s_2\right) z(t_1,t_2) \\
&= h_1(t_1,t_2) + h_2(t_1,t_2) z(t_1,t_2),
\end{aligned}$$

$(t_1,t_2) \in (\mathbb{R}_+ \cap \mathbb{T}_1) \times (\mathbb{R}_+ \cap \mathbb{T}_2)$, i. e.,

$$z_{t_1}^{\Delta_1}(t_1,t_2) \le h_1(t_1,t_2) + h_2(t_1,t_2) z(t_1,t_2),$$

$(t_1,t_2) \in (\mathbb{R}_+ \cap \mathbb{T}_1) \times (\mathbb{R}_+ \cap \mathbb{T}_2)$. From the last inequality and Lemma 2.1.1, we get

$$z(t_1,t_2) \le \int_0^{t_1} h_1(s_1,t_2) e_{\ominus h_2}(\sigma_1(s_1),t_1)\Delta_1 s_1,$$

$(t_1,t_2) \in (\mathbb{R}_+ \cap \mathbb{T}_1) \times (\mathbb{R}_+ \cap \mathbb{T}_2)$. Therefore

$$\begin{aligned}
u_{t_1 t_2}^{\Delta_1 \Delta_2}(t_1,t_2) &\le a(t_1,t_2) + z(t_1,t_2) \\
&\le a(t_1,t_2) + \int_0^{t_1} h_1(s_1,t_2) e_{\ominus h_2}(\sigma_1(s_1),t_1)\Delta_1 s_1 \\
&= h_3(t_1,t_2),
\end{aligned}$$

$$u_{t_1}^{\Delta_1}(t_1,t_2) - u_{t_1}^{\Delta_1}(t_1,0) \le \int_0^{t_2} h_3(t_1,s_2)\Delta_2 s_2,$$

$$u_{t_1}^{\Delta_1}(t_1,t_2) \le f(t_1) + \int_0^{t_2} h_3(t_1,s_2)\Delta_2 s_2,$$

$$u(t_1,t_2) - u(0,t_2) \le \int_0^{t_1}\left(f(s_1) + \int_0^{t_2} h_3(s_1,s_2)\Delta_2 s_2\right)\Delta_1 s_1,$$

$$u(t_1,t_2) \le g(t_2) + \int_0^{t_1}\left(f(s_1) + \int_0^{t_2} h_3(s_1,s_2)\Delta_2 s_2\right)\Delta_1 s_1,$$

$(t_1,t_2) \in (\mathbb{R}_+ \cap \mathbb{T}_1) \times (\mathbb{R}_+ \cap \mathbb{T}_2)$. This completes the proof. □

Theorem 8.2.2. *Let $u(t_1,t_2)$ be a nonnegative twice continuously-differentiable function for $(t_1,t_2) \in (\mathbb{R}_+ \cap \mathbb{T}_1) \times (\mathbb{R}_+ \cap \mathbb{T}_2)$ such that*

$$u_{t_1}^{\Delta_1}(t_1,t_2) \ge 0, \quad u_{t_2}^{\Delta_2}(t_1,t_2) \ge 0, \quad u_{t_1 t_2}^{\Delta_1\Delta_2}(t_1,t_2) \ge 0,$$

$(t_1,t_2) \in (\mathbb{R}_+ \cap \mathbb{T}_1) \times (\mathbb{R}_+ \cap \mathbb{T}_2)$, and

$$u_{t_1}^{\Delta_1}(t_1,0) = f_1(t_1), \quad u_{t_2}^{\Delta_2}(0,t_2) = f_2(t_2),$$
$$u(t_1,0) = f_3(t_1), \quad u(0,t_2) = f_4(t_2),$$
$$f_3(0) = f_4(0),$$

where $f_1(t_1)$ and $f_3(t_1)$ are nonnegative continuous functions for $t_1 \in \mathbb{R}_+ \cap \mathbb{T}_1$, $f_2(t_2)$ and $f_4(t_2)$ are nonnegative continuous functions for $t_2 \in \mathbb{R}_+ \cap \mathbb{T}_2$. Let also $b_1(t_1,t_2)$, $b_2(t_1,t_2)$, and $b_3(t_1,t_2)$ be nonnegative continuous functions for $(t_1,t_2) \in (\mathbb{R}_+ \cap \mathbb{T}_1) \times (\mathbb{R}_+ \cap \mathbb{T}_2)$. If

$$\begin{aligned} u_{t_1 t_2}^{\Delta_1\Delta_2}(t_1,t_2) \le \Bigg(&\int_0^{t_1}\int_0^{t_2} b_1(s_1,s_2)u(s_1,s_2)\Delta_2 s_2 \Delta_1 s_1 \\ &+ \int_0^{t_1}\int_0^{t_2} b_2(s_1,s_2)u_{t_1}^{\Delta_1}(s_1,s_2)\Delta_2 s_2 \Delta_1 s_1 \\ &+ \int_0^{t_1}\int_0^{t_2} b_3(s_1,s_2)u_{t_2}^{\Delta_2}(s_1,s_2)\Delta_2 s_2 \Delta_1 s_1 \Bigg), \end{aligned}$$

$(t_1,t_2) \in (\mathbb{R}_+ \cap \mathbb{T}_1) \times (\mathbb{R}_+ \cap \mathbb{T}_2)$, then

$$u_{t_1 t_2}^{\Delta_1\Delta_2}(t_1,t_2) \le \int_0^{t_1} h_5(s_1,t_2)e_{\ominus h_6}(\sigma_1(s_1),t_1)\Delta_1 s_1,$$

$$u_{t_1}^{\Delta_1}(t_1,t_2) \le f_1(t_1) + t_2\int_0^{t_1} h_5(s_1,t_2)e_{\ominus h_6}(\sigma_1(s_1),t_1)\Delta_1 s_1,$$

$$u_{t_2}^{\Delta_2}(t_1,t_2) \le f_2(t_2) + t_1 \int_0^{t_1} h_5(s_1,t_2) e_{\ominus h_6}(\sigma_1(s_1),t_1)\Delta_1 s_1,$$

$$u(t_1,t_2) \le f_4(t_2) + \int_0^{t_1} f_1(s_1)\Delta_1 s_1 + t_1 t_2 \int_0^{t_1} h_5(s_1,t_2) e_{\ominus h_6}(\sigma_1(s_1),t_1)\Delta_1 s_1,$$

$(t_1,t_2) \in (\mathbb{R}_+ \cap \mathbb{T}_1) \times (\mathbb{R}_+ \cap \mathbb{T}_2)$, *where*

$$F(t_1,t_2) = \max\left\{ \int_0^{t_2} \left(f_4(s_2) + \int_0^{t_1} f_1(s_1)\Delta_1 s_1 \right) b_1(t_1,s_2)\Delta_2 s_2, \int_0^{t_2} b_1(t_1,s_2) \left(f_3(t_1) + \int_0^{s_2} f_2(y_2)\Delta_2 y_2 \right) \Delta_2 s_2 \right\},$$

$$G(t_1,t_2) = t_1 \int_0^{t_2} b_1(t_1,s_2) s_2 \Delta_2 s_2,$$

$$h_1(t_1,t_2) = f_1(t_1) \int_0^{t_2} b_2(t_1,s_2)\Delta_2 s_2,$$

$$h_2(t_1,t_2) = \int_0^{t_2} b(t_1,s_2) s_2 \Delta_2 s_2,$$

$$h_3(t_1,t_2) = \int_0^{t_2} b_3(t_1,s_2) f_2(s_2)\Delta_2 s_2,$$

$$h_4(t_1,t_2) = t_1 \int_0^{t_2} b_3(t_1,s_2)\Delta_2 s_2,$$

$$h_5(t_1,t_2) = F(t_1,t_2) + h_1(t_1,t_2) + h_3(t_1,t_2),$$

$$h_6(t_1,t_2) = G(t_1,t_2) + h_2(t_1,t_2) + h_4(t_1,t_2),$$

$(t_1,t_2) \in (\mathbb{R}_+ \cap \mathbb{T}_1) \times (\mathbb{R}_+ \cap \mathbb{T}_2)$.

Proof. Let

$$z(t_1,t_2) = \int_0^{t_1} \int_0^{t_2} b_1(s_1,s_2) u(s_1,s_2)\Delta_2 s_2 \Delta_1 s_1$$

$$+\int_0^{t_1}\int_0^{t_2} b_2(s_1,s_2)u_{t_1}^{\Delta_1}(s_1,s_2)\Delta_2 s_2\Delta_1 s_1$$
$$+\int_0^{t_1}\int_0^{t_2} b_3(s_1,s_2)u_{t_2}^{\Delta_2}(s_1,s_2)\Delta_2 s_2\Delta_1 s_1,$$

$(t_1,t_2) \in (\mathbb{R}_+ \cap \mathbb{T}_1) \times (\mathbb{R}_+ \cap \mathbb{T}_2)$. Note that $z(t_1,t_2)$ is a nonnegative continuous nondecreasing function for $(t_1,t_2) \in (\mathbb{R}_+ \cap \mathbb{T}_1) \times (\mathbb{R}_+ \cap \mathbb{T}_2)$, and

$$\begin{aligned} z(0,t_2) &= 0, \\ z(t_1,0) &= 0, \\ u_{t_1t_2}^{\Delta_1\Delta_2}(t_1,t_2) &\le z(t_1,t_2), \end{aligned}$$

$(t_1,t_2) \in (\mathbb{R}_+ \cap \mathbb{T}_1) \times (\mathbb{R}_+ \cap \mathbb{T}_2)$. We have

$$u_{t_1}^{\Delta_1}(t_1,t_2) - u_{t_1}^{\Delta_1}(t_1,0) \le \int_0^{t_2} z(t_1,s_2)\Delta_2 s_2,$$

$$\begin{aligned} u_{t_1}^{\Delta_1}(t_1,t_2) &\le f_1(t_1) + \int_0^{t_2} z(t_1,s_2)\Delta_2 s_2 \\ &\le f_1(t_1) + t_2 z(t_1,t_2), \end{aligned}$$

$$u_{t_2}^{\Delta_2}(t_1,t_2) - u_{t_2}^{\Delta_2}(0,t_2) \le \int_0^{t_1} z(s_1,t_2)\Delta_1 s_1,$$

$$\begin{aligned} u_{t_2}^{\Delta_2}(t_1,t_2) &\le f_2(t_2) + \int_0^{t_1} z(s_1,t_2)\Delta_1 s_1 \\ &\le f_2(t_2) + t_1 z(t_1,t_2), \end{aligned}$$

$$u(t_1,t_2) - u(0,t_2) \le \int_0^{t_1} f_1(s_1)\Delta_1 s_1 + t_2\int_0^{t_1} z(s_1,t_2)\Delta_1 s_1,$$

$$\begin{aligned} u(t_1,t_2) &\le f_4(t_2) + \int_0^{t_1} f_1(s_1)\Delta_1 s_1 \\ &\quad + t_2\int_0^{t_1} z(s_1,t_2)\Delta_1 s_1 \\ &\le f_4(t_2) + \int_0^{t_1} f_1(s_1)\Delta_1 s_1 \\ &\quad + t_1 t_2 z(t_1,t_2), \end{aligned}$$

$$u(t_1,t_2) - u(t_1,0) \le \int_0^{t_2} f_2(s_2)\Delta_2 s_2 + t_1 \int_0^{t_2} z(t_1,s_2)\Delta_2 s_2,$$

$$\begin{aligned} u(t_1,t_2) &\le f_3(t_1) + \int_0^{t_2} f_2(s_2)\Delta_2 s_2 \\ &\quad + t_1 \int_0^{t_2} z(t_1,s_2)\Delta_2 s_2 \\ &\le f_3(t_1) + \int_0^{t_2} f_2(s_2)\Delta_2 s_2 \\ &\quad + t_1 t_2 z(t_1,t_2), \end{aligned}$$

$(t_1,t_2) \in (\mathbb{R}_+ \cap \mathbb{T}_1) \times (\mathbb{R}_+ \cap \mathbb{T}_2)$.

1. Let

$$u(t_1,t_2) \le f_4(t_2) + \int_0^{t_1} f_1(s_1)\Delta_1 s_1 + t_1 t_2 z(t_1,t_2),$$

   $(t_1,t_2) \in (\mathbb{R}_+ \cap \mathbb{T}_1) \times (\mathbb{R}_+ \cap \mathbb{T}_2)$. Then

$$\begin{aligned} \int_0^{t_2} b_1(t_1,s_2)u(t_1,s_2)\Delta_2 s_2 &\le \int_0^{t_2} b_1(t_1,s_2)\Bigg(f_4(s_2) + \int_0^{t_1} f_1(s_1)\Delta_1 s_1 \\ &\quad + t_1 s_2 z(t_1,s_2) \Bigg)\Delta_2 s_2 \\ &= \int_0^{t_2}\left(f_4(s_2) + \int_0^{t_1} f_1(s_1)\Delta_1 s_1 \right) b_1(t_1,s_2)\Delta_2 s_2 \\ &\quad + t_1 \int_0^{t_2} b_1(t_1,s_2) s_2 z(t_1,s_2)\Delta_2 s_2 \\ &\le F(t_1,t_2) \\ &\quad + \left(t_1 \int_0^{t_2} b_1(t_1,s_2) s_2 \Delta_2 s_2 \right) z(t_1,t_2) \\ &= F(t_1,t_2) + G(t_1,t_2) z(t_1,t_2), \end{aligned}$$

   $(t_1,t_2) \in (\mathbb{R}_+ \cap \mathbb{T}_1) \times (\mathbb{R}_+ \cap \mathbb{T}_2)$.

2. Let

$$u(t_1,t_2) \le f_3(t_1) + \int_0^{t_2} f_2(s_2)\Delta_2 s_2 + t_1 t_2 z(t_1,t_2),$$

$(t_1, t_2) \in (\mathbb{R}_+ \cap \mathbb{T}_1) \times (\mathbb{R}_+ \cap \mathbb{T}_2)$. Then

$$\begin{aligned}
\int_0^{t_2} b_1(t_1, s_2) u(t_1, s_2) \Delta_2 s_2 &\le \int_0^{t_2} b_1(t_1, s_2) \left(f_3(t_1) + \int_0^{s_2} f_2(y_2) \Delta_2 y_2 \right. \\
&\quad \left. + t_1 s_2 z(t_1, s_2) \right) \Delta_2 s_2 \\
&= \int_0^{t_2} b_1(t_1, s_2) \left(f_3(t_1) + \int_0^{s_2} f_2(y_2) \Delta_2 y_2 \right) \Delta_2 s_2 \\
&\quad + t_1 \int_0^{t_2} b_1(t_1, s_2) s_2 z(t_1, s_2) \Delta_2 s_2 \\
&\le F(t_1, t_2) + G(t_1, t_2) z(t_1, t_2),
\end{aligned}$$

$(t_1, t_2) \in (\mathbb{R}_+ \cap \mathbb{T}_1) \times (\mathbb{R}_+ \cap \mathbb{T}_2)$.

Next,

$$\begin{aligned}
\int_0^{t_2} b_2(t_1, s_2) u_{t_1}^{\Delta_1}(t_1, s_2) \Delta_2 s_2 &\le \int_0^{t_2} b_2(t_1, s_2)(f_1(t_1) + s_2 z(t_1, s_2)) \Delta_2 s_2 \\
&= f_1(t_1) \int_0^{t_2} b_2(t_1, s_2) \Delta_2 s_2 \\
&\quad + \int_0^{t_2} b_2(t_1, s_2) s_2 z(t_1, s_2) \Delta_2 s_2 \\
&\le h_1(t_1, t_2) \\
&\quad + \left(\int_0^{t_2} b_2(t_1, s_2) s_2 \Delta_2 s_2 \right) z(t_1, t_2) \\
&= h_1(t_1, t_2) + h_2(t_1, t_2) z(t_1, t_2), \\
\int_0^{t_2} b_3(t_1, s_2) u_{t_2}^{\Delta_2}(t_1, s_2) \Delta_2 s_2 &\le \int_0^{t_2} b_3(t_1, s_2)(f_2(s_2) + t_1 z(t_1, s_2)) \Delta_2 s_2 \\
&= \int_0^{t_2} b_3(t_1, s_2) f_2(s_2) \Delta_2 s_2 \\
&\quad + t_1 \int_0^{t_2} b_3(t_1, s_2) z(t_1, s_2) \Delta_2 s_2
\end{aligned}$$

$$
\begin{aligned}
&\leq h_3(t_1,t_2)\\
&\quad+\left(t_1\int_0^{t_2} b_3(t_1,s_2)\Delta_2 s_2\right)z(t_1,t_2)\\
&= h_3(t_1,t_2)+h_4(t_1,t_2)z(t_1,t_2),
\end{aligned}
$$

$(t_1,t_2)\in(\mathbb{R}_+\cap\mathbb{T}_1)\times(\mathbb{R}_+\cap\mathbb{T}_2)$. Then

$$
\begin{aligned}
z_{t_1}^{\Delta_1}(t_1,t_2) &= \int_0^{t_2} b_1(t_1,s_2)u(t_1,s_2)\Delta_2 s_2\\
&\quad+\int_0^{t_2} b_2(t_1,s_2)u_{t_1}^{\Delta_1}(t_1,s_2)\Delta_2 s_2\\
&\quad+\int_0^{t_2} b_3(t_1,s_2)u_{t_2}^{\Delta_2}(t_1,s_2)\Delta_2 s_2\\
&\leq F(t_1,t_2)+G(t_1,t_2)z(t_1,t_2)\\
&\quad+h_1(t_1,t_2)+h_2(t_1,t_2)z(t_1,t_2)\\
&\quad+h_3(t_1,t_2)+h_4(t_1,t_2)z(t_1,t_2)\\
&= h_5(t_1,t_2)+h_6(t_1,t_2)z(t_1,t_2),
\end{aligned}
$$

$(t_1,t_2)\in(\mathbb{R}_+\cap\mathbb{T}_1)\times(\mathbb{R}_+\cap\mathbb{T}_2)$. From the last inequality and Lemma 2.1.1, we obtain

$$
z(t_1,t_2)\leq\int_0^{t_1} h_5(s_1,t_2)e_{\ominus h_6}(\sigma_1(s_1),t_1)\Delta_1 s_1,
$$

$(t_1,t_2)\in(\mathbb{R}_+\cap\mathbb{T}_1)\times(\mathbb{R}_+\cap\mathbb{T}_2)$. Hence,

$$
\begin{aligned}
u_{t_1}^{\Delta_1}(t_1,t_2) &\leq f_1(t_1)+t_2 z(t_1,t_2)\\
&\leq f_1(t_1)\\
&\quad+t_2\int_0^{t_1} h_5(s_1,t_2)e_{\ominus h_6}(\sigma_1(s_1),t_1)\Delta_1 s_1,\\
u_{t_2}^{\Delta_2}(t_1,t_2) &\leq f_2(t_2)+t_1 z(t_1,t_2)\\
&\leq f_2(t_2)\\
&\quad+t_1\int_0^{t_1} h_5(s_1,t_2)e_{\ominus h_6}(\sigma_1(s_1),t_1)\Delta_1 s_1,\\
u(t_1,t_2) &\leq f_4(t_2)+\int_0^{t_1} f_1(s_1)\Delta_1 s_1\\
&\quad+t_1 t_2 z(t_1,t_2)
\end{aligned}
$$

$$\leq f_4(t_2) + \int_0^{t_1} f_1(s_1)\Delta_1 s_1 + t_1 t_2 \int_0^{t_1} h_5(s_1,t_2) e_{\ominus h_6}(\sigma_1(s_1),t_1)\Delta_1 s_1,$$

$(t_1,t_2) \in (\mathbb{R}_+ \cap \mathbb{T}_1) \times (\mathbb{R}_+ \cap \mathbb{T}_2)$. This completes the proof. □

Theorem 8.2.3. *Let $a(t_1,t_2)$, $b(t_1,t_2)$ and $c(t_1,t_2)$ be nonnegative continuous functions for $(t_1,t_2) \in (\mathbb{R}_+ \cap \mathbb{T}_1) \times (\mathbb{R}_+ \cap \mathbb{T}_2)$, $u(t_1,t_2)$ be a nonnegative twice continuously-differentiable function for $(t_1,t_2) \in (\mathbb{R}_+ \cap \mathbb{T}_1) \times (\mathbb{R}_+ \cap \mathbb{T}_2)$ such that*

$$u_{t_1}^{\Delta_1}(t_1,t_2) \geq 0, \quad u_{t_2}^{\Delta_2}(t_1,t_2) \geq 0,$$
$$u_{t_1t_2}^{\Delta_1\Delta_2}(t_1,t_2) \geq 0, \quad u_{t_2t_2}^{\Delta_2\Delta_2}(t_1,t_2) \geq 0,$$
$$u_{t_1}^{\Delta_1}(t_1,0) = f(t_1), \quad u_{t_2}^{\Delta_2}(0,t_2) = g(t_2),$$

$(t_1,t_2) \in (\mathbb{R}_+ \cap \mathbb{T}_1) \times (\mathbb{R}_+ \cap \mathbb{T}_2)$, where $f(t_1)$ is a nonnegative continuous function for $t_1 \in \mathbb{R}_+ \cap \mathbb{T}_1$, $g(t_2)$ is a nonnegative continuous function for $t_2 \in \mathbb{R}_+ \cap \mathbb{T}_2$. If

$$u_{t_1t_2}^{\Delta_1\Delta_2}(t_1,t_2) \leq a(t_1,t_2) + b(t_1,t_2)\Bigg(u_{t_2}^{\Delta_2}(t_1,t_2) + \int_0^{t_1}\int_0^{t_2} c(s_1,s_2)(u_{t_1}^{\Delta_1}(s_1,s_2) + u_{t_2}^{\Delta_2}(s_1,s_2))\Delta_2 s_2 \Delta_1 s_1 \Bigg),$$

$(t_1,t_2) \in (\mathbb{R}_+ \cap \mathbb{T}_1) \times (\mathbb{R}_+ \cap \mathbb{T}_2)$, then

$$u_{t_1t_2}^{\Delta_1\Delta_2}(t_1,t_2) \leq a(t_1,t_2) + b(t_1,t_2)g(t_2)e_{h_6}(t_1,0) + b(t_1,t_2)\int_0^{t_1} h_5(s_1,t_2)e_{\ominus h_6}(\sigma_1(s_1),t_1)\Delta_1 s_1,$$

$(t_1,t_2) \in (\mathbb{R}_+ \cap \mathbb{T}_1) \times (\mathbb{R}_+ \cap \mathbb{T}_2)$, where

$$h_1(t_1,t_2) = f(t_1) + \int_0^{t_2} a(t_1,s_2)\Delta_2 s_2,$$

$$h_2(t_1,t_2) = \int_0^{t_2} b(t_1,s_2)\Delta_2 s_2,$$

$$h_3(t_1,t_2) = g(t_2) + \int_0^{t_1} a(s_1,t_2)\Delta_1 s_1,$$

$$h_4(t_1,t_2) = \int_0^{t_1} b(s_1,t_2)\Delta_1 s_1,$$

$$h_5(t_1,t_2) = a(t_1,t_2) + \int_0^{t_2} c(t_1,s_2)h_1(t_1,s_2)\Delta_2 s_2 + \int_0^{t_2} c(t_1,s_2)h_3(t_1,s_2)\Delta_2 s_2,$$

$$h_6(t_1,t_2) = b(t_1,t_2) + \int_0^{t_2} c(t_1,s_2)h_2(t_1,s_2)\Delta_2 s_2 + \int_0^{t_2} c(t_1,s_2)h_4(t_1,s_2)\Delta_2 s_2,$$

$(t_1,t_2) \in (\mathbb{R}_+ \cap \mathbb{T}_1) \times (\mathbb{R}_+ \cap \mathbb{T}_2)$.

Proof. Let

$$z(t_1,t_2) = u_{t_2}^{\Delta_2}(t_1,t_2) + \int_0^{t_1}\int_0^{t_2} c(s_1,s_2)(u_{t_1}^{\Delta_1}(s_1,s_2) + u_{t_2}^{\Delta_2}(s_1,s_2))\Delta_2 s_2 \Delta_1 s_1,$$

$(t_1,t_2) \in (\mathbb{R}_+ \cap \mathbb{T}_1) \times (\mathbb{R}_+ \cap \mathbb{T}_2)$. Since

$$u_{t_1 t_2}^{\Delta_1\Delta_2}(t_1,t_2) \geq 0, \quad u_{t_2 t_2}^{\Delta_2\Delta_2}(t_1,t_2) \geq 0,$$

$(t_1,t_2) \in (\mathbb{R}_+ \cap \mathbb{T}_1) \times (\mathbb{R}_+ \cap \mathbb{T}_2)$, we have that $z(t_1,t_2)$ is a nonnegative nondecreasing continuous function for $(t_1,t_2) \in (\mathbb{R}_+ \cap \mathbb{T}_1) \times (\mathbb{R}_+ \cap \mathbb{T}_2)$. We have

$$u_{t_1 t_2}^{\Delta_1\Delta_2}(t_1,t_2) \leq a(t_1,t_2) + b(t_1,t_2)z(t_1,t_2),$$

$(t_1,t_2) \in (\mathbb{R}_+ \cap \mathbb{T}_1) \times (\mathbb{R}_+ \cap \mathbb{T}_2)$. Hence,

$$u_{t_1}^{\Delta_1}(t_1,t_2) - u_{t_1}^{\Delta_1}(t_1,0) \leq \int_0^{t_2} (a(t_1,s_2) + b(t_1,s_2)z(t_1,s_2))\Delta_2 s_2,$$

$(t_1,t_2) \in (\mathbb{R}_+ \cap \mathbb{T}_1) \times (\mathbb{R}_+ \cap \mathbb{T}_2)$, or

$$u_{t_1}^{\Delta_1}(t_1,t_2) \leq f(t_1) + \int_0^{t_2} (a(t_1,s_2) + b(t_1,s_2)z(t_1,s_2))\Delta_2 s_2$$

$$
\begin{aligned}
&= f(t_1) + \int_0^{t_2} a(t_1,s_2)\Delta_2 s_2 \\
&\quad + \int_0^{t_2} b(t_1,s_2) z(t_1,s_2)\Delta_2 s_2 \\
&\le f(t_1) + \int_0^{t_2} a(t_1,s_2)\Delta_2 s_2 \\
&\quad + \left(\int_0^{t_2} b(t_1,s_2)\Delta_2 s_2\right) z(t_1,t_2) \\
&= h_1(t_1,t_2) + h_2(t_1,t_2) z(t_1,t_2),
\end{aligned}
$$

$(t_1,t_2) \in (\mathbb{R}_+ \cap \mathbb{T}_1) \times (\mathbb{R}_+ \cap \mathbb{T}_2)$, and

$$
u_{t_2}^{\Delta_2}(t_1,t_2) - u_{t_2}^{\Delta_2}(0,t_2) \le \int_0^{t_1} (a(s_1,t_2) + b(s_1,t_2) z(s_1,t_2))\Delta_1 s_1,
$$

$(t_1,t_2) \in (\mathbb{R}_+ \cap \mathbb{T}_1) \times (\mathbb{R}_+ \cap \mathbb{T}_2)$, or

$$
\begin{aligned}
u_{t_2}^{\Delta_2}(t_1,t_2) &\le g(t_2) + \int_0^{t_1} (a(s_1,t_2) + b(s_1,t_2) z(s_1,t_2))\Delta_1 s_1 \\
&= g(t_2) + \int_0^{t_1} a(s_1,t_2)\Delta_1 s_1 \\
&\quad + \int_0^{t_1} b(s_1,t_2) z(s_1,t_2)\Delta_1 s_1 \\
&\le g(t_2) + \int_0^{t_1} a(s_1,t_2)\Delta_1 s_1 \\
&\quad + \left(\int_0^{t_1} b(s_1,t_2)\Delta_1 s_1\right) z(t_1,t_2) \\
&= h_3(t_1,t_2) + h_4(t_1,t_2) z(t_1,t_2),
\end{aligned}
$$

$(t_1,t_2) \in (\mathbb{R}_+ \cap \mathbb{T}_1) \times (\mathbb{R}_+ \cap \mathbb{T}_2)$. Now, using the definition of function $z(t_1,t_2)$, we get

$$
z_{t_1}^{\Delta_1}(t_1,t_2) = u_{t_1 t_2}^{\Delta_1 \Delta_2}(t_1,t_2) + \int_0^{t_2} c(t_1,s_2)(u_{t_1}^{\Delta_1}(t_1,s_2) + u_{t_2}^{\Delta_2}(t_1,s_2))\Delta_2 s_2
$$

$$
\begin{aligned}
&= u^{\Delta_1\Delta_2}_{t_1t_2}(t_1,t_2) + \int_0^{t_2} c(t_1,s_2)u^{\Delta_1}_{t_1}(t_1,s_2)\Delta_2 s_2 \\
&\quad + \int_0^{t_2} c(t_1,s_2)u^{\Delta_2}_{t_2}(t_1,s_2)\Delta_2 s_2 \\
&\le a(t_1,t_2) + b(t_1,t_2)z(t_1,t_2) \\
&\quad + \int_0^{t_2} c(t_1,s_2)(h_1(t_1,s_2) + h_2(t_1,s_2)z(t_1,s_2))\Delta_2 s_2 \\
&\quad + \int_0^{t_2} c(t_1,s_2)(h_3(t_1,s_2) + h_4(t_1,s_2)z(t_1,s_2))\Delta_2 s_2 \\
&= a(t_1,t_2) + b(t_1,t_2)z(t_1,t_2) \\
&\quad + \int_0^{t_2} c(t_1,s_2)h_1(t_1,s_2)\Delta_2 s_2 \\
&\quad + \int_0^{t_2} c(t_1,s_2)h_2(t_1,s_2)z(t_1,s_2)\Delta_2 s_2 \\
&\quad + \int_0^{t_2} c(t_1,s_2)h_3(t_1,s_2)\Delta_2 s_2 \\
&\quad + \int_0^{t_2} c(t_1,s_2)h_4(t_1,s_2)z(t_1,s_2)\Delta_2 s_2 \\
&\le a(t_1,t_2) + b(t_1,t_2)z(t_1,t_2) \\
&\quad + \int_0^{t_2} c(t_1,s_2)h_1(t_1,s_2)\Delta_2 s_2 \\
&\quad + \left(\int_0^{t_2} c(t_1,s_2)h_2(t_1,s_2)\Delta_2 s_2\right)z(t_1,t_2) \\
&\quad + \int_0^{t_2} c(t_1,s_2)h_3(t_1,s_2)\Delta_2 s_2 \\
&\quad + \left(\int_0^{t_2} c(t_1,s_2)h_4(t_1,s_2)\Delta_2 s_2\right)z(t_1,t_2) \\
&= h_5(t_1,t_2) + h_6(t_1,t_2)z(t_1,t_2),
\end{aligned}
$$

$(t_1, t_2) \in (\mathbb{R}_+ \cap \mathbb{T}_1) \times (\mathbb{R}_+ \cap \mathbb{T}_2)$, i. e.,

$$z_{t_1}^{\Delta_1}(t_1, t_2) \le h_5(t_1, t_2) + h_6(t_1, t_2) z(t_1, t_2),$$

$(t_1, t_2) \in (\mathbb{R}_+ \cap \mathbb{T}_1) \times (\mathbb{R}_+ \cap \mathbb{T}_2)$. From the last inequality and Lemma 2.1.1, we find

$$\begin{aligned} z(t_1, t_2) &\le z(0, t_2) e_{h_6}(t_1, 0) \\ &\quad + \int_0^{t_1} h_5(s_1, t_2) e_{\ominus h_6}(\sigma_1(s_1), t_1) \Delta_1 s_1 \\ &= g(t_2) e_{h_6}(t_1, 0) \\ &\quad + \int_0^{t_1} h_5(s_1, t_2) e_{\ominus h_6}(\sigma_1(s_1), t_1) \Delta_1 s_1, \end{aligned}$$

$(t_1, t_2) \in (\mathbb{R}_+ \cap \mathbb{T}_1) \times (\mathbb{R}_+ \cap \mathbb{T}_2)$. Consequently,

$$\begin{aligned} u_{t_1 t_2}^{\Delta_1 \Delta_2}(t_1, t_2) &\le a(t_1, t_2) + b(t_1, t_2) z(t_1, t_2) \\ &\le a(t_1, t_2) + b(t_1, t_2) g(t_2) e_{h_6}(t_1, 0) \\ &\quad + b(t_1, t_2) \int_0^{t_1} h_5(s_1, t_2) e_{\ominus h_6}(\sigma_1(s_1), t_1) \Delta_1 s_1, \end{aligned}$$

$(t_1, t_2) \in (\mathbb{R}_+ \cap \mathbb{T}_1) \times (\mathbb{R}_+ \cap \mathbb{T}_2)$. This completes the proof. □

9 Two-dimensional nonlinear integral inequalities

This chapter is devoted on Wendroff type and Pachpatte type two dimensional nonlinear integral inequalities. Some of the results in this chapter can be found in [8, 24] and [25].

Let $\mathbb{T}_1$ and $\mathbb{T}_2$ be time scales with forward jump operators and delta differentiation operators σ_1, σ_2 and Δ_1, Δ_2, respectively. Suppose that $0 \in \mathbb{T}_1$ and $0 \in \mathbb{T}_2$.

9.1 Wendroff's inequality

Theorem 9.1.1 (Wendroff's Inequality). *Let $u(t_1, t_2)$, $a(t_1, t_2)$, and $b(t_1, t_2)$ be nonnegative continuous functions for $(t_1, t_2) \in (\mathbb{R}_+ \cap \mathbb{T}_1) \times (\mathbb{R}_+ \cap \mathbb{T}_2)$. Let also*

$$L : (\mathbb{R}_+ \cap \mathbb{T}_1) \times (\mathbb{R}_+ \cap \mathbb{T}_2) \times \mathbb{R}_+ \to \mathbb{R}_+$$

be a continuous function such that

$$0 \le L(t_1, t_2, x) - L(t_1, t_2, y) \le M(t_1, t_2, y)(x - y)$$

for $(t_1, t_2) \in (\mathbb{R}_+ \cap \mathbb{T}_1) \times (\mathbb{R}_+ \cap \mathbb{T}_2)$ and $x \ge y \ge 0$, where

$$M : (\mathbb{R}_+ \cap \mathbb{T}_1) \times (\mathbb{R}_+ \cap \mathbb{T}_2) \times \mathbb{R}_+ \to \mathbb{R}_+$$

is a continuous function. If

$$\begin{aligned} u(t_1, t_2) &\le a(t_1, t_2) \\ &\quad + b(t_1, t_2) \int_0^{t_1} \int_0^{t_2} L(s_1, s_2, u(s_1, s_2)) \Delta_2 s_2 \Delta_1 s_1, \end{aligned}$$

$(t_1, t_2) \in (\mathbb{R}_+ \cap \mathbb{T}_1) \times (\mathbb{R}_+ \cap \mathbb{T}_2)$, then

$$\begin{aligned} u(t_1, t_2) &\le a(t_1, t_2) + b(t_1, t_2) h_1(t_1, t_2) \\ &\quad + b(t_1, t_2) A(t_1, t_2) e_B(t_2, 0), \end{aligned}$$

$(t_1, t_2) \in (\mathbb{R}_+ \cap \mathbb{T}_1) \times (\mathbb{R}_+ \cap \mathbb{T}_2)$, where

$$\begin{aligned} h_1(t_1, t_2) &= \int_0^{t_1} \int_0^{t_2} L(s_1, s_2, a(s_1, s_2)) \Delta_2 s_2 \Delta_1 s_1, \\ h_2(t_1, t_2) &= b(t_1, t_2) M(t_1, t_2, a(t_1, t_2)) h_1(t_1, t_2), \\ h_3(t_1, t_2) &= b(t_1, t_2) M(t_1, t_2, a(t_1, t_2)), \end{aligned}$$

https://doi.org/10.1515/9783110705553-009

$$A(t_1,t_2) = \int_0^{t_1}\int_0^{t_2} h_2(s_1,s_2)\Delta_2 s_2 \Delta_1 s_1,$$

$$B(t_1,t_2) = \int_0^{t_2} h_3(t_1,s_2)\Delta_2 s_2,$$

$(t_1,t_2) \in (\mathbb{R}_+ \cap \mathbb{T}_1) \times (\mathbb{R}_+ \cap \mathbb{T}_2)$.

Proof. Let

$$z(t_1,t_2) = \int_0^{t_1}\int_0^{t_2} L(s_1,s_2,u(s_1,s_2))\Delta_2 s_2 \Delta_1 s_1,$$

$(t_1,t_2) \in (\mathbb{R}_+ \cap \mathbb{T}_1) \times (\mathbb{R}_+ \cap \mathbb{T}_2)$. Then $z(t_1,t_2)$ is a nonnegative continuous function for $(t_1,t_2) \in (\mathbb{R}_+ \cap \mathbb{T}_1) \times (\mathbb{R}_+ \cap \mathbb{T}_2)$ and

$$u(t_1,t_2) \le a(t_1,t_2) + b(t_1,t_2)z(t_1,t_2),$$

$(t_1,t_2) \in (\mathbb{R}_+ \cap \mathbb{T}_1) \times (\mathbb{R}_+ \cap \mathbb{T}_2)$. Hence, using the conditions for the function L, we get

$$\begin{aligned} z(t_1,t_2) &\le \int_0^{t_1}\int_0^{t_2} L(s_1,s_2,a(s_1,s_2) + b(s_1,s_2)z(s_1,s_2))\Delta_2 s_2 \Delta_1 s_1 \\ &= \int_0^{t_1}\int_0^{t_2} L(s_1,s_2,a(s_1,s_2))\Delta_2 s_2 \Delta_1 s_1 \\ &\quad + \int_0^{t_1}\int_0^{t_2} (L(s_1,s_2,a(s_1,s_2) + b(s_1,s_2)z(s_1,s_2)) \\ &\quad - L(s_1,s_2,a(s_1,s_2)))\Delta_2 s_2 \Delta_1 s_1 \\ &\le h_1(t_1,t_2) \\ &\quad + \int_0^{t_1}\int_0^{t_2} b(s_1,s_2)M(s_1,s_2,a(s_1,s_2))z(s_1,s_2)\Delta_2 s_2 \Delta_1 s_1, \end{aligned}$$

$(t_1,t_2) \in (\mathbb{R}_+ \cap \mathbb{T}_1) \times (\mathbb{R}_+ \cap \mathbb{T}_2)$. Hence, by Corollary 6.2.2, we get

$$z(t_1,t_2) \le h_1(t_1,t_2) + A(t_1,t_2)e_B(t_2,0),$$

$(t_1,t_2) \in (\mathbb{R}_+ \cap \mathbb{T}_1) \times (\mathbb{R}_+ \cap \mathbb{T}_2)$. Therefore

$$\begin{aligned} u(t_1,t_2) &\le a(t_1,t_2) + b(t_1,t_2)z(t_1,t_2) \\ &\le a(t_1,t_2) + b(t_1,t_2)h_1(t_1,t_2) \\ &\quad + b(t_1,t_2)A(t_1,t_2)e_B(t_2,0), \end{aligned}$$

$(t_1,t_2) \in (\mathbb{R}_+ \cap \mathbb{T}_1) \times (\mathbb{R}_+ \cap \mathbb{T}_2)$. This completes the proof. □

9.2 Pachpatte's inequality

Theorem 9.2.1 (Pachpatte's Inequality). *Let functions $u(t_1,t_2)$, $a(t_1,t_2)$. and $L(t_1,t_2)$ satisfy all conditions of Theorem 9.1.1. Let also $F(u)$ be a continuous, strictly increasing, convex and submultiplicative function for $u > 0$,*

$$\lim_{u\to\infty} F(u) = \infty,$$

suppose F^{-1} denotes the inverse function of F, also let $\alpha(t_1,t_2)$ and $\beta(t_1,t_2)$ be continuous and positive functions for $(t_1,t_2) \in (\mathbb{R}_+ \cap \mathbb{T}_1) \times (\mathbb{R}_+ \cap \mathbb{T}_2)$ and

$$\alpha(t_1,t_2) + \beta(t_1,t_2) = 1,$$

$(t_1,t_2) \in (\mathbb{R}_+ \cap \mathbb{T}_1) \times (\mathbb{R}_+ \cap \mathbb{T}_2)$. If

$$\begin{aligned} u(t_1,t_2) &\le a(t_1,t_2) + b(t_1,t_2) \\ &\quad \times F^{-1}\left(\int_0^{t_1}\int_0^{t_2} L(s_1,s_2,F(u(s_1,s_2)))\Delta_2 s_2 \Delta_1 s_1\right), \end{aligned} \tag{9.1}$$

$(t_1,t_2) \in (\mathbb{R}_+ \cap \mathbb{T}_1) \times (\mathbb{R}_+ \cap \mathbb{T}_2)$, then

$$\begin{aligned} u(t_1,t_2) &\le a(t_1,t_2) + b(t_1,t_2) \\ &\quad \times F^{-1}\left(\int_0^{t_1}\int_0^{t_2} L(s_1,s_2,h_1(s_1,s_2))\Delta_2 s_2 \Delta_1 s_1 \right. \\ &\quad \left. + A(t_1,t_2)e_B(t_2,0)\right), \end{aligned}$$

$(t_1,t_2) \in (\mathbb{R}_+ \cap \mathbb{T}_1) \times (\mathbb{R}_+ \cap \mathbb{T}_2)$, where

$$\begin{aligned} h_1(t_1,t_2) &= \alpha(t_1,t_2)F\big(a(t_1,t_2)(\alpha(t_1,t_2))^{-1}\big), \\ h_2(t_1,t_2) &= \beta(t_1,t_2)F\big(b(t_1,t_2)(\beta(t_1,t_2))^{-1}\big), \\ h_3(t_1,t_2) &= \int_0^{t_1}\int_0^{t_2} L(s_1,s_2,h_1(s_1,s_2))\Delta_2 s_2\Delta_1 s_1, \\ h_4(t_1,t_2) &= h_2(t_1,t_2)M(t_1,t_2,h_1(t_1,t_2)), \\ h_5(t_1,t_2) &= h_2(t_1,t_2)M(t_1,t_2,h_1(t_1,t_2)), \\ A(t_1,t_2) &= \int_0^{t_1}\int_0^{t_2} h_4(s_1,s_2)\Delta_2 s_2\Delta_1 s_1, \\ B(t_1,t_2) &= \int_0^{t_2} h_5(t_1,s_2)\Delta_2 s_2, \end{aligned}$$

$(t_1,t_2) \in (\mathbb{R}_+ \cap \mathbb{T}_1) \times (\mathbb{R}_+ \cap \mathbb{T}_2)$.

Proof. We rewrite inequality (9.1) in the following form:

$$
\begin{aligned}
u(t_1,t_2) \le{}& \alpha(t_1,t_2)a(t_1,t_2)(\alpha(t_1,t_2))^{-1}\\
&+\beta(t_1,t_2)b(t_1,t_2)(\beta(t_1,t_2))^{-1}\\
&\times F^{-1}\left(\int_0^{t_1}\int_0^{t_2} L(s_1,s_2,F(u(s_1,s_2)))\Delta_2 s_2\Delta_1 s_1\right),
\end{aligned}
$$

$(t_1,t_2) \in (\mathbb{R}_+ \cap \mathbb{T}_1) \times (\mathbb{R}_+ \cap \mathbb{T}_2)$. Now, since F is strictly increasing, convex and submultiplicative, we get

$$
\begin{aligned}
F(u(t_1,t_2)) \le{}& F\Bigg(\alpha(t_1,t_2)a(t_1,t_2)(\alpha(t_1,t_2))^{-1}\\
&+\beta(t_1,t_2)b(t_1,t_2)(\beta(t_1,t_2))^{-1}\\
&\times F^{-1}\left(\int_0^{t_1}\int_0^{t_2} L(s_1,s_2,F(u(s_1,s_2)))\Delta_2 s_2\Delta_1 s_1\right)\Bigg)\\
\le{}& \alpha(t_1,t_2)F(a(t_1,t_2)(\alpha(t_1,t_2))^{-1})\\
&+\beta(t_1,t_2)F\Bigg(b(t_1,t_2)(\beta(t_1,t_2))^{-1}\\
&\times F^{-1}\left(\int_0^{t_1}\int_0^{t_2} L(s_1,s_2,F(u(s_1,s_2)))\Delta_2 s_2\Delta_1 s_1\right)\Bigg)\\
\le{}& \alpha(t_1,t_2)F(a(t_1,t_2)(\alpha(t_1,t_2))^{-1})\\
&+\beta(t_1,t_2)F(b(t_1,t_2)(\beta(t_1,t_2))^{-1})\\
&\times F\left(F^{-1}\left(\int_0^{t_1}\int_0^{t_2} L(s_1,s_2,F(u(s_1,s_2)))\Delta_2 s_2\Delta_1 s_1\right)\right)\\
={}& h_1(t_1,t_2)\\
&+h_2(t_1,t_2)\int_0^{t_1}\int_0^{t_2} L(s_1,s_2,F(u(s_1,s_2)))\Delta_2 s_2\Delta_1 s_1,
\end{aligned}
$$

$(t_1,t_2) \in (\mathbb{R}_+ \cap \mathbb{T}_1) \times (\mathbb{R}_+ \cap \mathbb{T}_2)$. Now we apply Theorem 9.1.1, and get

$$
\begin{aligned}
F(u(t_1,t_2)) \le{}& h_1(t_1,t_2) + h_2(t_1,t_2)h_3(t_1,t_2)\\
&+ h_2(t_1,t_2)A(t_1,t_2)e_B(t_2,0)\\
={}& \alpha(t_1,t_2)F(a(t_1,t_2)(\alpha(t_1,t_2))^{-1})\\
&+\beta(t_1,t_2)F(b(t_1,t_2)(\beta(t_1,t_2))^{-1})
\end{aligned}
$$

$$\times\left(\int_0^{t_1}\int_0^{t_2} L(s_1,s_2,h_1(s_1,s_2))\Delta_2 s_2\Delta_1 s_1 + A(t_1,t_2)e_B(t_2,0)\right),$$

$(t_1,t_2) \in (\mathbb{R}_+ \cap \mathbb{T}_1) \times (\mathbb{R}_+ \cap \mathbb{T}_2)$. Hence,

$$\begin{aligned}
u(t_1,t_2) &\le F^{-1}(h_1(t_1,t_2) + h_2(t_1,t_2)h_3(t_1,t_2) \\
&\quad + h_2(t_1,t_2)A(t_1,t_2)e_B(t_2,0)) \\
&= F^{-1}\Bigg(\alpha(t_1,t_2)F(a(t_1,t_2)(\alpha(t_1,t_2))^{-1}) \\
&\quad + \beta(t_1,t_2)F(b(t_1,t_2)(\beta(t_1,t_2))^{-1}) \\
&\quad \times\left(\int_0^{t_1}\int_0^{t_2} L(s_1,s_2,h_1(s_1,s_2))\Delta_2 s_2\Delta_1 s_1 \right. \\
&\quad \left. + A(t_1,t_2)e_B(t_2,0)\right)\Bigg) \\
&\le \alpha(t_1,t_2)F^{-1}(F(a(t_1,t_2)(\alpha(t_1,t_2))^{-1})) \\
&\quad + \beta(t_1,t_2)F^{-1}\Bigg(F(b(t_1,t_2)(\beta(t_1,t_2))^{-1}) \\
&\quad \times\left(\int_0^{t_1}\int_0^{t_2} L(s_1,s_2,h_1(s_1,s_2))\Delta_2 s_2\Delta_1 s_1 \right. \\
&\quad \left. + A(t_1,t_2)e_B(t_2,0)\right)\Bigg) \\
&\le \alpha(t_1,t_2)a(t_1,t_2)(\alpha(t_1,t_2))^{-1} \\
&\quad + \beta(t_1,t_2)F^{-1}(F(b(t_1,t_2)(\beta(t_1,t_2))^{-1})) \\
&\quad \times F^{-1}\left(\int_0^{t_1}\int_0^{t_2} L(s_1,s_2,h_1(s_1,s_2))\Delta_2 s_2\Delta_1 s_1 \right. \\
&\quad \left. + A(t_1,t_2)e_B(t_2,0)\right) \\
&= a(t_1,t_2) \\
&\quad + \beta(t_1,t_2)b(t_1,t_2)(\beta(t_1,t_2))^{-1} \\
&\quad \times F^{-1}\left(\int_0^{t_1}\int_0^{t_2} L(s_1,s_2,h_1(s_1,s_2))\Delta_2 s_2\Delta_1 s_1\right.
\end{aligned}$$

$$
\begin{aligned}
&\left.+A(t_1,t_2)e_B(t_2,0)\right)\\
&= a(t_1,t_2)+b(t_1,t_2)\\
&\quad\times F^{-1}\left(\int_0^{t_1}\int_0^{t_2} L(s_1,s_2,h_1(s_1,s_2))\Delta_2 s_2\Delta_1 s_1\right.\\
&\quad\left.+A(t_1,t_2)e_B(t_2,0)\right),
\end{aligned}
$$

$(t_1,t_2)\in(\mathbb{R}_+\cap\mathbb{T}_1)\times(\mathbb{R}_+\cap\mathbb{T}_2)$. This completes the proof. □

10 Delay integral inequalities

This chapter introduces delay integral inequalities with one and two independent variables. The material in this chapter is based on some results in [10, 14, 15, 19–22] and [30].

Let $\mathbb{T}$ be an unbounded time scale with forward jump operator and delta differentiation operator σ and Δ, respectively. Let also $t_0 \in \mathbb{T}$.

10.1 Linear Gronwall–Bellman-type delay integral inequalities

In this section we will investigate the inequality

$$x(t) \le a(t) + c(t) \int_{t_0}^{t} (f(s)x(\tau(s)) + g(s)x(s))\Delta s, \quad t \in [t_0, \infty), \tag{10.1}$$

with the initial conditions

$$\begin{aligned} x(t) &= \phi(t), \quad t \in [\alpha, t_0], \\ \phi(\tau(t)) &\le a(t), \quad t \in [t_0, \infty), \ \tau(t) \le t_0, \end{aligned} \tag{10.2}$$

where

(A1) $x, a, c, f, g \in C_{rd}([t_0, \infty))$ are nonnegative functions, a and c are nondecreasing functions on $[t_0, \infty)$.

(A2) $\tau : [t_0, \infty) \to \mathbb{T}$, $\tau(t) \le t$, $t \in [t_0, \infty)$,

$$-\infty < \alpha = \inf\{\tau(t) : t \in [t_0, \infty)\},$$

$\phi \in C_{rd}([\alpha, t_0])$ is a nonnegative function.

Theorem 10.1.1. *Suppose* (A1) *and* (A2) *hold. Then*

$$x(t) \le a(t) + c(t) \int_{t_0}^{t} e_{\ominus(f+g)c}(\sigma(s), t) a(s)(f(s) + g(s))\Delta s, \quad t \in [t_0, \infty).$$

Proof. Let $t^* \in [t_0, \infty)$. Define

$$z(t) = a(t^*) + c(t) \int_{t_0}^{t} (f(s)x(\tau(s)) + g(s)x(s))\Delta s, \quad t \in [t_0, t^*]. \tag{10.3}$$

Then z is a nondecreasing function on $[t_0, t^*]$ and

$$x(t) \le a(t^*) + c(t) \int_{t_0}^{t} (f(s)x(\tau(s)) + g(s)x(s))\Delta s$$

https://doi.org/10.1515/9783110705553-010

$$= z(t), \quad t \in [t_0, t^*].$$

Next,

$$\begin{aligned} x(\tau(t)) &\le z(\tau(t)) \\ &\le z(t), \quad t \in [t_0, t^*],\ \tau(t) \ge t_0, \end{aligned}$$

and

$$\begin{aligned} x(\tau(t)) &= \phi(\tau(t)) \\ &\le a(t) \\ &\le a(t^*) \\ &\le z(t), \quad t \in [t_0, t^*],\ \tau(t) \le t_0. \end{aligned}$$

Hence, by (10.3), we get

$$\begin{aligned} z(t) &\le a(t^*) + c(t) \int_{t_0}^{t} (f(s)z(s) + g(s)z(s))\Delta s \\ &= a(t^*) + c(t) \int_{t_0}^{t} (f(s) + g(s))z(s)\Delta s, \quad t \in [t_0, t^*]. \end{aligned}$$

Because $t^* \in [t_0, \infty)$ was arbitrarily chosen, by the last inequality, we obtain

$$z(t) \le a(t) + c(t) \int_{t_0}^{t} (f(s) + g(s))z(s)\Delta s, \quad t \in [t_0, \infty).$$

Then, applying Theorem 1.1.10, we arrive at

$$\begin{aligned} x(t) &\le z(t) \\ &\le a(t) + c(t) \int_{t_0}^{t} (f(s) + g(s))a(s)e_{\ominus(f+g)c}(\sigma(s), t)\Delta s, \quad t \in [t_0, \infty). \end{aligned}$$

This completes the proof. □

10.2 Nonlinear Gronwall–Bellman-type delay integral inequalities with one delay

Consider the integral inequality

$$(x(t))^p \le a(t) + c(t) \int_{t_0}^{t} (f(s)x(\tau(s)) + g(s))\Delta s, \quad t \in [t_0, \infty), \tag{10.4}$$

with the initial conditions

$$\begin{aligned} x(t) &= \phi(t), \quad t \in [\alpha, t_0], \\ \phi(\tau(t)) &\le (a(t))^{\frac{1}{p}}, \quad t \in [t_0, \infty),\ \tau(t) \le t_0, \end{aligned} \tag{10.5}$$

where

(A3) $x, a, c, f, g \in \mathcal{C}_{rd}([t_0, \infty))$ are nonnegative functions,

(A4) $p \ge 1$ is a constant, $\tau : [t_0, \infty) \to \mathbb{T}$, $\tau(t) \le t$, $t \in [t_0, \infty)$,

$$-\infty < \alpha = \inf\{\tau(t) : t \in [t_0, \infty)\},$$

$\phi \in \mathcal{C}_{rd}([t_0, \infty))$ is a nonnegative function.

We will start our investigations with the following useful lemma.

Lemma 10.2.1. *Let $p \ge q \ge 0$, $p \ne 0$, and $a \ge 0$. Then, for any $k > 0$, we have*

$$a^{\frac{q}{p}} \le \frac{q}{p} k^{\frac{q-p}{p}} a + \frac{p-q}{p} k^{\frac{q}{p}}.$$

Proof.

1. Let $a = 0$. Then

$$0 \le \frac{p-q}{p} k^{\frac{q}{p}}.$$

2. Let $a > 0$. Set

$$f(k) = \frac{q}{p} k^{\frac{q-p}{p}} a + \frac{p-q}{p} k^{\frac{q}{p}}, \quad k > 0.$$

   Then

$$\begin{aligned} f'(k) &= \frac{q(q-p)}{p^2} k^{\frac{q-2p}{p}} a + q \frac{p-q}{p^2} k^{\frac{q-p}{p}} \\ &= \frac{q(p-q)}{p^2} k^{\frac{q-2p}{p}} (k-a), \quad k > 0. \end{aligned}$$

   Hence, we conclude that

$$f(k) \ge f(a), \quad k > 0,$$

   and since

$$f(a) = \frac{q}{p} (a^{\frac{q-p}{p}}) a + \frac{p-q}{p} a^{\frac{q}{p}}$$

$$
\begin{aligned}
&= \frac{q}{p} a^{\frac{q}{p}} + \frac{p-q}{p} a^{\frac{q}{p}} \\
&= a^{\frac{q}{p}},
\end{aligned}
$$

we get the desired result. This completes the proof. □

Theorem 10.2.2. *Suppose* (A3), (A4) *hold and* a, c *are nondecreasing on* $[t_0, \infty)$. *Then inequality* (10.4) *with the initial data* (10.5) *implies*

$$
\begin{aligned}
x(t) &\le \left(a(t) + c(t) \left(h(t) + \int_{t_0}^{t} e_B(t, \sigma(s)) h(s) B(s) \Delta s \right) \right)^{\frac{1}{p}} \\
&\le \left(a(t) + c(t) h(t) e_B(t, t_0) \right)^{\frac{1}{p}}, \quad t \in [t_0, \infty),
\end{aligned}
$$

for any $k > 0$, *where*

$$
\begin{aligned}
h(t) &= \int_{t_0}^{t} \left(f(s) \left(\frac{p-1}{p} k^{\frac{1}{p}} + \frac{a(s)}{p k^{\frac{p-1}{p}}} \right) + g(s) \right) \Delta s, \\
B(t) &= \frac{c(t) f(t)}{p k^{\frac{p-1}{p}}}, \quad t \in [t_0, \infty).
\end{aligned}
$$

Proof. Take t^* arbitrarily. Define

$$
z(t) = \left(a(t^*) + c(t) \int_{t_0}^{t} (f(s) x(\tau(s)) + g(s)) \Delta s \right)^{\frac{1}{p}}, \quad t \in [t_0, t^*].
$$

We have that z is a nonnegative and nondecreasing function on $[t_0, t^*]$, and

$$
x(t) \le z(t), \quad t \in [t_0, t^*].
$$

If $t \in [t_0, t^*]$ and $\tau(t) \le t_0$, then

$$
\begin{aligned}
x(\tau(t)) &= \phi(\tau(t)) \\
&\le (a(t))^{\frac{1}{p}} \\
&\le (a(t^*))^{\frac{1}{p}} \\
&\le z(t).
\end{aligned}
$$

If $t \in [t_0, t^*]$ and $\tau(t) \ge t_0$, then

$$
x(\tau(t)) \le z(t).
$$

Therefore

$$(z(t))^p = a(t^*) + c(t)\int_{t_0}^{t}(f(s)x(\tau(s)) + g(s))\Delta s$$
$$\le a(t^*) + c(t)\int_{t_0}^{t}(f(s)z(s) + g(s))\Delta s, \quad t \in [t_0, t^*].$$

In particular,

$$(z(t^*))^p \le a(t^*) + c(t^*)\int_{t_0}^{t^*}(f(s)z(s) + g(s))\Delta s.$$

Therefore

$$(z(t))^p \le a(t) + c(t)\int_{t_0}^{t}(f(s)z(s) + g(s))\Delta s, \quad t \in [t_0, \infty), \tag{10.6}$$

and

$$x(t) \le z(t), \quad t \in [t_0, \infty). \tag{10.7}$$

Let

$$u(t) = \int_{t_0}^{t}(f(s)z(s) + g(s))\Delta s, \quad t \in [t_0, \infty).$$

Then, for any $k > 0$, applying Lemma 10.2.1, we get

$$z(t) \le (a(t) + c(t)u(t))^{\frac{1}{p}}$$
$$\le \frac{p-1}{p}k^{\frac{1}{p}} + \frac{a(t)}{pk^{\frac{p-1}{p}}} + \frac{c(t)u(t)}{pk^{\frac{p-1}{p}}}, \quad t \in [t_0, \infty).$$

Thus,

$$u(t) \le \int_{t_0}^{t}\left(f(s)\left(\frac{p-1}{p}k^{\frac{1}{p}} + \frac{a(s)}{pk^{\frac{p-1}{p}}} + \frac{c(s)u(s)}{pk^{\frac{p-1}{p}}}\right) + g(s)\right)\Delta s$$
$$= \int_{t_0}^{t}\left(f(s)\left(\frac{p-1}{p}k^{\frac{1}{p}} + \frac{a(s)}{pk^{\frac{p-1}{p}}}\right) + g(s)\right)\Delta s$$
$$+ \frac{1}{pk^{\frac{p-1}{p}}}\int_{t_0}^{t}f(s)c(s)u(s)\Delta s$$

$$= h(t) + \int_{t_0}^{t} B(s)u(s)\Delta s, \quad t \in [t_0, \infty).$$

Now, applying Theorem 1.1.10, we obtain

$$u(t) \le h(t) + \int_{t_0}^{t} B(s)h(s)e_B(\sigma(s), t)\Delta s, \quad t \in [t_0, \infty),$$

and

$$\begin{aligned} x(t) &\le z(t) \\ &\le (a(t) + c(t)u(t))^{\frac{1}{p}} \\ &\le \left(a(t) + c(t)h(t) + c(t) \int_{t_0}^{t} B(s)h(s)e_B(\sigma(s), t)\Delta s \right)^{\frac{1}{p}}, \end{aligned}$$

$t \in [t_0, \infty)$. Note that h is a nonnegative and nondecreasing function on $[t_0, \infty)$. Then

$$\begin{aligned} x(t) &\le \left(a(t) + c(t)h(t) \left(1 + \int_{t_0}^{t} B(s)e_B(\sigma(s), t)\Delta s \right) \right)^{\frac{1}{p}} \\ &= \left(a(t) + c(t)h(t) \left(1 - \int_{t_0}^{t} e_B^{\Delta_s}(t, s)\Delta s \right) \right)^{\frac{1}{p}} \\ &= (a(t) + c(t)h(t)(1 - e_B(t, t) + e_B(t, t_0)))^{\frac{1}{p}} \\ &= (a(t) + c(t)h(t)e_B(t, t_0))^{\frac{1}{p}}, \quad t \in [t_0, \infty). \end{aligned}$$

This completes the proof. □

Next, consider the inequality

$$\begin{aligned} \eta(x(t)) \le a(t) &+ \int_{t_0}^{t} (f(t, s)\psi(x(\tau(s)))w(x(\tau(s))) + g(t, s)\psi(x(\tau(s))))\Delta s \\ &+ \int_{t_0}^{t} \int_{t_0}^{s} h(\xi)\psi(x(\tau(\xi)))\Delta\xi\Delta s \end{aligned} \tag{10.8}$$

with the initial conditions

$$\begin{aligned} \eta(x(t) &= \phi(t), \quad t \in [\alpha, t_0], \\ \phi(\tau(t)) &\le a(t), \quad t \in [t_0, \infty), \ \tau(t) \le t_0, \end{aligned} \tag{10.9}$$

where

(A6) $x, a \in C_{rd}([t_0,\infty))$ are nonnegative and nondecreasing functions, $f(t,s), f_t^{\Delta}(t,s) \in C_{rd}([t_0,\infty)\times[t_0,\infty))$ are nonnegative functions, $w \in C([t_0,\infty))$ is a nonnegative and nondecreasing function, $p > 0$ is a constant, $\tau : [t_0,\infty) \to \mathbb{T}$, $\tau(t) \leq t$, $t \in [t_0,\infty)$,

$$-\infty < \alpha = \inf\{\tau(t) : t \in [t_0,\infty)\},$$

$\phi \in C_{rd}([\alpha, t_0])$ is a nonnegative function.

Theorem 10.2.3. *Suppose* (A6) *holds. Let also* $g(t,s), g_t^{\Delta}(t,s) \in C_{rd}([t_0,\infty))$ *be nonnegative functions,* $h \in C_{rd}([t_0,\infty))$ *be a nonnegative function,* $\psi, \eta \in C([t_0,\infty))$ *be nonnegative functions, and suppose* ψ *is nondecreasing, while* η *is strictly increasing. If* x *satisfies inequality* (10.8) *with the initial data* (10.9), *then*

$$x(t) \leq \eta^{-1}\left(\widetilde{G}^{-1}\left(\widetilde{H}^{-1}\left(\widetilde{H}\left(\widetilde{G}(a(t)) + \int_{t_0}^{t}\left(g(t,s) + \int_{t_0}^{s} h(\xi)\Delta\xi\right)\Delta s\right)\right.\right.\right.$$

$$\left.\left.\left. + \int_{t_0}^{t} f(t,s)\Delta s\right)\right)\right), \quad t \in [t_0,\infty),$$

where $\widetilde{G}$ *and* $\widetilde{H}$ *are increasing, bijective functions, and*

$$\left(\widetilde{G}(v(t))\right)^{\Delta} = \frac{v^{\Delta}(t)}{\psi(\eta^{-1}(v(t)))},$$

$$\left(\widetilde{H}(z(t))\right)^{\Delta} = \frac{z^{\Delta}(t)}{w(\eta^{-1}(\widetilde{G}^{-1}(z(t))))}, \quad t \in [t_0,\infty).$$

Proof. Let $t^* \in [t_0,\infty)$ be arbitrarily chosen. Define

$$z(t) = a(t^*) + \int_{t_0}^{t} (f(t,s)\psi(x(\tau(s)))w(x(\tau(s))) + g(t,s)\psi(x(\tau(s))))\Delta s$$

$$+ \int_{t_0}^{t}\int_{t_0}^{s} h(\xi)\psi(x(\tau(\xi)))\Delta\xi\Delta s, \quad t \in [t_0,\infty).$$

Then z is a nondecreasing and nonnegative function on $[t_0, t^*]$. Next,

$$\eta(x(t)) \leq a(t) + \int_{t_0}^{t} (f(t,s)\psi(x(\tau(s)))w(x(\tau(s))) + g(t,s)\psi(x(\tau(s))))\Delta s$$

$$+ \int_{t_0}^{t}\int_{t_0}^{s} h(\xi)\psi(x(\tau(\xi)))\Delta\xi\Delta s$$

$$\begin{aligned}
&\leq a(t^*) + \int_{t_0}^{t} (f(t,s)\psi(x(\tau(s)))w(x(\tau(s))) + g(t,s)\psi(x(\tau(s))))\Delta s \\
&\quad + \int_{t_0}^{t}\int_{t_0}^{s} h(\xi)\psi(x(\tau(\xi)))\Delta\xi\Delta s \\
&= z(t), \quad t \in [t_0, t^*],
\end{aligned}$$

whereupon

$$x(t) \leq \eta^{-1}(z(t)), \quad t \in [t_0, t^*].$$

If $t \in [t_0, t^*]$ and $\tau(t) \leq t_0$, then

$$\begin{aligned}
\eta(x(\tau(t))) &= \phi(\tau(t)) \\
&\leq a(t) \\
&\leq a(t^*) \\
&\leq z(t).
\end{aligned}$$

If $t \in [t_0, t^*]$ and $\tau(t) \geq t_0$, then

$$\begin{aligned}
\eta(x(\tau(t))) &\leq z(\tau(t)) \\
&\leq z(t).
\end{aligned}$$

Therefore

$$\eta(x(\tau(t))) \leq z(t), \quad t \in [t_0, t^*],$$

and

$$x(\tau(t)) \leq \eta^{-1}(z(t)), \quad t \in [t_0, t^*].$$

Next,

$$\begin{aligned}
z^{\Delta}(t) &= \int_{t_0}^{t} (f_t^{\Delta}(t,s)\psi(x(\tau(s)))w(x(\tau(s))) + g_t^{\Delta}(t,s)\psi(x(\tau(s))))\Delta s \\
&\quad + f(\sigma(t),t)\psi(x(\tau(t)))w(x(\tau(t))) \\
&\quad + g(\sigma(t),t)\psi(x(\tau(t))) + \int_{t_0}^{t} h(\xi)\psi(x(\tau(\xi)))\Delta\xi \\
&\leq \Bigg(\int_{t_0}^{t} (f_t^{\Delta}(t,s)w(\eta^{-1}(z(s))) + g_t^{\Delta}(t,s))\Delta s
\end{aligned}$$

$$+f(\sigma(t),t)w(\eta^{-1}(z(t)))+g(\sigma(t),t)+\int_{t_0}^{t}h(\xi)\Delta\xi\Bigg)$$

$$\times\psi(\eta^{-1}(z(t))),\quad t\in[t_0,t^*],$$

and

$$\frac{z^{\Delta}(t)}{\psi(\eta^{-1}(z(t)))}\le\int_{t_0}^{t}(f_t^{\Delta}(t,s)w(\eta^{-1}(z(s)))+g_t^{\Delta}(t,s))\Delta s$$

$$+f(\sigma(t),t)w(\eta^{-1}(z(t)))+g(\sigma(t),t)+\int_{t_0}^{t}h(\xi)\Delta\xi$$

$$=\left(\int_{t_0}^{t}\left(f(t,s)w(\eta^{-1}(z(s)))+g(t,s)+\int_{t_0}^{s}h(\xi)\Delta\xi\right)\Delta s\right)^{\Delta},$$

$t\in[t_0,t^*]$. Now we integrate the last inequality from t_0 to t and get

$$\widetilde{G}(z(t))-\widetilde{G}(z(t_0))\le\int_{t_0}^{t}\Bigg(f(t,s)w(\eta^{-1}(z(s)))+g(t,s)$$

$$+\int_{t_0}^{s}h(\xi)\Delta\xi\Bigg)\Delta s,\quad t\in[t_0,t^*].$$

Hence, using that $z(t_0)=a(t^*)$, we find

$$\widetilde{G}(z(t))\le\widetilde{G}(a(t^*))+\int_{t_0}^{t}\Bigg(f(t,s)w(\eta^{-1}(z(s)))+g(t,s)$$

$$+\int_{t_0}^{s}h(\xi)\Delta\xi\Bigg)\Delta s,\quad t\in[t_0,t^*],$$

and

$$z(t)\le\widetilde{G}^{-1}\Bigg(\widetilde{G}(a(t^*))+\int_{t_0}^{t}\Bigg(f(t,s)w(\eta^{-1}(z(s)))+g(t,s)$$

$$+\int_{t_0}^{s}h(\xi)\Delta\xi\Bigg)\Delta s\Bigg),\quad t\in[t_0,t^*].$$

Let

$$v(t) = \widetilde{G}(a(t^*)) + \int_{t_0}^{t^*} \Bigg(f(t,s)w(\eta^{-1}(z(s))) + g(t,s) + \int_{t_0}^{s} h(\xi)\Delta\xi \Bigg)\Delta s + \int_{t_0}^{t} f(t,s)w(\eta^{-1}(z(s)))\Delta s, \quad t \in [t_0, t^*].$$

Then

$$z(t) \le \widetilde{G}^{-1}(v(t)), \quad t \in [t_0, t^*],$$

and

$$\begin{aligned} v^{\Delta}(t) &= \int_{t_0}^{t} f_t^{\Delta}(t,s)w(\eta^{-1}(z(s)))\Delta s + f(\sigma(t),t)w(\eta^{-1}(z(t))) \\ &\le \Bigg(\int_{t_0}^{t} f_t^{\Delta}(t,s)\Delta s + f(\sigma(t),t) \Bigg) w(\eta^{-1}(\widetilde{G}^{-1}(v(t)))), \end{aligned}$$

$t \in [t_0, t^*]$, and

$$\begin{aligned} \frac{v^{\Delta}(t)}{w(\eta^{-1}(\widetilde{G}^{-1}(v(t))))} &\le \int_{t_0}^{t} f_t^{\Delta}(t,s)\Delta s + f(\sigma(t),t) \\ &= \Bigg(\int_{t_0}^{t} f(t,s)\Delta s \Bigg)^{\Delta}, \quad t \in [t_0, t^*]. \end{aligned}$$

Integrating the last inequality from t_0 to t, $t \in [t_0, t^*]$, we get

$$\widetilde{H}(v(t)) - \widetilde{H}(v(t_0)) \le \int_{t_0}^{t} f(t,s)\Delta s, \quad t \in [t_0, t^*],$$

and

$$\begin{aligned} v(t) &\le \widetilde{H}^{-1}\Bigg(\widetilde{H}(v(t_0)) + \int_{t_0}^{t} f(t,s)\Delta s \Bigg) \\ &= \widetilde{H}^{-1}\Bigg(\widetilde{H}\Bigg(\widetilde{G}(a(t^*)) + \int_{t_0}^{t^*} \Bigg(f(t,s)w(\eta^{-1}(z(s))) + g(t,s) \end{aligned}$$

$$+\int_{t_0}^{s} h(\xi)\Delta\xi \Big)\Delta s\Big)$$

$$+\int_{t_0}^{t} f(t,s)\Delta s\Big), \quad t\in[t_0,t^*].$$

Hence,

$$\begin{aligned}x(t) &\le \eta^{-1}(z(t))\\ &\le \eta^{-1}(\widetilde{G}^{-1}(v(t)))\\ &\le \eta^{-1}\Big(\widetilde{G}^{-1}\Big(\widetilde{H}^{-1}\Big(\widetilde{H}\Big(\widetilde{G}(a(t^*))+\int_{t_0}^{t^*}\Big(f(t,s)w(\eta^{-1}(z(s)))+g(t,s)\\ &\quad+\int_{t_0}^{s} h(\xi)\Delta\xi\Big)\Delta s\Big)+\int_{t_0}^{t} f(t,s)\Delta s\Big)\Big)\Big),\end{aligned}$$

$t\in[t_0,t^*]$. In particular, for $t=t^*$, we find

$$\begin{aligned}x(t^*) \le \eta^{-1}\Big(\widetilde{G}^{-1}\Big(\widetilde{H}^{-1}\Big(\widetilde{H}\Big(\widetilde{G}(a(t^*))+\int_{t_0}^{t^*}\Big(f(t,s)w(\eta^{-1}(z(s)))+g(t,s)\\ +\int_{t_0}^{s} h(\xi)\Delta\xi\Big)\Delta s\Big)+\int_{t_0}^{t^*} f(t^*,s)\Delta s\Big)\Big)\Big).\end{aligned}$$

Because $t^*\in[t_0,\infty)$ was arbitrarily chosen, we get the desired result. This completes the proof. □

Next, consider the delay integral inequality

$$(x(t))^p \le a(t)+\int_{t_0}^{t} b(s)(x(s))^p\Delta s + c(t)\int_{t_0}^{t}(f(s)(x(\tau(s)))^q+g(s)(x(s))^r)\Delta s, \quad t\in[t_0,\infty), \tag{10.10}$$

with the initial data

$$\begin{aligned}x(t) &= \phi(t), \quad t\in[\alpha,t_0],\\ \phi(\tau(t)) &\le (a(t))^{\frac{1}{p}}, \quad t\in[t_0,\infty), \quad \tau(t)\le t_0,\end{aligned} \tag{10.11}$$

where

(A7) p, q, r are constants, $p\neq 0$, $p\ge q\ge 0$, $p\ge r\ge 0$, $\tau:[t_0,\infty)\to\mathbb{T}$, $\tau(t)\le t$, $t\in[t_0,\infty)$,

$$-\infty<\alpha=\inf\{\tau(t):t\in[t_0,\infty)\},$$

$\phi\in C_{rd}([\alpha,t_0])$ is a nonnegative function,

(A8) $x, a, b, c, f, g \in \mathcal{C}_{rd}([t_0, \infty))$ are nonnegative functions, a and c are nondecreasing functions on $[t_0, \infty)$.

Theorem 10.2.4. *If x satisfies* (10.10) *with the initial data* (10.11), *then*

$$x(t) \le \left(e_b(t, t_0)\left(a(t) + c(t)\left(F(t) + \int_{t_0}^{t} e_G(t, \sigma(s))F(s)G(s)\Delta s \right)\right)\right)^{\frac{1}{p}}$$
$$\le (e_b(t, t_0)(a(t) + c(t)F(t)e_G(t, t_0)))^{\frac{1}{p}}, \quad t \in [t_0, \infty),$$

for any $k > 0$, where

$$F(t) = \int_{t_0}^{t} \left(\frac{f(s)(e_b(s, t_0))^{\frac{q}{p}}(k(p-q) + qa(s))}{pk^{\frac{p-q}{p}}} + \frac{g(s)(e_b(s, t_0))^{\frac{r}{p}}(k(p-r) + ra(s))}{pk^{\frac{p-q}{p}}} \right)\Delta s,$$
$$G(t) = c(t)\left(\frac{qf(t)(e_b(t, t_0))^{\frac{q}{p}}}{pk^{\frac{p-q}{p}}} + \frac{rg(t)(e_b(t, t_0))^{\frac{r}{p}}}{pk^{\frac{p-q}{p}}} \right),$$

$t \in [t_0, \infty)$.

Proof. Define the function

$$z(t) = \left(a(t) + \int_{t_0}^{t} b(s)(x(s))^p \Delta s + c(t) \int_{t_0}^{t} (f(s)(x(\tau(s)))^q + g(s)(x(s))^r)\Delta s \right)^{\frac{1}{p}},$$

$t \in [t_0, \infty)$. We have that z is a nonnegative and nondecreasing function on $[t_0, \infty)$ and

$$x(t) \le z(t), \quad t \in [t_0, \infty).$$

Also, if $t \in [t_0, \infty)$, $\tau(t) > t_0$, then

$$x(\tau(t)) \le z(\tau(t))$$
$$\le z(t).$$

If $t \in [t_0, \infty)$ and $\tau(t) \le t_0$, then

$$x(\tau(t)) = \phi(\tau(t))$$
$$\le (a(t))^{\frac{1}{p}}$$
$$\le z(t).$$

Therefore

$$x(\tau(t)) \le z(t), \quad t \in [t_0, \infty).$$

From here,

$$\begin{aligned}(z(t))^p &= a(t) + \int_{t_0}^{t} b(s)(x(s))^p \Delta s + c(t) \int_{t_0}^{t} (f(s)(x(\tau(s)))^q + g(s)(x(s))^r)\Delta s \\ &\le a(t) + \int_{t_0}^{t} b(s)(z(s))^p \Delta s + c(t) \int_{t_0}^{t} (f(s)(z(s))^q + g(s)(z(s))^r)\Delta s,\end{aligned}$$

$t \in [t_0, \infty)$. Define

$$\begin{aligned}u(t) &= \int_{t_0}^{t} (f(s)(z(s))^q + g(s)(z(s))^r)\Delta s, \\ w(t) &= a(t) + c(t)u(t), \quad t \in [t_0, \infty).\end{aligned}$$

Then

$$(z(t))^p \le w(t) + \int_{t_0}^{t} b(s(z(s))^p \Delta s, \quad t \in [t_0, \infty).$$

Hence, by Theorem 1.1.10, we arrive at

$$(z(t))^p \le w(t) + \int_{t_0}^{t} e_b(t, \sigma(s)) w(s) b(s) \Delta s, \quad t \in [t_0, \infty).$$

Note that w is a nondecreasing function on $[t_0, \infty)$. Then

$$\begin{aligned}(z(t))^p &\le w(t) + w(t) \int_{t_0}^{t} e_b(t, \sigma(s)) b(s) \Delta s \\ &= w(t)\left(1 - \int_{t_0}^{t} e_b^{\Delta_s}(t, s)\Delta s\right) \\ &= w(t)(1 - e_b(t, t) + e_b(t, t_0)) \\ &= w(t) e_b(t, t_0), \quad t \in [t_0, \infty).\end{aligned} \tag{10.12}$$

Therefore

$$(z(t))^p \le (a(t) + c(t)u(t)) e_b(t, t_0), \quad t \in [t_0, \infty).$$

Taking $k > 0$ arbitrarily and applying Lemma 10.2.1, we get

$$\begin{aligned}
(z(t))^q &= (e_b(t,t_0))^{\frac{q}{p}}(a(t)+c(t)u(t))^{\frac{q}{p}}\\
&\le (e_b(t,t_0))^{\frac{q}{p}}\left(\frac{q}{p}k^{\frac{q-p}{p}}(a(t)+c(t)u(t))+\frac{p-q}{p}k^{\frac{q}{p}}\right)\\
&= (e_b(t,t_0))^{\frac{q}{p}}\left(\frac{qa(t)+qc(t)u(t)}{pk^{\frac{p-q}{p}}}+\frac{(p-q)k}{pk^{\frac{p-q}{p}}}\right)\\
&= (e_b(t,t_0))^{\frac{q}{p}}\left(\frac{\gamma(p-q)+qa(t)}{pk^{\frac{p-q}{p}}}+\frac{qc(t)u(t)}{pk^{\frac{p-q}{p}}}\right),
\end{aligned}$$

$t \in [t_0, \infty)$, and

$$\begin{aligned}
(z(t))^r &\le (e_b(t,t_0))^{\frac{r}{p}}(a(t)+c(t)u(t))^{\frac{r}{p}}\\
&\le (e_b(t,t_0))^{\frac{r}{p}}\left(\frac{k(p-r)+ra(t)}{pk^{\frac{p-q}{p}}}+\frac{rc(t)u(t)}{pk^{\frac{p-q}{p}}}\right),
\end{aligned}$$

$t \in [t_0, \infty)$. By the last two inequalities and the definition of function u, we find

$$\begin{aligned}
u(t) &= \int_{t_0}^{t} f(s)(z(s))^q\Delta s + \int_{t_0}^{t} g(s)(z(s))^r\Delta s\\
&\le \int_{t_0}^{t} f(s)(e_b(s,t_0))^{\frac{q}{p}}\left(\frac{k(p-q)+qa(s)}{pk^{\frac{p-q}{p}}}+\frac{qc(s)u(s)}{pk^{\frac{p-q}{p}}}\right)\Delta s\\
&\quad + \int_{t_0}^{t} g(s)(e_b(s,t_0))^{\frac{r}{p}}\left(\frac{k(p-r)+ra(s)}{pk^{\frac{p-q}{p}}}+\frac{rc(s)u(s)}{pk^{\frac{p-q}{p}}}\right)\Delta s\\
&= \int_{t_0}^{t}\left(f(s)(e_b(s,t_0))^{\frac{q}{p}}\frac{k(p-q)+qa(s)}{pk^{\frac{p-q}{p}}}+g(s)(e_b(s,t_0))^{\frac{r}{p}}\frac{r(p-r)+ra(s)}{pk^{\frac{p-q}{p}}}\right)\Delta s\\
&\quad + \int_{t_0}^{t}\left(f(s)(e_b(s,t_0))^{\frac{q}{p}}\frac{qc(s)}{pk^{\frac{p-q}{p}}}+g(s)(e_b(s,t_0))^{\frac{r}{p}}\frac{rc(s)}{pk^{\frac{p-q}{p}}}\right)u(s)\Delta s\\
&= F(t) + \int_{t_0}^{t} G(s)u(s)\Delta s, \quad t \in [t_0,\infty).
\end{aligned}$$

Applying Theorem 1.1.10, we find

$$u(t) \le F(t) + \int_{t_0}^{t} e_G(t,\sigma(s))F(s)G(s)\Delta s, \quad t \in [t_0,\infty),$$

and then

$$\begin{aligned} w(t) &= a(t) + c(t)u(t) \\ &\leq a(t) + c(t)\left(F(t) + \int_{t_0}^{t} e_G(t,\sigma(s))F(s)G(s)\Delta s \right), \quad t \in [t_0,\infty), \end{aligned}$$

and, using (10.12), we obtain

$$\begin{aligned} (x(t))^p &\leq (z(t))^p \\ &\leq w(t)e_b(t,t_0) \\ &\leq e_b(t,t_0)\left(a(t) + c(t)\left(F(t) + \int_{t_0}^{t} e_G(t,\sigma(s))F(s)G(s)\Delta s \right)\right) \\ &\leq e_b(t,t_0)\left(a(t) + c(t)F(t)\left(1 + \int_{t_0}^{t} e_G(t,\sigma(s))G(s)\Delta s \right)\right) \\ &= e_b(t,t_0)\left(a(t) + c(t)F(t)\left(1 - \int_{t_0}^{t} e_G^{\Delta_s}(t,s)\Delta s \right)\right) \\ &= e_b(t,t_0)(a(t) + c(t)F(t)(1 - e_G(t,t) + e_G(t,t_0))) \\ &= e_b(t,t_0)(a(t) + c(t)F(t)e_G(t,t_0)), \quad t \in [t_0,\infty). \end{aligned}$$

This completes the proof. □

Now, we consider a special case of the inequality (10.10)

$$(x(t))^p \leq c + \int_{t_0}^{t} b(s)(x(s))^p \Delta s + \int_{t_0}^{t} f(s)(x(\tau(s)))^{p-1}\Delta s, \quad t \in [t_0,\infty), \tag{10.13}$$

with initial data

$$\begin{aligned} x(t) &= \phi(t), \quad t \in [\alpha, t_0], \\ \phi(\tau(t)) &\leq c^{\frac{1}{p}}, \quad t \in [t_0,\infty), \quad \tau(t) \leq t_0, \end{aligned} \tag{10.14}$$

where

(A9) c is a positive constant, $p > 0$, τ, α and ϕ are as in (A7).

(A10) $x, b, f \in C_{rd}([t_0,\infty))$ be nonnegative functions.

Theorem 10.2.5. *Suppose* (A9), (A10) *hold. If x satisfies* (10.13) *with initial data* (10.14), *then*

$$x(t) \leq c^{\frac{1}{p}} e_{\frac{b}{p}}(t,t_0) + \frac{1}{p}\int_{t_0}^{t} e_{\frac{b}{p}}(t,\sigma(s))f(s)\Delta s, \quad t \in [t_0,\infty).$$

Proof. Define the function

$$(z(t))^p = c + \int_{t_0}^{t} b(s)(x(s))^p \Delta s + \int_{t_0}^{t} f(s)(x(\tau(s)))^{p-1} \Delta s, \quad t \in [t_0, \infty).$$

As in the proof of Theorem 10.2.4, we have

$$\begin{aligned} x(t) &\le z(t), \\ x(\tau(t)) &\le z(t), \quad t \in [t_0, \infty). \end{aligned}$$

By Pötzsche's chain rule, for any $t \in [t_0, \infty)$, there exists a $\theta \in [t, \sigma(t)]$ such that

$$\begin{aligned} p(z(\theta))^{p-1} z^{\Delta}(t) &= b(t)(x(t))^p + f(t)(x(\tau(t)))^{p-1} \\ &\le b(t)(z(t))^p + f(t)(z(t))^{p-1}, \quad t \in [t_0, \infty). \end{aligned}$$

Since z is a nondecreasing function and $\theta \in [t, \sigma(t)]$ for any $t \in [t_0, \infty)$, we have

$$p(z(t))^{p-1} z^{\Delta}(t) \le b(t)(z(t))^p + f(t)(z(t))^{p-1}, \quad t \in [t_0, \infty),$$

whereupon

$$z^{\Delta}(t) \le \frac{b(t)}{p} z(t) + \frac{f(t)}{p}, \quad t \in [t_0, \infty).$$

Hence, using the Gronwall inequality, we get

$$\begin{aligned} x(t) &\le z(t) \\ &\le c^{\frac{1}{p}} e_{\frac{b}{p}}(t, t_0) + \int_{t_0}^{t} e_{\frac{b}{p}}(t, \sigma(s)) \frac{f(s)}{p} \Delta s, \quad t \in [t_0, \infty). \end{aligned}$$

This completes the proof. □

Next, consider the delay integral inequality of the form

$$(x(t))^p \le a(t) + c(t) \int_{t_0}^{t} (f(s)(x(s))^q + L(s, x(\tau(s)))) \Delta s, \quad t \in [t_0, \infty), \tag{10.15}$$

subject to the initial condition (10.11), where q, τ, α satisfy (A7), x, a, c, f satisfy (A8) and

(A11) $L \in C([t_0, \infty) \times \mathbb{R})$ is a nonnegative function such that

$$0 \le L(t, x) - L(t, y) \le K(t, y)(x - y),$$

$t \in [t_0, \infty)$, $x \ge y \ge 0$, $K \in C([t_0, \infty) \times \mathbb{R})$.

Theorem 10.2.6. *If x is a solution of inequality* (10.15) *with the initial data* (10.11), *then*

$$\begin{aligned} x(t) &\le \left(a(t) + c(t)\left(H(t) + \int_{t_0}^{t} e_J(t,\sigma(s))H(s)J(s)\Delta s\right)\right)^{\frac{1}{p}} \\ &\le a(t) + c(t)H(t)e_J(t,t_0), \quad t \in [t_0,\infty), \end{aligned}$$

where

$$\begin{aligned} H(t) &= \int_{t_0}^{t}\left(\frac{f(s)(k(p-q)+qa(s))}{pk^{\frac{p-q}{p}}} + L\left(s, \frac{p-1}{p} + \frac{a(s)}{p}\right)\right)\Delta s, \\ J(t) &= \frac{qc(t)f(t)}{pk^{\frac{p-q}{p}}} + K\left(t, \frac{p-1}{p} + \frac{a(t)}{p}\right)\frac{c(t)}{p}, \quad t \in [t_0,\infty). \end{aligned}$$

Proof. Define the function

$$z(t) = \int_{t_0}^{t} (f(s)(x(s))^q + L(s, x(\tau(s))))\Delta s, \quad t \in [t_0,\infty).$$

We have that z is a nonnegative and nondecreasing function on $[t_0,\infty)$ and then, using Lemma 10.2.1 for $k = 1$, obtain

$$\begin{aligned} x(t) &\le (a(t) + c(t)z(t))^{\frac{1}{p}} \\ &\le \frac{p-1}{p} + \frac{a(t)}{p} + \frac{c(t)z(t)}{p}, \quad t \in [t_0,\infty). \end{aligned}$$

If $t \in [t_0,\infty)$, $\tau(t) > t_0$, then

$$\begin{aligned} x(\tau(t)) &\le \frac{p-1}{p} + \frac{a(\tau(t))}{p} + \frac{c(\tau(t))z(\tau(t))}{p} \\ &\le \frac{p-1}{p} + \frac{a(t)}{p} + \frac{c(t)z(t)}{p}. \end{aligned}$$

If $t \in [t_0,\infty)$ and $\tau(t) \le t_0$, then

$$\begin{aligned} x(\tau(t)) &= \phi(\tau(t)) \\ &\le (a(t))^{\frac{1}{p}} \\ &\le x(t) \\ &\le \frac{p-1}{p} + \frac{a(t)}{p} + \frac{c(t)z(t)}{p}. \end{aligned}$$

Thus,

$$x(\tau(t)) \le \frac{p-1}{p} + \frac{a(t)}{p} + \frac{c(t)z(t)}{p}, \quad t \in [t_0,\infty).$$

Therefore, applying Lemma 10.2.1 for $k > 0$, we have

$$
\begin{aligned}
z(t) &= \int_{t_0}^{t} (f(s)(x(s))^q + L(s, x(\tau(s))))\Delta s \\
&\le \int_{t_0}^{t} f(s)(a(s) + c(s)z(s))^{\frac{q}{p}} \Delta s \\
&\quad + \int_{t_0}^{t} L\left(s, \frac{p-1}{p} + \frac{a(s)}{p} + \frac{c(s)z(s)}{p}\right)\Delta s \\
&\le \int_{t_0}^{t} f(s)\left(\frac{k(p-q) + qa(s)}{pk^{\frac{p-q}{p}}} + \frac{qc(s)z(s)}{pk^{\frac{p-q}{p}}}\right)\Delta s \\
&\quad + \int_{t_0}^{t} \left(L\left(s, \frac{p-1}{p} + \frac{a(s)}{p} + \frac{c(s)z(s)}{p}\right) - L\left(s, \frac{p-1}{p} + \frac{a(s)}{p}\right)\right. \\
&\quad \left. + L\left(s, \frac{p-1}{p} + \frac{a(s)}{p}\right)\right)\Delta s \\
&\le \int_{t_0}^{t} \left(\frac{f(s)(k(p-q) + qa(s))}{pk^{\frac{p-q}{p}}} + L\left(s, \frac{p-1}{p} + \frac{a(s)}{p}\right)\right)\Delta s \\
&\quad + \int_{t_0}^{t} \left(\frac{qc(s)f(s)}{pk^{\frac{p-q}{p}}} + K\left(s, \frac{p-1}{p} + \frac{a(s)}{p}\right)\frac{c(s)}{p}\right)z(s)\Delta s \\
&= H(t) + \int_{t_0}^{t} J(s)z(s)\Delta s, \quad t \in [t_0, \infty).
\end{aligned}
$$

Hence, by Theorem 1.1.10, we get

$$
\begin{aligned}
z(t) &\le H(t) + \int_{t_0}^{t} e_J(t, \sigma(s))H(s)J(s)\Delta s \\
&\le H(t)\left(1 + \int_{t_0}^{t} e_J(t, \sigma(s))J(s)\Delta s\right) \\
&= H(t)\left(1 - \int_{t_0}^{t} e_J^{\Delta_s}(t, s)\Delta s\right) \\
&= H(t)(1 - e_J(t, t) + e_J(t, t_0)) \\
&= H(t)e_J(t, t_0), \quad t \in [t_0, \infty),
\end{aligned}
$$

and

$$\begin{aligned} x(t) &\leq (a(t) + c(t)z(t))^{\frac{1}{p}} \\ &\leq \left(a(t) + c(t)\left(H(t) + \int_{t_0}^{t} e_J(t, \sigma(s))H(s)J(s)\Delta s \right)\right)^{\frac{1}{p}} \\ &\leq (a(t) + c(t)H(t)e_J(t, t_0))^{\frac{1}{p}}, \quad t \in [t_0, \infty). \end{aligned}$$

This completes the proof. □

10.3 Gronwall–Bellman-type nonlinear delay integral inequalities with several delays

Consider the inequality

$$(x(t))^p \leq c + \int_{t_0}^{t} (f(s)(x(\tau_1(s)))^q + g(s)(x(\tau_2(s)))^q w(x(\tau_2(s))))\Delta s, \quad t \in [t_0, \infty), \tag{10.16}$$

with the initial conditions

$$\begin{aligned} x(t) &= \phi(t), \quad t \in [\alpha, t_0], \\ \phi(\tau_i(t)) &\leq (a(t))^{\frac{1}{p}}, \quad t \in [t_0, \infty), \quad \tau_i(t) \leq t_0, \quad i \in \{1, 2\}, \end{aligned} \tag{10.17}$$

where

(A12) $a, x, f, g \in \mathcal{C}_{rd}([t_0, \infty))$ are nonnegative and nondecreasing functions, $w \in \mathcal{C}([0, \infty))$ is a nonnegative function,

(A13) $\tau_i \in \mathcal{C}([t_0, \infty))$, $\tau_i(t) \leq t$, $t \in [t_0, \infty)$, $i \in \{1, 2\}$,

$$-\infty < \alpha = \inf\left\{ \min_{i \in \{1,2\}} \tau_i(t) : t \in [t_0, \infty) \right\},$$

$\phi \in \mathcal{C}_{rd}([\alpha, t_0])$, p, q, c are positive constants, $p > q$.

Theorem 10.3.1. *Suppose* (A12), (A13) *hold. If x satisfies* (10.16) *with the initial condition* (10.17), *then*

$$x(t) \leq \left(G^{-1}\left(H^{-1}\left(H\left(G(c) + \int_{t_0}^{t} f(s)\Delta s \right) + \int_{t_0}^{t} g(s)\Delta s \right)\right)\right)^{\frac{1}{p}}, \quad t \in [t_0, \infty),$$

where G and H are increasing, bijective functions and

$$G(v) = \int_{1}^{v} \frac{1}{r^{\frac{q}{p}}} dr, \quad v > 0,$$

$$H(z) = \int_1^z \frac{1}{w((G^{-1}(r))^{\frac{1}{p}})} dr, \quad z > 0, \quad H(\infty) = \infty.$$

Proof. Let

$$z(t) = c + \int_{t_0}^{t} (f(s)(x(\tau_1(s)))^q + g(s)(x(\tau_2(s)))^q w(x(\tau_2(s))))\Delta s, \quad t \in [t_0, \infty).$$

Then

$$x(t) \le (z(t))^{\frac{1}{p}}, \quad t \in [t_0, \infty), \tag{10.18}$$

and, as in the proof of Theorem 10.2.6, we have

$$x(\tau_i(t)) \le (z(t))^{\frac{1}{p}}, \quad t \in [t_0, \infty), \quad i \in \{1, 2\}.$$

Next,

$$\begin{aligned} z^\Delta(t) &= f(t)(x(\tau_1(t)))^q + g(t)(x(\tau_2(t)))^q w(x(\tau_2(t))) \\ &\le f(t)(z(t))^{\frac{q}{p}} + g(t)(z(t))^{\frac{q}{p}} w((z(t))^{\frac{1}{p}}), \quad t \in [t_0, \infty). \end{aligned}$$

On the other hand, if $t \in [t_0, \infty)$ and $\sigma(t) > t$, then

$$\begin{aligned} (G(z(t)))^\Delta &= \frac{G(z(\sigma(t))) - G(z(t))}{\sigma(t) - t} \\ &= \frac{1}{\sigma(t) - t} \int_{z(t)}^{z(\sigma(t))} \frac{1}{r^{\frac{q}{p}}} dr \\ &\le \frac{z(\sigma(t)) - z(t)}{\sigma(t) - t} \frac{1}{(z(t))^{\frac{q}{p}}} \\ &= \frac{z^\Delta(t)}{(z(t))^{\frac{q}{p}}}. \end{aligned}$$

If $t \in [t_0, \infty)$, $\sigma(t) = t$, then

$$\begin{aligned} (G(z(t)))^\Delta &= \lim_{s \to t} \frac{G(z(t)) - G(z(s))}{t - s} \\ &= \lim_{s \to t} \frac{1}{t - s} \int_{z(s)}^{z(t)} \frac{1}{r^{\frac{q}{p}}} dr \\ &= \lim_{s \to t} \frac{z(t) - z(s)}{t - s} \frac{1}{(z(\xi))^{\frac{q}{p}}} \\ &= \frac{z^\Delta(t)}{(z(t))^{\frac{q}{p}}}. \end{aligned}$$

Thus,

$$\begin{aligned}(G(z(t)))^{\Delta} &\leq \frac{z^{\Delta}(t)}{(z(t))^{\frac{q}{p}}}\\ &\leq f(t)+g(t)w((z(t))^{\frac{1}{p}}),\quad t\in[t_0,\infty).\end{aligned}$$

Hence,

$$G(z(t))-G(z(t_0))\leq\int_{t_0}^{t}(f(s)+g(s)w((z(s))^{\frac{1}{p}}))\Delta s,\quad t\in[t_0,\infty),$$

or

$$G(z(t))\leq G(c)+\int_{t_0}^{t}(f(s)+g(s)w((z(s))^{\frac{1}{p}}))\Delta s,\quad t\in[t_0,\infty).$$

Therefore

$$z(t)\leq G^{-1}\left(G(c)+\int_{t_0}^{t}(f(s)+g(s)w((z(s))^{\frac{1}{p}}))\Delta s\right),\quad t\in[t_0,\infty).$$

Let $t^*\in[t_0,\infty)$ be arbitrarily chosen. Set

$$v(t)=G(c)+\int_{t_0}^{t^*}f(s)\Delta s+\int_{t_0}^{t}g(s)w((z(s))^{\frac{1}{p}})\Delta s,\quad t\in[t_0,t^*].$$

Then

$$\begin{aligned}z(t)&\leq G^{-1}\left(G(c)+\int_{t_0}^{t}(f(s)+g(s)w((z(s))^{\frac{q}{p}}))\Delta s\right)\\ &\leq G^{-1}\left(G(c)+\int_{t_0}^{t^*}f(s)\Delta s+\int_{t_0}^{t}g(s)w((z(s))^{\frac{1}{p}})\Delta s\right)\\ &=G^{-1}(v(t)),\quad t\in[t_0,t^*].\end{aligned}\tag{10.19}$$

Next,

$$\begin{aligned}v^{\Delta}(t)&=g(t)w((z(t))^{\frac{1}{p}})\\ &\leq g(t)w((G^{-1}(v(t)))^{\frac{1}{p}}),\quad t\in[t_0,t^*].\end{aligned}$$

Thus,

$$\frac{v^{\Delta}(t)}{w((G^{-1}(v(t)))^{\frac{1}{p}})} \le g(t), \quad t \in [t_0, t^*].$$

If $t \in [t_0, t^*]$, $\sigma(t) > t$, then

$$\begin{aligned}
(H(v(t)))^{\Delta} &= \frac{H(v(\sigma(t))) - H(v(t))}{\sigma(t) - t} \\
&= \frac{1}{\sigma(t) - t} \int_{v(t)}^{v(\sigma(t))} \frac{1}{w((G^{-1}(v(r)))^{\frac{1}{p}})} dr \\
&= \frac{v(\sigma(t)) - v(t)}{\sigma(t) - t} \frac{1}{w((G^{-1}(v(t)))^{\frac{1}{p}})} \\
&= \frac{v^{\Delta}(t)}{w((G^{-1}(v(t)))^{\frac{1}{p}})}.
\end{aligned}$$

If $t \in [t_0, t^*]$, $\sigma(t) = t$, then

$$\begin{aligned}
(H(v(t)))^{\Delta} &= \lim_{s \to t} \frac{H(v(t)) - H(v(s))}{t - s} \\
&= \lim_{s \to t} \frac{1}{t - s} \int_{v(s)}^{v(t)} \frac{1}{w((G^{-1}(v(r)))^{\frac{1}{p}})} dr \\
&= \lim_{s \to t} \frac{v(t) - v(s)}{t - s} \frac{1}{w((G^{-1}(v(\xi)))^{\frac{1}{p}})} \\
&= \frac{v^{\Delta}(t)}{w((G^{-1}(v(t)))^{\frac{1}{p}})}.
\end{aligned}$$

Therefore

$$\begin{aligned}
(H(v(t)))^{\Delta} &\le \frac{v^{\Delta}(t)}{w((G^{-1}(v(t)))^{\frac{1}{p}})} \\
&\le g(t), \quad t \in [t_0, t^*],
\end{aligned}$$

and

$$H(v(t)) - H(v(t_0)) \le \int_{t_0}^{t} g(s)\Delta s, \quad t \in [t_0, t^*],$$

or

$$H(v(t)) \le H\left(G(c) + \int_{t_0}^{t^*} f(s)\Delta s \right) + \int_{t_0}^{t} g(s)\Delta s, \quad t \in [t_0, t^*],$$

or

$$v(t) \le H^{-1}\left(H\left(G(c) + \int_{t_0}^{t^*} f(s)\Delta s\right) + \int_{t_0}^{t} g(s)\Delta s\right), \quad t \in [t_0, t^*].$$

By the last inequality and (10.18), (10.19), we get

$$\begin{aligned} x(t) &\le (z(t))^{\frac{1}{p}} \\ &\le (G^{-1}(v(t)))^{\frac{1}{p}} \\ &\le \left(G^{-1}\left(H^{-1}\left(H\left(G(c) + \int_{t_0}^{t^*} f(s)\Delta s\right) + \int_{t_0}^{t} g(s)\Delta s\right)\right)\right)^{\frac{1}{p}}, \end{aligned}$$

$t \in [t_0, t^*]$. In particular,

$$x(t^*) \le \left(G^{-1}\left(H^{-1}\left(H\left(G(c) + \int_{t_0}^{t^*} f(s)\Delta s\right) + \int_{t_0}^{t^*} g(s)\Delta s\right)\right)\right)^{\frac{1}{p}}.$$

Because $t^* \in [t_0, \infty)$ was arbitrarily chosen, we conclude that

$$x(t) \le \left(G^{-1}\left(H^{-1}\left(H\left(G(c) + \int_{t_0}^{t} f(s)\Delta s\right) + \int_{t_0}^{t} g(s)\Delta s\right)\right)\right)^{\frac{1}{p}}, \quad t \in [t_0, \infty).$$

This completes the proof. □

10.4 Delay integral inequalities in two independent variables

Let $x_0, y_0 \in \mathbb{T}$, $\hat{\mathbb{T}} = [x_0, \infty)$, $\widetilde{\mathbb{T}} = [y_0, \infty)$. Consider the integral inequality

$$\begin{aligned} \eta(u(x,y)) \le a(x,y) + b(x,y) \sum_{j=1}^{n} \int_{x_0}^{x}\int_{y_0}^{y} \Bigg(& f(s,t)(u(\tau_{1j}(s), \tau_{2j}(t)))^q w(u(\tau_{1j}(s), \tau_{2j}(t))) \\ & + g_j(s,t)(u(\tau_{1j}(s), \tau_{2j}(t)))^q \\ & + \int_{x_0}^{s}\int_{y_0}^{t} h_j(\xi,\zeta)(u(\tau_{1j}(\xi), \tau_{2j}(\zeta)))^q \Delta\zeta\Delta\xi \Bigg)\Delta t\Delta s, \end{aligned} \tag{10.20}$$

$(x,y) \in \hat{\mathbb{T}} \times \widetilde{\mathbb{T}}$, subject to the initial condition

$$\begin{aligned} \eta(u(x,y)) &= \phi(x,y), \quad (x,y) \in [\alpha, x_0] \times [\beta, y_0], \\ \phi(\tau_{1j}(x), \tau_{2j}(y)) &\le a(x,y) \end{aligned} \tag{10.21}$$

if $\tau_{1j}(x) \le x_0$ or $\tau_{2j}(y) \le y_0$, $(x,y) \in \hat{\mathbb{T}} \times \widetilde{\mathbb{T}}$, $j \in \{1, \ldots, n\}$, where

(A14) $u, a, b, f_j, g_j, h_j \in \mathcal{C}_{rd}(\hat{\mathbb{T}} \times \tilde{\mathbb{T}})$ are nonnegative functions, $j \in \{1, \dots, n\}$, a and b are nondecreasing, $\eta \in \mathcal{C}([0, \infty))$ is a nonnegative strictly increasing function, $q > 0$,

(A15)

$$-\infty < \alpha = \inf\Big\{\min_{j\in\{1,\dots,n\}} \tau_{1j}(x) : x \in \hat{\mathbb{T}}\Big\},$$

$$-\infty < \beta = \inf\Big\{\min_{j\in\{1,\dots,n\}} \tau_{2j}(y) : y \in \tilde{\mathbb{T}}\Big\},$$

$\tau_{1j} \in \mathcal{C}_{rd}([\alpha, x_0])$, $\tau_{2j} \in \mathcal{C}_{rd}([\beta, y_0])$, $j \in \{1, \dots, n\}$, $\tau_{1j}(x) \le x$, $x \in \hat{\mathbb{T}}$, $\tau_{2j}(y) \le y$, $y \in \tilde{\mathbb{T}}$, $j \in \{1, \dots, n\}$.

Theorem 10.4.1. *Suppose* (A14) *and* (A15) *hold. If* u *satisfies the integral inequality* (10.20) *with the initial condition* (10.21), *then*

$$\begin{aligned} u(x,y) \le \eta^{-1}\Bigg(G^{-1}\Bigg(H^{-1}\Bigg(H\Bigg(G(a(x,y)) \\ + b(x,y)\sum_{j=1}^{n}\int_{x_0}^{x}\int_{y_0}^{y}\Bigg(g_j(s,t) + \int_{x_0}^{s}\int_{y_0}^{t} h_j(\xi,\zeta)\Delta\zeta\Delta\xi\Bigg)\Delta t\Delta s\Bigg) \\ + b(x,y)\sum_{j=1}^{n}\int_{x_0}^{x}\int_{y_0}^{y} f_j(s,t)\Delta t\Delta s\Bigg)\Bigg)\Bigg), \quad (x,y) \in \hat{\mathbb{T}} \times \tilde{\mathbb{T}}, \end{aligned} \tag{10.22}$$

where G *and* H *are increasing, bijective functions and*

$$(G(v(x,y)))_x^{\Delta} = \frac{v_x^{\Delta}(x,y)}{(\eta^{-1}(v(x,y)))^q},$$

$$(H(z(x,y)))_x^{\Delta} = \frac{z_x^{\Delta}(x,y)}{w(\eta^{-1}(G^{1}(z(x,y))))}, \quad (x,y) \in \hat{\mathbb{T}} \times \tilde{\mathbb{T}}.$$

Proof. Let $X \in \hat{\mathbb{T}}$ and $Y \in \tilde{\mathbb{T}}$ be arbitrarily chosen. Define

$$\begin{aligned} v(x,y) = a(X,Y) + b(X,Y)\sum_{j=1}^{n}\int_{x_0}^{x}\int_{y_0}^{y}\Bigg(f_j(s,t)(u(\tau_{1j}(s),\tau_{2j}(t)))^q w(u(\tau_{1j}(s),\tau_{2j}(t))) \\ + g_j(s,t)(u(\tau_{1j}(s),\tau_{2j}(t)))^q \\ + \int_{x_0}^{s}\int_{y_0}^{t} h_j(\xi,\zeta)(u(\tau_{1j}(\xi),\tau_{2j}(\zeta)))^q\Delta\zeta\Delta\xi\Bigg)\Delta t\Delta s, \end{aligned}$$

$(x,y) \in [x_0, X] \times [y_0, Y]$. Then

$$u(x,y) \le \eta^{-1}\Bigg(a(x,y) + b(x,y)\sum_{j=1}^{n}\int_{x_0}^{x}\int_{y_0}^{y}\Bigg(f_j(s,t)(u(\tau_{1j}(s),\tau_{2j}(t)))^q w(u(\tau_{1j}(s),\tau_{2j}(t)))$$

$$
\begin{aligned}
&+ g_j(s,t)(u(\tau_{1j}(s),\tau_{2j}(t)))^q \\
&+ \int_{x_0}^{s}\int_{y_0}^{t} h_j(\xi,\zeta)(u(\tau_{1j}(\xi),\tau_{2j}(\zeta)))^q \Delta\zeta\Delta\xi \Bigg)\Delta t\Delta s \Bigg) \\
\le{}& \eta^{-1}\Bigg(a(X,Y) + b(X,Y)\sum_{j=1}^{n}\int_{x_0}^{x}\int_{y_0}^{y}\Bigg(f_j(s,t)(u(\tau_{1j}(s),\tau_{2j}(t)))^q w(u(\tau_{1j}(s),\tau_{2j}(t))) \\
&+ g_j(s,t)(u(\tau_{1j}(s),\tau_{2j}(t)))^q \\
&+ \int_{x_0}^{s}\int_{y_0}^{t} h_j(\xi,\zeta)(u(\tau_{1j}(\xi),\tau_{2j}(\zeta)))^q \Delta\zeta\Delta\xi \Bigg)\Delta t\Delta s \Bigg) \\
={}& \eta^{-1}(v(x,y)), \quad (x,y) \in [x_0,X]\times[y_0,Y].
\end{aligned}
$$

As in the proof of Theorem 10.2.6, we have that

$$
u(\tau_{1j}(x),\tau_{2j}(y)) \le \eta^{-1}(v(x,y)), \quad x \in [x_0,X],\ y \in [y_0,Y].
$$

Next,

$$
\begin{aligned}
v_x^{\Delta}(x,y) ={}& b(X,Y)\sum_{j=1}^{n}\int_{y_0}^{y}\Bigg(f_j(x,t)(u(\tau_{1j}(x),\tau_{2j}(t)))^q w(u(\tau_{1j}(x),\tau_{2j}(t))) \\
&+ g_j(x,t)(u(\tau_{1j}(x),\tau_{2j}(t)))^q \\
&+ \int_{x_0}^{x}\int_{y_0}^{t} h_j(\xi,\zeta)(u(\tau_{1j}(\xi),\tau_{2j}(\zeta)))^q \Delta\zeta\Delta\xi \Bigg)\Delta t \\
\le{}& \Bigg(b(X,Y)\sum_{j=1}^{n}\int_{y_0}^{y}\Bigg(f_j(x,t)w(\eta^{-1}(v(x,t))) + g_j(x,t) \\
&+ \int_{x_0}^{x}\int_{y_0}^{t} h_j(\xi,\zeta)\Delta\zeta\Delta\xi \Bigg)\Delta t \Bigg)(\eta^{-1}(v(x,y)))^q,
\end{aligned}
$$

$(x,y) \in [x_0,X]\times[y_0,Y]$, which implies that

$$
\begin{aligned}
\frac{v_x^{\Delta}(x,y)}{(\eta^{-1}(v(x,y)))^q} \le{}& b(X,Y)\sum_{j=1}^{n}\int_{y_0}^{y}\Bigg(f_j(x,t)w(\eta^{-1}(v(x,t))) + g_j(x,t) \\
&+ \int_{x_0}^{x}\int_{y_0}^{t} h_j(\xi,\zeta)\Delta\zeta\Delta\xi \Bigg)\Delta t,
\end{aligned}
$$

$(x,y) \in [x_0,X]\times[y_0,Y]$. As in the proof of Theorem 10.3.1, we get

$$
(G(v(x,y)))_x^{\Delta} = \frac{v_x^{\Delta}(x,y)}{(\eta^{-1}(v(x,y)))^q}, \quad (x,y) \in [x_0,X]\times[y_0,Y].
$$

Thus,

$$(G(v(x,y)))_x^{\Delta} \le \left(b(X,Y) \sum_{j=1}^{n} \int_{y_0}^{y} \left(f_j(x,t) w(\eta^{-1}(v(x,t))) + g_j(x,t) \right.\right.$$
$$\left.\left. + \int_{x_0}^{x} \int_{y_0}^{t} h_j(\xi,\zeta) \Delta\zeta \Delta\xi \right) \Delta t \right), \quad (x,y) \in [x_0,X] \times [y_0,Y],$$

whereupon

$$G(v(x,y)) - G(v(x_0,y)) \le b(X,Y) \sum_{j=1}^{n} \int_{x_0}^{x} \int_{y_0}^{y} \left(f_j(s,t) w(\eta^{-1}(v(s,t))) + g_j(s,t) \right.$$
$$\left. + \int_{x_0}^{s} \int_{y_0}^{t} h_j(\xi,\zeta) \Delta\zeta \Delta\xi \right) \Delta t \Delta s, \quad (x,y) \in [x_0,X] \times [y_0,Y],$$

or

$$G(v(x,y)) \le G(a(X,Y)) + b(X,Y) \sum_{j=1}^{n} \int_{x_0}^{x} \int_{y_0}^{y} \left(f_j(s,t) w(\eta^{-1}(v(s,t))) + g_j(s,t) \right.$$
$$\left. + \int_{x_0}^{s} \int_{y_0}^{t} h_j(\xi,\zeta) \Delta\zeta \Delta\xi \right) \Delta t \Delta s, \quad (x,y) \in [x_0,X] \times [y_0,Y],$$

or

$$v(x,y) \le G^{-1}\left(G(a(X,Y)) + \left(b(X,Y) \sum_{j=1}^{n} \int_{x_0}^{x} \int_{y_0}^{y} \left(f_j(s,t) w(\eta^{-1}(v(s,t))) + g_j(s,t) \right.\right.\right.$$
$$\left.\left.\left. + \int_{x_0}^{s} \int_{y_0}^{t} h_j(\xi,\zeta) \Delta\zeta \Delta\xi \right) \Delta t \Delta s \right) \right), \quad (x,y) \in [x_0,X] \times [y_0,Y].$$

Let

$$z(x,y) = G(a(X,Y))$$
$$+ b(X,Y) \sum_{j=1}^{n} \int_{x_0}^{x} \int_{y_0}^{y} f_j(s,t) w(\eta^{-1}(v(s,t))) \Delta t \Delta s$$
$$+ b(X,Y) \sum_{j=1}^{n} \int_{x_0}^{X} \int_{y_0}^{Y} \left(g_j(s,t) \right.$$
$$\left. + \sum_{j=1}^{n} \int_{x_0}^{s} \int_{y_0}^{t} h_j(\xi,\zeta) \Delta\zeta \Delta\xi \right) \Delta t \Delta s, \quad (x,y) \in [x_0,X] \times [y_0,Y].$$

We have

$$v(x,y) \le G^{-1}(z(x,y)), \quad (x,y) \in [x_0, X] \times [y_0, Y],$$

and furthermore,

$$\begin{aligned} z^{\Delta}(x,y) &\le b(X,Y) \sum_{j=1}^{n} \int_{y_0}^{y} f_j(x,t) w(\eta^{-1}(v(x,t)))\Delta t \\ &\le \left(b(X,Y) \sum_{j=1}^{n} \int_{y_0}^{y} f_j(x,t)\Delta t \right) w(\eta^{-1}(G^{-1}(z(x,y)))), \end{aligned}$$

$(x,y) \in [x_0, X] \times [y_0, Y]$, which is followed by

$$\frac{z^{\Delta}(x,y)}{w(\eta^{-1}(G^{-1}(z(x,y))))} \le b(X,Y) \sum_{j=1}^{n} \int_{y_0}^{y} f_j(x,t)\Delta t,$$

$(x,y) \in [x_0, X] \times [y_0, Y]$. As in the proof of Theorem 10.3.1, we have

$$\begin{aligned} (H(z(x,y)))_x^{\Delta} &= \frac{z_x^{\Delta}(x,y)}{w(\eta^{-1}(G^{-1}(z(x,y))))} \\ &\le b(X,Y) \sum_{j=1}^{n} \int_{y_0}^{y} f_j(x,t)\Delta t, \end{aligned}$$

$(x,y) \in [x_0, X] \times [y_0, Y]$. Hence,

$$H(z(x,y)) - H(z(x_0,y)) \le b(X,Y) \sum_{j=1}^{n} \int_{x_0}^{x} \int_{y_0}^{y} f_j(x,t)\Delta t \Delta s,$$

$(x,y) \in [x_0, X] \times [y_0, Y]$, whereupon

$$z(x,y) \le H^{-1}\left(H(z(x_0,y)) + b(X,Y) \sum_{j=1}^{n} \int_{x_0}^{x} \int_{y_0}^{y} f_j(x,t)\Delta t \Delta s \right),$$

$(x,y) \in [x_0, X] \times [y_0, Y]$. Then

$$\begin{aligned} u(x,y) &\le \eta^{-1}(v(x,y)) \\ &\le \eta^{-1}(G^{-1}(z(x,y))) \\ &\le \eta^{-1}\Bigg(G^{-1}\Bigg(H^{-1}\Bigg(H(z(x_0,y)) \\ &\quad + b(X,Y) \sum_{j=1}^{n} \int_{x_0}^{x} \int_{y_0}^{y} f_j(x,t)\Delta t \Delta s \Bigg)\Bigg)\Bigg), \end{aligned}$$

$(x,y) \in [x_0, X] \times [y_0, Y]$. In particular,

$$u(X,Y) \le \eta^{-1}\Bigg(G^{-1}\Bigg(H^{-1}\Bigg(H(z(x_0,Y)) + b(X,Y)\sum_{j=1}^{n}\int_{x_0}^{X}\int_{y_0}^{Y} f_j(x,t)\Delta t\Delta s\Bigg)\Bigg)\Bigg).$$

Because $(X,Y) \in \hat{\mathbb{T}}\times\tilde{\mathbb{T}}$ was arbitrarily chosen, from the last inequality, we get inequality (10.22). This completes the proof. □

10.5 Delay Volterra–Fredholm-type integral inequalities

Let $T \in [t_0,\infty)$, $I = [t_0, T]$. In this section we will investigate the inequality

$$\begin{aligned}(u(t))^l &\le a(t) + b(t)\int_{\alpha(t_0)}^{\alpha(t)} f_1(s)\Bigg((u(s))^{q_1} + f_2(s)\int_{\alpha(t_0)}^{s} g_1(\tau)(u(\tau))^{r_1}\Delta\tau\Bigg)^{\theta_1}\Delta s\\ &+ \int_{\beta(t_0)}^{\beta(T)} f_3(s)\Bigg((u(s))^{q_2} + f_4(s)\int_{\beta(t_0)}^{s} q_2(\tau)(u(\tau))^{r_2}\Delta\tau\Bigg)^{\theta_2}\Delta s\\ &+ \int_{\gamma(t_0)}^{\gamma(T)} f_5(s)\Bigg((u(s))^{q_3} + f_6(s)\int_{\gamma(t_0)}^{s} q_3(\tau)(u(\tau))^{r_3}\Delta\tau\Bigg)^{\theta_3}\Delta s, \quad t \in I,\end{aligned} \tag{10.23}$$

where

(A16) $\alpha : I \to \mathbb{T}$ is continuous and strictly increasing such that $\alpha(t) \le t$, $t \in I$, $\alpha^\Delta \in C_{rd}(I)$,

(A17) $\beta, \gamma \in C(I)$, $\beta(t) \le t$, $\gamma(t) \le t$, $t \in I$,

(A18) $u, a, b \in C_{rd}(I)$ are nonnegative functions, a is nondecreasing and $b^\Delta \ge 0$ on I,

(A19) $f_i, g_i \in C_{rd}(I)$, $i \in \{1,\dots,6\}$,

(A20) $0 \le q_i \le l$, $0 \le r_0 \le l$, $l \ne 0$, $0 \le \theta_i \le 1$, $i \in \{1,2,3\}$.

Define

$$\begin{aligned}\widetilde{K} = \int_{\beta(t_0)}^{\beta(T)}\Bigg(&\theta_2 k_4^{\theta_2-1} f_3(s)\Bigg(\frac{l-q_2}{l}k_5^{q_2} + f_4(s)\int_{\beta(t_0)}^{s} g_2(\tau)\Bigg(\frac{l-r_2}{l}k_6^{r_2}\Bigg)\Delta\tau\Bigg)\\ &+ (1-\theta_2)k_4^{\theta_2} f_3(s)\Bigg)\Delta s\end{aligned}$$

$$+\int_{\gamma(t_0)}^{\gamma(T)}\left(\theta_3 k_7^{\theta_3-1}f_5(s)\left(\frac{l-q_3}{l}k_8^{q_3}+f_6(s)\int_{\beta(t_0)}^{s}g_3(\tau)\left(\frac{l-r_3}{l}k_7^{r_3}\right)\Delta\tau\right)\right.$$

$$\left.+(1-\theta_3)k_7^{\theta_3}f_5(s)\right)\Delta s,$$

$$\widetilde{V}(t)=a(t)+b(t)\int_{\alpha(t_0)}^{\alpha(t)}\left(\theta_1 k_1^{\theta_1-1}f_1(s)\left(\frac{l-q_1}{l}k_2^{q_1}\right.\right.$$

$$\left.\left.+f_2(s)\int_{\alpha(t_0)}^{s}g_1(\tau)\left(\frac{l-r_1}{l}k_1^{r_1}\right)\Delta\tau\right)+(1-\theta_1)k_1^{\theta_1}f_1(s)\right)\Delta s,$$

$$A(t)=b^{\Delta}(t)\int_{\alpha(t_0)}^{\alpha(\sigma(t))}\left(\theta_1 k_1^{\theta_1-1}\widetilde{f}_1(s)\left(1+f_2(s)\int_{\alpha(t_0)}^{s}\widetilde{g}_1(\tau)\Delta\tau\right)\right)\Delta s,$$

$$B(t)=\frac{A(t)}{1-\mu(t)A(t)},$$

$$C(t)=b(t)\left(\theta_1 k_1^{\theta-1}\widetilde{f}_1(t)\left(1+f_2(t)\int_{\alpha(t_0)}^{t}\widetilde{g}_1(\tau)\Delta\tau\right)\right)\alpha^{\Delta}(t),$$

$$\widetilde{f}_1(t)=\frac{q_1}{l}k_2^{q_1-l}f_1(t),$$

$$\widetilde{f}_3(t)=\frac{q_2}{l}k_3^{q_2-l}f_3(t),$$

$$\widetilde{f}_5(t)=\frac{q_3}{l}k_8^{q_3-l}f_5(t),$$

$$\widetilde{g}_1(t)=\frac{r_1}{l}k_2^{l-q_1}k_3^{r_1-l}g_1(t),$$

$$\widetilde{g}_2(t)=\frac{r_2}{l}k_5^{l-q_2}k_6^{r_2-l}g_2(t),$$

$$\widetilde{g}_3(t)=\frac{r_3}{l}k_8^{l-q_3}k_9^{r_3-l}g_3(t),\quad t\in I.$$

Suppose that

(A21) there exist positive constants k_i, $i\in\{1,\dots,9\}$, such that

$$\lambda=\int_{\beta(t_0)}^{\beta(T)}\left(\theta_2 k_4^{\theta_2-1}f_3(s)\left(e_{B\oplus C}(s,t_0)+f_4(s)\int_{\beta(t_0)}^{s}g_2(\tau)e_{B\oplus C}(\tau,t_0)\Delta\tau\right)\right)\Delta s$$

$$+\int_{\gamma(t_0)}^{\gamma(T)}\left(\theta_3 k_7^{\theta_3-1}f_5(s)\left(e_{B\oplus C}(s,t_0)+f_6(s)\int_{\beta(t_0)}^{s}g_3(\tau)e_{B\oplus C}(\tau,t_0)\Delta\tau\right)\right)\Delta s$$

$$<1,$$

(A22) $\mu(t)A(t)<1$, $t\in I$.

Theorem 10.5.1. *Assume* (A16)–(A22) *hold. If u satisfies inequality* (10.23), *then*

$$u(t) \leq \left(\frac{\widetilde{K} + \widetilde{V}(T)}{1-\lambda} e_{B\oplus C}(t, t_0)\right)^{\frac{1}{l}}, \quad t \in I.$$

Proof. Let

$$z(t) = a(t) + b(t) \int_{\alpha(t_0)}^{\alpha(t)} f_1(s)\left((u(s))^{q_1} + f_2(s) \int_{\alpha(t_0)}^{s} g_1(\tau)(u(\tau))^{r_1}\Delta\tau\right)^{\theta_1} \Delta s$$
$$+ \int_{\beta(t_0)}^{\beta(T)} f_3(s)\left((u(s))^{q_2} + f_4(s) \int_{\beta(t_0)}^{s} q_2(\tau)(u(\tau))^{r_2}\Delta\tau\right)^{\theta_2} \Delta s$$
$$+ \int_{\gamma(t_0)}^{\gamma(T)} f_5(s)\left((u(s))^{q_3} + f_6(s) \int_{\gamma(t_0)}^{s} q_3(\tau)(u(\tau))^{r_3}\Delta\tau\right)^{\theta_3} \Delta s, \quad t \in I.$$

Then z is a nondecreasing function on I and

$$(u(t))^l \leq z(t), \quad t \in I.$$

Furthermore, using Lemma 10.2.1, we get

$$z(t) \leq a(t) + b(t) \int_{\alpha(t_0)}^{\alpha(t)} \left(\theta_1 k_1^{\theta_1 - 1} f_1(s)\left(\frac{q_1}{l} k_2^{q_1 - l}(u(s))^l\right.\right.$$
$$+ \frac{l - q_1}{l} k_2^{q_1} + f_2(s) \int_{\alpha(t_0)}^{s} g_1(\tau)\left(\frac{r_1}{l} k_3^{r_1 - l}(u(\tau))^l + \frac{l - r_1}{l} k_3^{r_1}\right)\Delta\tau\Bigg)$$
$$\left. + (1 - \theta_1) k_1^{\theta_1} f_1(s)\right)\Delta s$$
$$+ \int_{\beta(t_0)}^{\beta(T)} \left(\theta_2 k_4^{\theta_2 - 1} f_3(s)\left(\frac{q_2}{l} k_3^{q_2 - l}(u(s))^l + \frac{l - q_2}{l} k_5^{q_2}\right.\right.$$
$$+ f_4(s) \int_{\beta(t_0)}^{s} q_2(\tau)\left(\frac{r_2}{l} k_6^{r_2 - l}(u(\tau))^l + \frac{l - r_2}{l} k_6^{r_2}\right)\Delta\tau\Bigg)$$
$$\left. + (1 - \theta_2) k_4^{\theta_2} f_3(s)\right)\Delta s$$
$$+ \int_{\gamma(t_0)}^{\gamma(T)} \left(\theta_3 k_7^{\theta_3 - 1} f_5(s)\left(\frac{q_3}{l} k_8^{q_3 - l}(u(s))^l + \frac{l - q_3}{l} k_8^{q_3}\right.\right.$$

$$+f_6(s)\int_{\gamma(t_0)}^{s} g_3(\tau)\left(\frac{r_3}{l}k_9^{r_3-l}(u(\tau))^l+\frac{l-r_3}{l}k_9^{r_3}\right)\Delta\tau\Bigg)$$

$$+(1-\theta_3)k_7^{\theta_3}f_5(s)\Bigg)\Delta s$$

$$\begin{aligned}
&=\widetilde{K}+\widetilde{V}(t)\\
&\quad+b(t)\int_{\alpha(t_0)}^{\alpha(t)}\left(\theta_1 k_1^{\theta_1-1}\widetilde{f}_1(s)\left(z(s)+f_2(s)\int_{\alpha(t_0)}^{s}\widetilde{g}_1(\tau)z(\tau)\Delta\tau\right)\right)\Delta s\\
&\quad+\int_{\beta(t_0)}^{\beta(T)}\left(\theta_2 k_4^{\theta_2-1}\widetilde{f}_3(s)\left(z(s)+f_4(s)\int_{\beta(t_0)}^{s}\widetilde{g}_2(\tau)z(\tau)\Delta\tau\right)\right)\Delta s\\
&\quad+\int_{\gamma(t_0)}^{\gamma(T)}\left(\theta_3 k_7^{\theta_3-1}\widetilde{f}_5(s)\left(z(s)+f_6(s)\int_{\gamma(t_0)}^{s}\widetilde{g}_3(\tau)z(\tau)\Delta\tau\right)\right)\Delta s,\quad t\in I.
\end{aligned}$$

Since $\widetilde{V}$ is nondecreasing, from the last inequality, we get

$$\begin{aligned}
z(t)&\le\widetilde{K}+\widetilde{V}(T)\\
&\quad+b(t)\int_{\alpha(t_0)}^{\alpha(t)}\left(\theta_1 k_1^{\theta_1-1}\widetilde{f}_1(s)\left(z(s)+f_2(s)\int_{\alpha(t_0)}^{s}\widetilde{g}_1(\tau)z(\tau)\Delta\tau\right)\right)\Delta s\\
&\quad+\int_{\beta(t_0)}^{\beta(T)}\left(\theta_2 k_4^{\theta_2-1}\widetilde{f}_3(s)\left(z(s)+f_4(s)\int_{\beta(t_0)}^{s}\widetilde{g}_2(\tau)z(\tau)\Delta\tau\right)\right)\Delta s\\
&\quad+\int_{\gamma(t_0)}^{\gamma(T)}\left(\theta_3 k_7^{\theta_3-1}\widetilde{f}_5(s)\left(z(s)+f_6(s)\int_{\gamma(t_0)}^{s}\widetilde{g}_3(\tau)z(\tau)\Delta\tau\right)\right)\Delta s,\quad t\in I.
\end{aligned}$$

Let

$$\begin{aligned}
M&=\widetilde{K}+\widetilde{V}(T)\\
&\quad+\int_{\beta(t_0)}^{\beta(T)}\left(\theta_2 k_4^{\theta_2-1}\widetilde{f}_3(s)\left(z(s)+f_4(s)\int_{\beta(t_0)}^{s}\widetilde{g}_2(\tau)z(\tau)\Delta\tau\right)\right)\Delta s\\
&\quad+\int_{\gamma(t_0)}^{\gamma(T)}\left(\theta_3 k_7^{\theta_3-1}\widetilde{f}_5(s)\left(z(s)+f_6(s)\int_{\gamma(t_0)}^{s}\widetilde{g}_3(\tau)z(\tau)\Delta\tau\right)\right)\Delta s.
\end{aligned}$$

Therefore

$$z(t)\le M+b(t)\int_{\alpha(t_0)}^{\alpha(t)}\left(\theta_1 k_1^{\theta_1-1}\widetilde{f}_1(s)\left(z(s)+f_2(s)\int_{\alpha(t_0)}^{s}\widetilde{g}_1(\tau)z(\tau)\Delta\tau\right)\right)\Delta s,\quad t\in I.$$

Let

$$w(t) = M + b(t) \int_{\alpha(t_0)}^{\alpha(t)} \left(\theta_1 k_1^{\theta_1 - 1} \tilde{f}_1(s) \left(z(s) + f_2(s) \int_{\alpha(t_0)}^{s} \tilde{g}_1(\tau) z(\tau) \Delta\tau \right) \right) \Delta s \quad t \in I.$$

Then w is nondecreasing on I and

$$\begin{aligned} w^\Delta(t) &= b^\Delta(t) \int_{\alpha(t_0)}^{\alpha(\sigma(t))} \left(\theta_1 k_1^{\theta_1 - 1} \tilde{f}_1(s) \left(z(s) + f_2(s) \int_{\alpha(t_0)}^{s} \tilde{g}_1(\tau) z(\tau) \Delta\tau \right) \right) \Delta s \\ &\quad + b(t) \left(\theta_1 k_1^{\theta_1 - 1} \tilde{f}_1(t) \left(z(t) + f_2(t) \int_{\alpha(t_0)}^{t} \tilde{g}_1(\tau) z(\tau) \Delta\tau \right) \right) \alpha^\Delta(t) \\ &\le b^\Delta(t) \int_{\alpha(t_0)}^{\alpha(\sigma(t))} \left(\theta_1 k_1^{\theta_1 - 1} \tilde{f}_1(s) \left(w(s) + f_2(s) \int_{\alpha(t_0)}^{s} \tilde{g}_1(\tau) w(\tau) \Delta\tau \right) \right) \Delta s \\ &\quad + b(t) \left(\theta_1 k_1^{\theta_1 - 1} \tilde{f}_1(t) \left(w(t) + f_2(t) \int_{\alpha(t_0)}^{t} \tilde{g}_1(\tau) w(\tau) \Delta\tau \right) \right) \alpha^\Delta(t) \\ &\le w(\sigma(t)) b^\Delta(t) \int_{\alpha(t_0)}^{\alpha(\sigma(t))} \left(\theta_1 k_1^{\theta_1 - 1} \tilde{f}_1(s) \left(1 + f_2(s) \int_{\alpha(t_0)}^{s} \tilde{g}_1(\tau) \Delta\tau \right) \right) \Delta s \\ &\quad + w(t) b(t) \left(\theta_1 k_1^{\theta_1 - 1} \tilde{f}_1(t) \left(1 + f_2(t) \int_{\alpha(t_0)}^{t} \tilde{g}_1(\tau) \Delta\tau \right) \right) \alpha^\Delta(t) \\ &= A(t) w(\sigma(t)) + w(t) C(t) \\ &= \frac{B(t)}{1 + \mu(t) B(t)} w(\sigma(t)) + w(t) C(t) \\ &= \frac{B(t)}{1 + \mu(t) B(t)} (w(t) + \mu(t) w^\Delta(t)) + w(t) C(t), \quad t \in I. \end{aligned}$$

Hence,

$$\frac{w^\Delta(t)}{1 + \mu(t) B(t)} \le \left(\frac{B(t)}{1 + \mu(t) B(t)} + C(t) \right) w(t), \quad t \in I,$$

or

$$\begin{aligned} w^\Delta(t) &\le (B(t) + C(t) + \mu(t) B(t) C(t)) w(t) \\ &= (B \oplus C)(t) w(t), \quad t \in I. \end{aligned}$$

Now we apply the Gronwall inequality and obtain

$$\begin{aligned} w(t) &\le w(t_0) e_{B \oplus C}(t, t_0) \\ &= M e_{B \oplus C}(t, t_0), \quad t \in I, \end{aligned}$$

and

$$z(t) \le M e_{B\oplus C}(t, t_0), \quad t \in I.$$

Now, by the definition of M, we get

$$\begin{aligned}
M &\le \widetilde{K} + \widetilde{V}(T) + M\Bigg(\int\limits_{\beta(t_0)}^{\beta(T)} \Bigg(\theta_2 k_4^{\theta_2-1} f_3(s)\Bigg(e_{B\oplus C}(s, t_0) + f_4(s) \int\limits_{\beta(t_0)}^{s} g_2(\tau) e_{B\oplus C}(\tau, t_0)\Delta\tau \Bigg)\Delta s \\
&\quad + \int\limits_{\gamma(t_0)}^{\gamma(T)} \Bigg(\theta_3 k_7^{\theta_3-1} f_5(s)\Bigg(e_{B\oplus C}(s, t_0) + f_6(s) \int\limits_{\beta(t_0)}^{s} g_3(\tau) e_{B\oplus C}(\tau, t_0)\Delta\tau \Bigg)\Delta s \Bigg)\Bigg)\Bigg) \\
&= \widetilde{K} + \widetilde{V}(T) + M\lambda,
\end{aligned}$$

whereupon

$$M \le \frac{\widetilde{K} + \widetilde{V}(T)}{1 - \lambda}.$$

Consequently,

$$\begin{aligned}
(u(t))^l &\le z(t) \\
&\le \frac{\widetilde{K} + \widetilde{V}(T)}{1 - \lambda} e_{B\oplus C}(t, t_0), \quad t \in I.
\end{aligned}$$

This completes the proof. □

11 Applications

This chapter deals with some applications of some linear integral inequalities and some linear integro-dynamic inequalities. They are investigated the existence and uniqueness of the solutions of first order dynamic equations, continuous dependence on initial conditions of the solutions of first order dynamic equations. They are given applications for some second order integro-dynamic equations and Volterra integral equations. THey are deducted some bounds of the solutions of delay dynamic equations. Some of the results in this chapter can be found in [6, 7, 11, 12, 20, 21] and [23].

Let $\mathbb{T}$ be a time scale with forward jump operator and delta differentiation operator σ and Δ, respectively. Let also $t_0, a \in \mathbb{T}$, $a > t_0$, $J = [t_0, a]$.

11.1 Existence and uniqueness of the solution of first order dynamic equations

Consider the following IVP:

$$y^{\Delta}(t) = f(t, y(t)), \quad t \in J, \tag{11.1}$$

$$y(t_0) = y_0, \tag{11.2}$$

where y_0 is a given constant.

Theorem 11.1.1. *Let $f : J \times \mathbb{R} \to \mathbb{R}$, $f \in \mathcal{C}(J \times \mathbb{R})$,*

$$|f(t, y(t))| \le M \quad \text{for any } t \in J,\ y \in \mathcal{C}(J),$$

$$|f(t, y_1(t)) - f(t, y_2(t))| \le L|y_1(t) - y_2(t)|, \quad t \in J,\ y_1, y_2 \in \mathcal{C}(J),$$

where L and M are some positive constants. Then the problem (11.1), (11.2) *has a unique solution $y \in \mathcal{C}(J)$.*

Proof. Let $y(t)$, $t \in J$, be a solution of the IVP (11.1), (11.2). We integrate equation (11.1) from t_0 to t and, using (11.2), get

$$y(t) = y_0 + \int_{t_0}^{t} f(\tau, y(\tau))\Delta\tau, \quad t \in J. \tag{11.3}$$

Let now $y \in \mathcal{C}(J)$ be a solution of the integral equation (11.3). We put $t = t_0$ in (11.3) and get

$$y(t_0) = y_0.$$

Differentiating equation (11.3) with respect to t, we obtain that $y(t)$, $t \in J$, satisfies equation (11.1). Therefore $y(t)$, $t \in J$, is a solution of the problem (11.1), (11.2). Hence,

https://doi.org/10.1515/9783110705553-011

the IVP (11.1), (11.2) is equivalent to the integral equation (11.3). Define the sequence $\{y_l(t)\}_{l\in\mathbb{N}_0}$ as follows:

$$\begin{aligned}
y_0(t) &= y_0,\\
y_l(t) &= y_0 + \int_{t_0}^{t} f(\tau, y_{l-1}(\tau))\Delta\tau, \quad l \in \mathbb{N},\ t \in J.
\end{aligned}$$

We have

$$\begin{aligned}
|y_1(t) - y_0(t)| &= \left| y_0 + \int_{t_0}^{t} f(\tau, y_0(\tau))\Delta\tau - y_0 \right|\\
&= \left| \int_{t_0}^{t} f(\tau, y_0(\tau))\Delta\tau \right|\\
&\le \int_{t_0}^{t} |f(\tau, y_0(\tau))|\Delta\tau\\
&\le M(t - t_0)\\
&= Mh_1(t, t_0), \quad t \in J.
\end{aligned}$$

Assume that

$$|y_l(t) - y_{l-1}(t)| \le ML^{l-1}h_l(t, t_0), \quad t \in J, \tag{11.4}$$

for some $l \in \mathbb{N}$. We will prove that

$$|y_{l+1}(t) - y_l(t)| \le ML^l h_{l+1}(t, t_0), \quad t \in J.$$

In fact, we have

$$\begin{aligned}
|y_{l+1}(t) - y_l(t)| &= \left| y_0 + \int_{t_0}^{t} f(\tau, y_l(\tau))\Delta\tau \right.\\
&\qquad \left. - y_0 - \int_{t_0}^{t} f(\tau, y_{l-1}(\tau))\Delta\tau \right|\\
&= \left| \int_{t_0}^{t} (f(\tau, y_l(\tau)) - f(\tau, y_{l-1}(\tau)))\Delta\tau \right|\\
&\le \int_{t_0}^{t} |f(\tau, y_l(\tau)) - f(\tau, y_{l-1}(\tau))|\Delta\tau
\end{aligned}$$

$$\begin{aligned}
&\leq L \int_{t_0}^{t} |y_l(\tau) - y_{l-1}(\tau)| \Delta\tau \\
&\leq ML^l \int_{t_0}^{t} h_l(\tau, t_0) \Delta\tau \\
&= ML^l h_{l+1}(t, t_0), \quad t \in J.
\end{aligned}$$

Consequently, (11.4) holds for any $l \in \mathbb{N}$. Note that

$$\begin{aligned}
\lim_{l\to\infty} |y_l(t) - y_0(t)| &= \left| \sum_{r=1}^{\infty} (y_r(t) - y_{r-1}(t)) \right| \\
&\leq \sum_{r=1}^{\infty} |y_r(t) - y_{r-1}(t)| \\
&\leq M \sum_{r=1}^{\infty} L^{r-1} h_r(t, t_0) \\
&\leq \frac{M}{L} e_L(t, t_0), \quad t \in J.
\end{aligned}$$

Therefore $y_l(t)$ converges uniformly on J to a solution $y(t)$, $t \in J$, of the IVP (11.1), (11.2). Suppose that the problem (11.1), (11.2) has two solutions $y(t)$ and $z(t)$, $t \in J$. For $z(t)$, $t \in J$, we have the following representation:

$$z(t) = y_0 + \int_{t_0}^{t} f(\tau, z(\tau)) \Delta\tau, \quad t \in J.$$

Then

$$\begin{aligned}
|y(t) - z(t)| &= \left| y_0 + \int_{t_0}^{t} f(\tau, y(\tau)) \Delta\tau - y_0 - \int_{t_0}^{t} f(\tau, z(\tau)) \Delta\tau \right| \\
&= \left| \int_{t_0}^{t} (f(\tau, y(\tau)) - f(\tau, z(\tau))) \Delta\tau \right| \\
&\leq \int_{t_0}^{t} |f(\tau, y(\tau)) - f(\tau, z(\tau))| \Delta\tau \\
&\leq L \int_{t_0}^{t} |y(\tau) - z(\tau)| \Delta\tau, \quad t \in J.
\end{aligned}$$

From the last inequality and Theorem 1.1.10, we get

$$|y(t) - z(t)| = 0, \quad t \in J,$$

i. e., $y = z$ on J. This completes the proof. □

11.2 Continuous dependence on initial conditions of the solutions of first order dynamic equations

Let $D \subset J \times \mathbb{R}$, $t_1 \in J$, $t_1 \geq t_0$, $(t_0, y_0), (t_1, z_1) \in D$.

Theorem 11.2.1. *Let $f \in \mathcal{C}(D)$, $|f(t,y)| \leq M$, $(t,y) \in D$,*

$$|f(t,y_1(t)) - f(t,y_2(t))| \leq L|y_1(t) - y_2(t)|, \quad t \in J, \quad y_1, y_2 \in \mathcal{C}(J).$$

Let also $g \in \mathcal{C}(D)$ and

$$|g(t,z)| \leq M_1, \quad (t,z) \in D,$$

where M, M_1, and L are some positive constants. If y and z are solutions of the IVP (11.1), (11.2), *and*

$$z^{\Delta}(t) = f(t,z(t)) + g(t,z(t)), \quad z(t_1) = z_1,$$

respectively, then

$$\begin{aligned} |y(t) - z(t)| &\leq |y_0 - z_1| + M(t_1 - t_0) + M_1|t - t_1| \\ &\quad + L\int_{t_0}^{t} (|y_0 - z_1| + M(t_1 - t_0) + M_1|s - t_1|) e_{\ominus L}(\sigma(s), t)\Delta s, \quad t \in J. \end{aligned}$$

Proof. We have

$$\begin{aligned} y(t) &= y_0 + \int_{t_0}^{t} f(\tau, y(\tau))\Delta\tau, \\ z(t) &= z_1 + \int_{t_1}^{t} f(\tau, z(\tau))\Delta\tau + \int_{t_1}^{t} g(\tau, z(\tau))\Delta\tau \\ &= z_1 + \int_{t_0}^{t} f(\tau, z(\tau))\Delta\tau - \int_{t_0}^{t_1} f(\tau, z(\tau))\Delta\tau \\ &\quad + \int_{t_1}^{t} g(\tau, z(\tau))\Delta\tau, \\ |y(t) - z(t)| &= \Bigg| y_0 + \int_{t_0}^{t} f(\tau, y(\tau))\Delta\tau - z_1 - \int_{t_0}^{t} f(\tau, z(\tau))\Delta\tau \\ &\quad + \int_{t_0}^{t_1} f(\tau, z(\tau))\Delta\tau - \int_{t_1}^{t} g(\tau, z(\tau))\Delta\tau \Bigg| \end{aligned}$$

$$
\begin{aligned}
&= \left| y_0 - z_1 + \int_{t_0}^{t_1} f(\tau, z(\tau))\Delta\tau - \int_{t_1}^{t} g(\tau, z(\tau))\Delta\tau \right. \\
&\quad \left. + \int_{t_0}^{t} (f(\tau, y(\tau)) - f(\tau, z(\tau)))\Delta\tau \right| \\
&\leq |y_0 - z_1| + \left| \int_{t_0}^{t_1} f(\tau, z(\tau))\Delta\tau \right| + \left| \int_{t_1}^{t} g(\tau, z(\tau))\Delta\tau \right| \\
&\quad + \left| \int_{t_0}^{t} (f(\tau, y(\tau)) - f(\tau, z(\tau)))\Delta\tau \right| \\
&\leq |y_0 - z_1| + \int_{t_0}^{t_1} |f(\tau, z(\tau))|\Delta\tau + \left| \int_{t_1}^{t} |g(\tau, z(\tau))|\Delta\tau \right| \\
&\quad + \int_{t_0}^{t} |f(\tau, y(\tau)) - f(\tau, z(\tau))|\Delta\tau \\
&\leq |y_0 - z_1| + M(t_1 - t_0) + M_1|t - t_1| \\
&\quad + L \int_{t_0}^{t} |y(\tau) - z(\tau)|\Delta\tau, \quad t \in J.
\end{aligned}
$$

Hence, by Theorem 1.1.10, we get

$$
\begin{aligned}
|y(t) - z(t)| &\leq |y_0 - z_1| + M(t_1 - t_0) + M_1|t - t_1| \\
&\quad + L \int_{t_0}^{t} (|y_0 - z_1| + M(t_1 - t_0) + M_1|s - t_1|) e_{\ominus L}(\sigma(s), t)\Delta s, \quad t \in J.
\end{aligned}
$$

This completes the proof. □

11.3 Second order integro-dynamic equations

Let $f, g \in \mathcal{C}(J)$, $k \in \mathcal{C}(J \times J \times \mathbb{R})$, $h \in \mathcal{C}(J \times \mathbb{R} \times \mathbb{R})$. Suppose that

$$
f(t) \neq 0, \quad t \in J, \quad f^{\sigma}(a) \neq 0. \tag{11.5}
$$

Consider the following second order integro-dynamic equation:

$$
(f(t)x^{\Delta})^{\Delta} + g(t)x = h\left(t, x, \int_{t_0}^{t} k(t, s, x(s))\Delta s \right), \quad t \in J,
$$

which we can rewrite in the form

$$f(t)x^{\Delta^2}(t) + f^{\Delta}(t)x^{\Delta\sigma}(t) + g(t)x(t) = h\left(t, x(t), \int_0^t k(t,s,x(s))\Delta s\right), \quad t \in J. \tag{11.6}$$

Let $x_1(t)$ and $x_2(t)$, $t \in J$, be two linearly independent solutions of the equation

$$f^{\sigma}(t)x^{\Delta^2}(t) + f^{\Delta}(t)x^{\Delta}(t) + g(t)x(t) = 0, \quad t \in J. \tag{11.7}$$

Then

$$\begin{aligned} &f^{\sigma}(t)x_1^{\Delta^2}(t) + f^{\Delta}(t)x_1^{\Delta}(t) + g(t)x_1(t) = 0, \\ &f^{\sigma}(t)x_2^{\Delta^2}(t) + f^{\Delta}(t)x_2^{\Delta}(t) + g(t)x_2(t) = 0, \quad t \in J. \end{aligned} \tag{11.8}$$

We will search for a solution $x(t)$ of equation (11.6) in the form

$$x(t) = c_1(t)x_1(t) + c_2(t)x_2(t), \quad t \in J,$$

where $c_1, c_2 \in \mathcal{C}^1(J)$ will be determined below. We have

$$\begin{aligned} x^{\Delta}(t) &= c_1^{\Delta}(t)x_1^{\sigma}(t) + c_1(t)x_1^{\Delta}(t) \\ &\quad + c_2^{\Delta}(t)x_2^{\sigma}(t) + c_2(t)x_2^{\Delta}(t), \quad t \in J. \end{aligned}$$

We want

$$c_1^{\Delta}(t)x_1^{\sigma}(t) + c_2^{\Delta}(t)x_2^{\sigma}(t) = 0, \quad t \in J. \tag{11.9}$$

Then

$$x^{\Delta}(t) = c_1(t)x_1^{\Delta}(t) + c_2(t)x_2^{\Delta}(t), \quad t \in J,$$

and

$$\begin{aligned} x^{\Delta^2}(t) &= c_1^{\Delta}(t)x_1^{\Delta\sigma}(t) + c_1(t)x_1^{\Delta^2}(t) \\ &\quad + c_2^{\Delta}(t)x_2^{\Delta\sigma}(t) + c_2(t)x_2^{\Delta^2}(t), \quad t \in J. \end{aligned}$$

Hence, using (11.6) and (11.8), we obtain

$$\begin{aligned} &f^{\sigma}(t)x^{\Delta^2}(t) + f^{\Delta}(t)x^{\Delta}(t) + g(t)x(t) \\ &\quad = f^{\sigma}(t)c_1^{\Delta}(t)x_1^{\Delta\sigma}(t) + f^{\sigma}(t)c_1(t)x_1^{\Delta^2}(t) \\ &\qquad + f^{\sigma}(t)c_2^{\Delta}(t)x_2^{\Delta\sigma}(t) + f^{\sigma}(t)c_2(t)x_2^{\Delta^2}(t) \\ &\qquad + f^{\Delta}(t)c_1(t)x_1^{\Delta}(t) + f^{\Delta}(t)c_2(t)x_2^{\Delta}(t) \\ &\qquad + g(t)c_1(t)x_1(t) + g(t)c_2(t)x_2(t) \\ &\quad = f^{\sigma}(t)(c_1^{\Delta}(t)x_1^{\Delta\sigma}(t) + c_2^{\Delta}(t)x_2^{\Delta\sigma}(t)) \\ &\quad = h\left(t, x(t), \int_{t_0}^t k(t,s,x(s))\Delta s\right), \quad t \in J, \end{aligned}$$

whereupon

$$c_1^\Delta(t)x_1^{\Delta\sigma}(t) + c_2^\Delta(t)x_2^{\Delta\sigma}(t) = \frac{1}{f^\sigma(t)}h\left(t, x(t), \int_{t_0}^{t} k(t,s,x(s))\Delta s\right), \quad t \in J.$$

From the last equation and (11.9), we get the following system:

$$\begin{cases} c_1^\Delta(t)x_1^\sigma(t) + c_2^\Delta(t)x_2^\sigma(t) = 0, \\ c_1^\Delta(t)x_1^{\Delta\sigma}(t) + c_2^\Delta(t)x_2^{\Delta\sigma}(t) = \dfrac{1}{f^\sigma(t)}h\left(t, x(t), \displaystyle\int_{t_0}^{t} k(t,s,x(s))\Delta s\right), \quad t \in J. \end{cases}$$

Hence,

$$\begin{cases} c_1^\Delta(t)(x_1^{\Delta\sigma}(t)x_2^\sigma(t) - x_1^\sigma(t)x_2^{\Delta\sigma}(t)) = \dfrac{x_2^\sigma(t)}{f^\sigma(t)}h\left(t, x(t), \displaystyle\int_{t_0}^{t} k(t,s,x(s))\Delta s\right), \\ c_2^\Delta(t)(-x_1^{\Delta\sigma}(t)x_2^\sigma(t) + x_1^\sigma(t)x_2^{\Delta\sigma}(t)) = \dfrac{x_1^\sigma(t)}{f^\sigma(t)}h\left(t, x(t), \displaystyle\int_{t_0}^{t} k(t,s,x(s))\Delta s\right), \quad t \in J, \end{cases}$$

or

$$\begin{cases} c_1^\Delta(t) = \dfrac{x_2^\sigma(t)}{f^\sigma(t)(x_1^{\Delta\sigma}(t)x_2^\sigma(t) - x_1^\sigma(t)x_2^{\Delta\sigma}(t))}h\left(t, x(t), \displaystyle\int_{t_0}^{t} k(t,s,x(s))\Delta s\right), \\ c_2^\Delta(t) = \dfrac{x_1^\sigma(t)}{f^\sigma(t)(-x_1^{\Delta\sigma}(t)x_2^\sigma(t) + x_1^\sigma(t)x_2^{\Delta\sigma}(t))}h\left(t, x(t), \displaystyle\int_{t_0}^{t} k(t,s,x(s))\Delta s\right), \quad t \in J. \end{cases}$$

We take

$$\begin{cases} c_1(t) = \displaystyle\int_{t_0}^{t} \dfrac{x_2^\sigma(y)}{f^\sigma(y)(x_1^{\Delta\sigma}(y)x_2^\sigma(y) - x_1^\sigma(y)x_2^{\Delta\sigma}(y))} \\ \qquad \times h\left(y, x(y), \displaystyle\int_{t_0}^{y} k(y,s,x(s))\Delta s\right)\Delta y, \\ c_2(t) = \displaystyle\int_{t_0}^{t} \dfrac{x_1^\sigma(y)}{f^\sigma(y)(-x_1^{\Delta\sigma}(y)x_2^\sigma(y) + x_1^\sigma(y)x_2^{\Delta\sigma}(y))} \\ \qquad \times h\left(y, x(y), \displaystyle\int_{t_0}^{y} k(y,s,x(s))\Delta s\right)\Delta y, \quad t \in J. \end{cases}$$

Suppose that

$$f^\sigma(t)(x_1^{\Delta\sigma}(t)x_2^\sigma(t) - x_1^\sigma(t)x_2^{\Delta\sigma}(t)) = 1, \quad t \in J. \tag{11.10}$$

Then

$$\begin{cases} c_1(t) = \int\limits_{t_0}^{t} x_2^\sigma(y)h\left(y, x(y), \int\limits_{t_0}^{y} k(y,s,x(s))\Delta s\right)\Delta y, \\ c_2(t) = -\int\limits_{t_0}^{t} x_1^\sigma(y)h\left(y, x(y), \int\limits_{t_0}^{y} k(y,s,x(s))\Delta s\right)\Delta y, \quad t \in J, \end{cases}$$

and

$$\begin{aligned} x(t) &= c_1x_1(t) + c_2x_2(t) + c_1(t)x_1(t) + c_2(t)x_2(t) \\ &= c_1x_1(t) + c_2x_2(t) \\ &\quad + \int\limits_{t_0}^{t} x_1(t)x_2^\sigma(y)h\left(y, x(y), \int\limits_{t_0}^{y} k(y,s,x(s))\Delta s\right)\Delta y \\ &\quad - \int\limits_{t_0}^{t} x_2(t)x_1^\sigma(y)h\left(y, x(y), \int\limits_{t_0}^{y} k(y,s,x(s))\Delta s\right)\Delta y \\ &= c_1x_1(t) + c_2x_2(t) \\ &\quad + \int\limits_{t_0}^{t} (x_1(t)x_2^\sigma(y) - x_2(t)x_1^\sigma(y))h\left(y, x(y), \int\limits_{t_0}^{y} k(y,s,x(s))\Delta s\right)\Delta y, \quad t \in J, \end{aligned}$$

is a solution of equation (11.6). Below, using Pachpatte's inequalities, we will deduct some of the properties of the solution of equation (11.6).

Theorem 11.3.1. *Let $f, g \in \mathcal{C}(\mathbb{T})$, $k \in \mathcal{C}(\mathbb{T} \times \mathbb{T} \times \mathbb{R})$, $h \in \mathcal{C}(\mathbb{T} \times \mathbb{R} \times \mathbb{R})$,*

$$\begin{aligned} f(t) &\neq 0, \quad t \in \mathbb{T}, \\ |k(t,s,x)| &\leq p(s)|x|, \quad (t,s,x) \in \mathbb{T} \times \mathbb{T} \times \mathbb{R}, \\ |h(t,x,z)| &\leq q(t)(|x| + |z|), \quad (t,x,z) \in \mathbb{T} \times \mathbb{R} \times \mathbb{R}, \end{aligned}$$

where $p, q \in \mathcal{C}(\mathbb{T})$ are nonnegative functions such that

$$\int\limits_{t_0}^{a} p(s)\Delta s < \infty, \quad \int\limits_{t_0}^{a} q(s)\Delta s < \infty.$$

If $x_1(t)$ and $x_2(t)$ are bounded solutions of equation (11.7) *so that* (11.10) *holds, then the corresponding solution $x(t)$ of equation* (11.6) *is bounded on J.*

Proof. Let $c_1, c_2 \in \mathbb{R}$ be arbitrarily chosen. Let also $c > 0$ and $M > 0$ be chosen so that

$$|c_1x_1(t) + c_2x_2(t)| \leq c \quad \text{for any } t \in J,$$

and

$$|x_1(t)x_2^\sigma(y) - x_2(t)x_1^\sigma(y)| \le M \quad \text{for any } t, y \in J.$$

Then

$$\begin{aligned}
|x(t)| &= \left| c_1x_1(t) + c_2x_2(t) \right.\\
&\quad \left. + \int_{t_0}^{t} (x_1(t)x_2^\sigma(y) - x_2(t)x_1^\sigma(y))h\left(y, x(y), \int_{t_0}^{y} k(y,s,x(s))\Delta s\right)\Delta y \right|\\
&\le |c_1x_1(t) + c_2x_2(t)|\\
&\quad + \int_{t_0}^{t} |x_1(t)x_2^\sigma(y) - x_2(t)x_1^\sigma(y)| \left| h\left(y, x(y), \int_{t_0}^{y} k(y,s,x(s))\Delta s\right)\right| \Delta y\\
&\le c + M\int_{t_0}^{t} q(y)\left(|x(y)| + \left|\int_{t_0}^{y} k(y,s,x(s))\Delta s\right|\right)\Delta y\\
&= c + M\int_{t_0}^{t} q(y)|x(y)|\Delta y\\
&\quad + M\int_{t_0}^{t} q(y)\int_{t_0}^{y} |k(y,s,x(s))|\Delta s\Delta y\\
&\le c + M\int_{t_0}^{t} q(y)|x(y)|\Delta y\\
&\quad + M\int_{t_0}^{t} q(y)\int_{t_0}^{y} p(s)|x(s)|\Delta s\Delta y, \quad t \in J.
\end{aligned}$$

Hence, by Theorem 1.5.1, we get

$$|x(t)| \le c\left(1 + M\int_{t_0}^{t} q(y)e_{Mq+p}(y,0)\Delta y\right), \quad t \in J.$$

From here we conclude that $x(t)$ is bounded on J. This completes the proof. □

11.4 Volterra integral equations

Consider the following Volterra integral equation:

$$x(t) = f(t) + \int_{t_0}^{t} F(t, s, x(s))\Delta s, \quad t \in J, \tag{11.11}$$

where $f \in \mathcal{C}(J)$, $F \in \mathcal{C}(J \times J \times \mathbb{R})$ are given functions, x is unknown.

Theorem 11.4.1. *Let $f \in \mathcal{C}(J)$, $F \in \mathcal{C}(J \times J \times \mathbb{R})$,*

$$|F(t, s, x(s))| \le g(t)(h(s) + |x(s)|), \quad t, s \in J,$$

where $g, h \in \mathcal{C}(J)$ are nonnegative functions which are bounded on J. If $x(t)$ is a solution of equation (11.11)*, then it is bounded on J.*

Proof. We have

$$\begin{aligned}
|x(t)| &= \left| f(t) + \int_{t_0}^{t} F(t, s, x(s))\Delta s \right| \\
&\le |f(t)| + \left| \int_{t_0}^{t} F(t, s, x(s))\Delta s \right| \\
&\le |f(t)| + \int_{t_0}^{t} |F(t, s, x(s))|\Delta s \\
&\le |f(t)| + g(t) \int_{t_0}^{t} (h(s) + |x(s)|)\Delta s \\
&= |f(t)| + g(t) \int_{t_0}^{t} h(s)\Delta s + g(t) \int_{t_0}^{t} |x(s)|\Delta s, \quad t \in J.
\end{aligned}$$

Let

$$p(t) = |f(t)| + g(t) \int_{t_0}^{t} h(s)\Delta s, \quad t \in J.$$

Then

$$|x(t)| \le p(t) + g(t) \int_{t_0}^{t} |x(s)|\Delta s, \quad t \in J.$$

Hence, by Theorem 1.1.10, we obtain

$$|x(t)| \le p(t) + g(t)\int_{t_0}^{t} p(s)e_{\ominus g}(\sigma(s),t)\Delta s, \quad t \in J.$$

Therefore $x(t)$ is bounded on J. This completes the proof. □

11.5 Applications to nonlinear delay dynamic equations

Suppose that the time scale $\mathbb{T}$ is unbounded above. Consider the following IVP:

$$((x(t))^p)^\Delta = M(t, x(\tau(t))), \quad t \in [t_0, \infty), \tag{11.12}$$

with the initial data

$$\begin{aligned} x(t) &= \phi(t), \quad t \in [\alpha, t_0], \\ \phi(\tau(t)) &= c^{\frac{1}{p}}, \quad t \in [t_0, \infty),\ \tau(t) \le t_0, \end{aligned} \tag{11.13}$$

where
(A1) $M : [t_0, \infty) \times \mathbb{R} \to \mathbb{R}$ is a continuous function,
(A2) $p \ge 1$, $c = (x(t_0))^p$ are constants, $\tau : [t_0, \infty) \to \mathbb{T}$, $\tau(t) \le t$, $t \in [t_0, \infty)$,

$$-\infty < \alpha = \inf\{\tau(t) : t \in [t_0, \infty)\},$$

$\phi \in \mathcal{C}_{rd}([t_0, \infty))$.

Theorem 11.5.1. *Suppose* (A1), (A2) *and*

$$|M(t,x)| \le f(t)|x| + g(t), \quad t \in [t_0, \infty),\ x \in \mathbb{R},$$

where $f, g \in \mathcal{C}_{rd}([t_0, \infty))$ *are nonnegative functions. If x is a solution of the IVP* (11.12), (11.13), *then*

$$\begin{aligned} |x(t)| &\le \left(|c| + \overline{h}(t)\int_{t_0}^{t} e_{\overline{B}}(t, \sigma(s))\overline{h}(s)\overline{B}(s)\Delta s\right)^{\frac{1}{p}} \\ &\le \left(|c| + \overline{h}(t)e_{\overline{B}}(t, t_0)\right)^{\frac{1}{p}}, \quad t \in [t_0, \infty), \end{aligned} \tag{11.14}$$

for any $k > 0$, where

$$\begin{aligned} \overline{h}(t) &= \int_{t_0}^{t}\left(f(s)\left(\frac{p-1}{p}k^{\frac{1}{p}} + \frac{|c|}{pk^{\frac{p-1}{p}}}\right) + g(s)\right)\Delta s, \\ \overline{B}(t) &= \frac{f(t)}{pk^{\frac{p-1}{p}}}, \quad t \in [t_0, \infty). \end{aligned}$$

Proof. The solution x of the IVP (11.12), (11.13) satisfies the following integral equation:

$$|x(t)|^p = c + \int_{t_0}^{t} M(s, x(\tau(s)))\Delta s, \quad t \in [t_0, \infty),$$

whereupon

$$|x(t)|^p \le |c| + \int_{t_0}^{t} (|f(s)||x(\tau(s))| + g(s))\Delta s, \quad t \in [t_0, \infty).$$

Hence, by Theorem 10.2.2, we get inequality (11.14). This completes the proof. □

Consider the following delay dynamic equation:

$$x^{\Delta}(t) = F\left(t, x(\tau(t)), \int_{t_0}^{t} M(\xi, x(\tau(\xi)))\Delta\xi\right), \quad t \in [t_0, \infty), \tag{11.15}$$

with initial data

$$\begin{aligned} x(t_0) &= c, \\ x(t) &= \phi(t), \quad t \in [\alpha, t_0], \\ \phi(\tau(t)) &\le |c|, \quad t \in [t_0, \infty),\ \tau(t) \le t_0, \end{aligned} \tag{11.16}$$

where

(A3) c is a constant, $c \neq 0$, $\phi \in C_{rd}([\alpha, t_0])$ is a nonnegative function, $\tau : [t_0, \infty) \to \mathbb{T}$, $\tau \in C([t_0, \infty))$, $\tau(t) \le t$, $t \in [t_0, \infty)$,

$$-\infty < \alpha = \inf\{\tau(t) : t \in [t_0, \infty)\},$$

(A4) $F \in C([t_0, \infty) \times \mathbb{R} \times \mathbb{R})$, $M \in C([t_0, \infty) \times \mathbb{R})$.

Theorem 11.5.2. *Assume* (A3), (A4) *and*

$$\begin{aligned} |F(t, u, v)| &\le f(t)\psi(|u|)w(|u|) + g(t)\psi(|u|) + |v|, \\ |M(t, u)| &\le h(t)\psi(|u|), \quad t \in [t_0, \infty),\ u, v \in \mathbb{R}, \end{aligned}$$

where $f, g, h \in C_{rd}([t_0, \infty))$ are nonnegative functions with f not equivalent to zero, $w, \psi \in C([t_0, \infty))$ are nonnegative and nondecreasing functions. Then the solution x of the IVP (11.15), (11.16) *satisfies the following estimate:*

$$|x(t)| \le \widetilde{G}^{-1}\left(\widetilde{H}^{-1}\left(\widetilde{H}\left(\widetilde{G}\left(|c| + \int_{t_0}^{t}\left(g(s) + \int_{t_0}^{s} h(\xi)\Delta\xi\right)\Delta s\right) + \int_{t_0}^{t} f(s)\Delta s\right)\right)\right), \quad t \in [t_0, \infty),$$

where $\widetilde{G}$ and $\widetilde{H}$ are increasing and bijective functions and

$$\left(\widetilde{G}(v(t))\right)^{\Delta} = \frac{v^{\Delta}(t)}{\psi(v(t))},$$

$$\left(\widetilde{H}(z(t))\right)^{\Delta} = \frac{z^{\Delta}(t)}{w(\widetilde{G}^{-1}(z(t)))}, \quad t \in [t_0, \infty).$$

Proof. We have that the solution of the IVP (11.15), (11.16) satisfies the following integral equation:

$$x(t) = c + \int_{t_0}^{t} F\left(s, x(\tau(s)), \int_{t_0}^{s} M(\xi, x(\tau(\xi)))\Delta\xi\right)\Delta s, \quad t \in [t_0, \infty).$$

Then

$$\begin{aligned}
|x(t)| &\leq |c| + \int_{t_0}^{t} \left|F\left(s, x(\tau(s)), \int_{t_0}^{s} M(\xi, x(\tau(\xi)))\Delta\xi\right)\right|\Delta s \\
&\leq |c| + \int_{t_0}^{t} \Bigg(f(s)\psi(|x(\tau(s))|)w(|x(\tau(s))|) + g(s)\psi(|x(\tau(s))|) \\
&\quad + \left|\int_{t_0}^{s} M(\xi, x(\tau(\xi)))\Delta\xi\right| \Bigg)\Delta s \\
&\leq |c| + \int_{t_0}^{t} (f(s)\psi(|x(\tau(s))|)w(|x(\tau(s))|) + g(s)\psi(|x(\tau(s))|))\Delta s \\
&\quad + \int_{t_0}^{t}\int_{t_0}^{s} |M(\xi, x(\tau(\xi)))|\Delta\xi\Delta s \\
&\leq |c| + \int_{t_0}^{t} (f(s)\psi(|x(\tau(s))|)w(|x(\tau(s))|) + g(s)\psi(|x(\tau(s))|))\Delta s \\
&\quad + \int_{t_0}^{t}\int_{t_0}^{s} h(\xi)\psi(|x(\tau(\xi))|)\Delta\xi\Delta s, \quad t \in [t_0, \infty).
\end{aligned}$$

Hence, by Theorem 10.2.3, it follows that

$$\begin{aligned}
|x(t)| \leq \widetilde{G}^{-1}\Bigg(\widetilde{H}^{-1}\Bigg(\widetilde{H}\Bigg(\widetilde{G}\Bigg(|c| + \int_{t_0}^{t}\left(g(s) + \int_{t_0}^{s} h(\xi)\Delta\xi\right)\Delta s\Bigg) \\
+ \int_{t_0}^{t} f(s)\Delta s\Bigg)\Bigg)\Bigg), \quad t \in [t_0, \infty).
\end{aligned}$$

This completes the proof. □

Now, consider the delay dynamic equation

$$((x(t))^p)^\Delta = M(t, x(t), x(\tau(t))), \quad t \in [t_0, \infty), \tag{11.17}$$

subject to the initial condition

$$\begin{aligned} x(t) &= \phi(t), \quad t \in [\alpha, t_0], \\ \phi(\tau(t)) &= c^{\frac{1}{p}}, \quad t \in [t_0, \infty), \quad \tau(t) \le t_0, \end{aligned} \tag{11.18}$$

where
(A5) $p > 0$, $c = (x(t_0))^p$, $\tau \in \mathcal{C}_{rd}([t_0, \infty))$, $\tau(t) \le t$, $t \in [t_0, \infty)$,

$$-\infty < \alpha = \inf\{\tau(t) : t \in [t_0, \infty)\}.$$

(A6) $M \in \mathcal{C}([t_0, \infty) \times \mathbb{R}^2)$ and

$$|M(t, u, v)| \le f(t)|v|^q + g(t)|u|^r,$$

$t \in [t_0, \infty)$, $u, v \in \mathbb{R}$, q and r are constants such that $p \ge q \ge 0$ and $p \ge r \ge 0$.

Theorem 11.5.3. *Assume* (A5), (A6). *Then the solution x of the IVP* (11.17), (11.18) *satisfies the following estimate:*

$$\begin{aligned} |x(t)| &\le \left(|c| + \overline{F}(t) + \int_{t_0}^{t} e_{\overline{G}}(t, \sigma(s)) \overline{F}(s) \overline{G}(s) \Delta s \right)^{\frac{1}{p}} \\ &\le (|c| + \overline{F}(t) e_{\overline{G}}(t, t_0))^{\frac{1}{p}}, \quad t \in [t_0, \infty), \end{aligned} \tag{11.19}$$

for any $k > 0$, where

$$\begin{aligned} \overline{F}(t) &= \int_{t_0}^{t} \left(\frac{f(s)(k(p-q) + q|c|)}{pk^{\frac{p-q}{p}}} + \frac{g(s)(k(p-r) + r|c|)}{pk^{\frac{p-q}{p}}} \right) \Delta s, \\ \overline{G}(t) &= \frac{qf(t)}{pk^{\frac{p-q}{p}}} + \frac{rg(t)}{pk^{\frac{p-q}{p}}}, \quad t \in [t_0, \infty). \end{aligned}$$

Proof. We have that the solution x of the IVP (11.17), (11.18) satisfies the integral equation

$$(x(t))^p = c + \int_{t_0}^{t} M(s, x(s), x(\tau(s))) \Delta s, \quad t \in [t_0, \infty).$$

Hence,

$$|x(t)|^p = \left| c + \int_{t_0}^{t} M(s, x(s), x(\tau(s))) \Delta s \right|$$

$$\leq |c| + \left|\int_{t_0}^{t} M(s, x(s), x(\tau(s)))\Delta s\right|$$

$$\leq |c| + \int_{t_0}^{t} |M(s, x(s), x(\tau(s)))|\Delta s$$

$$\leq |c| + \int_{t_0}^{t} (f(s)|x(\tau(s))|^q + g(s)|x(s)|^r)\Delta s, \quad t \in [t_0, \infty).$$

By the last inequality and Theorem 10.2.4, we get the estimate (11.19). This completes the proof. □

Consider the IVP

$$((x(t))^p)^\Delta = F\left(t, x(\tau_1(t)), \int_{t_0}^{t} M(\xi, x(\tau_2(\xi)))\Delta\xi\right), \quad t \in [t_0, \infty), \tag{11.20}$$

subject to the initial condition

$$\begin{aligned} x(t) &= \phi(t), \quad t \in [\alpha, t_0], \\ |\phi(\tau_i(t))| &\leq |c|^{\frac{1}{p}}, \quad t \in [t_0, \infty),\ \tau_i(t) \leq t_0,\ i \in \{1, 2\}, \end{aligned} \tag{11.21}$$

where

(A6) $x \in \mathcal{C}_{rd}([t_0, \infty))$, $c = (x(t_0))^p$, $p \geq 1$, $\tau_i \in \mathcal{C}([t_0, \infty))$, $\tau_i(t) \leq t$, $t \in [t_0, \infty)$, $i \in \{1, 2\}$,

$$-\infty < \alpha = \inf\left\{\min_{i\in\{1,2\}} \tau_i(t) : t \in [t_0, \infty)\right\},$$

$\phi \in \mathcal{C}_{rd}([t_0, \infty))$,

(A7) $F \in \mathcal{C}([t_0, \infty) \times \mathbb{R}^2)$, $M \in \mathcal{C}([t_0, \infty) \times \mathbb{R})$,

$$\begin{aligned} |F(t, u, v)| &\leq f(t)|u| + |v|, \\ |M(t, u)| &\leq h(t)|u|, \quad t \in [t_0, \infty),\ u, v \in \mathbb{R}, \end{aligned}$$

$f, h \in \mathcal{C}_{rd}([t_0, \infty))$ are nonnegative functions.

Theorem 11.5.4. *If x is a solution to the IVP (11.20), (11.21), then*

$$|x(t)| \leq \left(G^{-1}\left(G(|c|) + \int_{t_0}^{t}\left(f(s) + \int_{t_0}^{s} h(\xi)\Delta\xi\right)\Delta s\right)\right)^{\frac{1}{p}}, \quad t \in [t_0, \infty), \tag{11.22}$$

where G is an increasing, bijective function and

$$G(v) = \int_1^v \frac{1}{r^{\frac{q}{p}}} dr, \quad v > 0.$$

Proof. Every solution of the IVP (11.20), (11.21), satisfies the integral equation

$$(x(t))^p = c + \int_{t_0}^{t} F\left(s, x(\tau_1(s)), \int_{t_0}^{s} M(\xi, x(\tau_2(\xi)))\Delta\xi\right)\Delta s, \quad t \in [t_0, \infty).$$

Then

$$\begin{aligned}
|x(t)|^p &= \left|c + \int_{t_0}^{t} F\left(s, x(\tau_1(s)), \int_{t_0}^{s} M(\xi, x(\tau_2(\xi)))\Delta\xi\right)\Delta s\right| \\
&\le |c| + \left|\int_{t_0}^{t} F\left(s, x(\tau_1(s)), \int_{t_0}^{s} M(\xi, x(\tau_2(\xi)))\Delta\xi\right)\Delta s\right| \\
&\le |c| + \int_{t_0}^{t} \left|F\left(s, x(\tau_1(s)), \int_{t_0}^{s} M(\xi, x(\tau_2(\xi)))\Delta\xi\right)\right|\Delta s \\
&\le |c| + \int_{t_0}^{t} \left(f(s)|x(\tau_1(s))| + \left|\int_{t_0}^{s} M(\xi, x(\tau_2(\xi)))\Delta\xi\right|\right)\Delta s \\
&\le |c| + \int_{t_0}^{t} \left(f(s)|x(\tau_1(s))| + \int_{t_0}^{s} |M(\xi, x(\tau_2(\xi)))|\Delta\xi\right)\Delta s \\
&\le |c| + \int_{t_0}^{t} \left(f(s)|x(\tau_1(s))| + \int_{t_0}^{s} h(\xi)|x(\tau_2(\xi))|\Delta\xi\right)\Delta s,
\end{aligned}$$

$t \in [t_0, \infty)$. Hence, by Theorem 10.3.1, we get the estimate (11.22). This completes the proof. □

Bibliography

[1] R. Agarwal, M. Bohner and A. Peterson. Inequalities on time scales: A survey, Math. Inequal. Appl. 4 (4) (2001), 535–557.

[2] E. Akin-Bohner, M. Bohner and F. Akin. Pachpatte inequalities on time scales, JIPAM. J. Inequal. Pure Appl. Math. 6 (1) (2005), 1–23.

[3] M. Ammi, R. Ferreira and D. Torres. Diamond-α Jensen's inequality on time scales, J. Inequal. Appl. 2018.

[4] G. Anastassiou. Principles of delta fractional calculus on time scales and inequalities, Math. Comput. Model. 52 (2016), 556–566.

[5] M. Bohner and S. Georgiev. Multivariable Dynamic Calculus on Time Scales, Springer, 2017.

[6] M. Bohner and A. Peterson. Dynamic Equations on Time Scales: An Introduction with Applications, Birkhäuser, Boston, 2001.

[7] M. Bohner and A. Peterson. Advances in Dynamic Equations on Time Scales, Birkhäuser, Boston, 2003.

[8] K. Boukerrioua, D. Diabi and T. Chineb. Further nonlinear inetgral inequalities in two independent variables on time scales and their applications, Malaya J. Mat. 5 (1) (2017), 109–114.

[9] L. Du and R. Xu. Some new Pachpatte type inequalities on time scales and their applications, J. Math. Inequal. 6 (2) (2012), 229–240.

[10] A. El-Deeb, H. El-Sennary and Z. Khan. Some Steffensen-type dynamic inequalities on time scales, Adv. Differ. Equ. (2019).

[11] A. A. El-Deeb and W.-S. Cheung. A variety of dynamic inequalities on time scales with retardation, J. Nonlinear Sci. Appl. 11 (2018), 1185–1206.

[12] A. A. El-Deeb, H. Xu, A. Abdeldaim and G. Wang. Some dynamic inequalities on time scales and their applications, Adv. Differ. Equ. (2019).

[13] Q. Feng, F. Meng, Y. Zhang, B. Zheng and J. Zhou. Some nonlinear delay integral inequalities on time scales arising in the theory of dynamic equations, J. Inequal. Appl. (2011).

[14] Q. Feng, F. Meng, Y. Zhang, J. Zhou and B. Zheng. Some nonlinear delay integral inequalities on time scales and their applications in the theory of dynamic equations, Abstr. Appl. Anal. (2012).

[15] Q. Feng, F. Meng and B. Zheng. Gronwall–Bellman type nonlinear delay integral inequalities on time scales, J. Math. Anal. Appl. 382 (2011), 772–784.

[16] W. Li. Some new dynamic inequalities on time scales, J. Math. Anal. Appl. 319, 2 (2006), 802–814.

[17] W. Li. Bounds for certain delay integral inequalities on time scales, J. Inequal. Appl. (2008).

[18] W. Li and W. Sheng. Some nonlinear integral inequalities on time scales, J. Inequal. Appl. 2007 (2007), 70465, 15 pages. doi:10.1155/2007/70465.

[19] W. N. Li, Some Pachpatte type inequalities on time scales, Comput. Math. Appl. 57 (2009), 275–282.

[20] W. N. Li, Some delay integral inequalities on time scales, Comput. Math. Appl. 59 (2010), 1929–1936. doi:10.1016/j.camwa.2009.11.006.

[21] H. Liu. On some nonlinear retarded Volterra–Fredholm type integral inequalities on time scales and their applications, J. Inequal. Appl. (2018).

[22] Y. Mi. A generalized Gronwall–Bellman type delay integral inequality with two independent variables on time scales, J. Math. Inequal. 11 (4) (2017), 1151–1160.

[23] U. Ozkan and H. Yildirim. Steffenesen's integral inequalities on time scales, J. Inequal. Appl. (2007).

[24] B. G. Pachpatte. Inequalities for Differential and Integral Equations, volume 197 of

https://doi.org/10.1515/9783110705553-012

Mathematics in Science and Engineering, Academic Press Inc., San Diego, CA, 1998.
[25] B. G. Pachpatte. Inequalities for Finite Difference Equations, volume 247 of Monographs and Textbooks in Pure and Applied Mathematics, Marcel Dekker Inc., New York, 2002.
[26] D. B. Pachpatte. Explicit estimates on integral inequalities with time scales, JIPAM. J. Inequal. Pure Appl. Math. 7 (4) (2006), article 143.
[27] M. Sahir. Dynamic fractional inequalities amplified on time scale calculus revealing coalition disctretness and continuity, Fractal Fract. 2 (2008), 25.
[28] D. Ucar, V. Hatipoglu and A. Akincali. Fractional integral inequalities on time scales, Open J. Math. Sci. 2 (1) (2018), 361–370.
[29] F. Wong, C. C. Yeh and C. H. Hong. Gronwall inequalities on time scales, Math. Inequal. Appl. 9 (2006), 75–86.
[30] B. Zheng, Y. Zhang and Q. Feng. Some new delay integral inequalities in two independent variables on time scales, J. Appl. Math. (2011).

Index

www.ingramcontent.com/pod-product-compliance
Lightning Source LLC
Chambersburg PA
CBHW080931220326
41598CB00034B/5755
9783110705508